Collins

Revision

NEW GCSE SCIENCE

Science A and Additional Science A

OCR

Twenty First Century Science

Authors: John Beeby
            Michael Brimicombe
            Brian Cowie
            Sarah Mansel
            Ann Tiernan

Revision Guide +
Exam Practice Workbook

Contents

Contents

What genes do

What genes are and how they work

- **Genes** carry the instructions that control how you develop and function. They do this by telling the cells to make the **proteins** needed for your body to work.

- Each gene is a section of a very long molecule of a chemical called **DNA** (deoxyribonucleic acid).

- Lengths of DNA are coiled and packed into structures called **chromosomes**. These are found in the nuclei of the body's cells. We have between 20 000 and 25 000 genes on our chromosomes.

- Strands of DNA are made up of four chemicals called bases, as well as phosphate groups and sugar molecules.

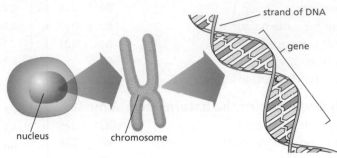

A gene is a length of DNA that codes for a particular protein.

- The order of the bases in a DNA strand determines the order of amino acids in a protein.

- Proteins fall into two groups:

  1 **Functional proteins** enable the body to function. Examples include **enzymes**, antibodies and hormones.

  2 **Structural proteins** give the body structure, rigidity and strength. Examples include collagen in ligaments and keratin in skin.

- The Human Genome Project has identified the location of all the genes on human chromosomes. We call the complete gene set of an organism its genome.

- The project will help us to understand how genes control our characteristics and development, and can lead to certain diseases.

- The project has ethical implications, e.g. some drug companies want to patent or 'own' genes. They could then charge other scientists money to investigate the genes, which would restrict research.

We're all different

- Our characteristics are controlled by our genes (e.g. dimples), our **environment** (e.g. the presence of scars or dyed hair), or a combination of these (e.g. our body weight).

- We inherit genes from each of our parents, so we're similar, but not identical to, each parent.

- Differences in genes produce **variation** in offspring.

- Some characteristics are controlled by several genes working together. These characteristics will show **continuous variation** across a population, e.g. the continuous range of eye colours and different heights.

- For a particular characteristic, you can describe a person by their **genotype** or **phenotype**.

- A person's genotype is his or her genetic make-up. It can refer to the whole of an individual's genes or (more often) the genes for a particular characteristic, such as if we have dimples. The genotype is usually written as two letters, for example DD.

- A person's phenotype describes a his or her observable physical features, e.g. build, or a single characteristic. The phenotype will depend on the person's genes, but may also be affected by how these interact with the environment.

Studies on twins

- Identical twins have identical genotypes because they develop after a fertilised egg splits into two.

- Studies of identical twins, especially those that have been separated, can help us to understand the effect of the environment on a person's genotype.

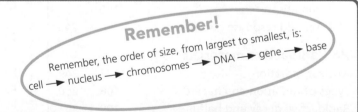

Remember!

Remember, the order of size, from largest to smallest, is:

cell → nucleus → chromosomes → DNA → gene → base

Improve your grade

Studies on twins

Higher: How do studies of identical twins help us to understand the effect of the environment on the phenotype for a characteristic?

AO2 [3 marks]

Genes working together and variation

How genes and chromosomes are organised

- Chromosomes can be arranged into pairs. Human cells have 23 pairs of chromosomes (a total of 46 chromosomes).

- Sex cells are eggs (ova) and sperm. These have 23 chromosomes, one from each pair.

- At **fertilisation**, an egg and sperm join together to produce a **zygote**. The zygote has a full set of 46 chromosomes – 23 from the mother and 23 from the father.

- The human baby that develops has a combination of genes from its mother and father.

- Pairs of chromosomes have genes for the same characteristic at identical positions on each chromosome of the pair.

- Changes to our DNA sometimes occur, causing a **mutation**. This can take place when sex cells are being made, or after fertilisation.

- One type of mutation is a chromosome mutation. This results in an individual having extra chromosomes. For example, a person with an extra chromosome 21 will have Down's syndrome.

Variation in offspring

- The combination of chromosomes in an egg or sperm will always be different. For example, in an egg, chromosome 1 could have been inherited from the mother, while chromosomes 2 and 3 could have been inherited from the father, etc. So, the combination of chromosomes (and therefore genes) will be unique to that person – unless he or she is an identical twin.

- Environmental effects will also add to the variation.

Pairs of alleles

- **Alleles** are the different forms in which the genes controlling a characteristic can occur. So, for dimples: one allele is for the presence of dimples and one allele is for a lack of dimples.

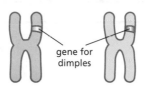

alternative alleles: ☐ dimples ☐ no dimples

The alleles for having dimples or for not having dimples are located on the same part of each chromosome in a pair. This individual has one of each allele.

- If the two alleles of a gene are identical, the person is said to be **homozygous** (for that characteristic).

- If the two alleles are different, the person is said to be **heterozygous** (for that characteristic).

> **Remember!**
> Two combinations of alleles, e.g. DD and dd, are possible in an organism that is homozygous for a characteristic. Only one combination, i.e. Dd, is possible for an organism that is heterozygous for a characteristic.

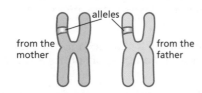

This pair of chromosomes shows the same pair of alleles, so this individual is homozygous. In the diagram above the individual is heterozygous.

Higher

Ideas about science

You should be able to:

- distinguish between questions that could be answered using a scientific approach (e.g. determining which alleles are dominant and which are recessive) and questions which are difficult to answer or could not be answered (e.g. the combination of alleles that occurs when sex cells are produced and at fertilisation).

Improve your grade

Variation in offspring

Foundation: Explain how variation occurs in the offspring produced by humans. *AO1* [5 marks]

Genetic crosses and sex determination

Genetic traits

- Traits are passed on from parents to their offspring through genes on chromosomes.
- Genes for a particular trait are found at the same place on each chromosome of the chromosome pair.
- The different forms of a gene that control a certain trait are called alleles.
- In most cases, alleles for a trait can be **dominant** or **recessive**. For example, the allele for hairy toes is dominant to the allele for hairless toes.
- Dominant alleles are written with upper-case letters in genetic diagrams, e.g. H for hairy toes.
- Recessive alleles are written with lower-case letters in genetic diagrams, e.g. h for hairless toes.
- If at least one dominant allele is present (e.g. Hh), the trait shown will be the dominant one (e.g. hairy toes).
- For the recessive trait (e.g. hairless toes) to be shown there must be two recessive alleles present (e.g. hh).

Genetic diagrams

- You can use a **Punnett square** to:
  - show genetics crosses
  - find the probability of two parents producing different types of offspring.
- You can show the inheritance of a trait in a family over several generations using a **family tree diagram**.
- A family tree diagram is very useful when tracing a genetic disorder, such as Huntingdon's disease, over generations.

	mother	
	X	X
X (father)	X X female	X X female
Y	X Y male	X Y male

This Punnett square shows the possible outcomes for the sex of a baby. You need to be able to draw a Punnett square like this one.

Sex determination

- One of the 23 pairs of chromosomes – the 23rd pair – determines our sex.
- A female has two X chromosomes, written XX.
- A male has an X and a Y chromosome, written XY.

Remember!
The X chromosome is always listed first when writing combinations of sex chromosomes.

- Eggs and sperm have 23 chromosomes.
  - Each egg cell produced by a female will have an X chromosome.
  - Half the sperm cells produced by a male will have an X chromosome and half will have a Y chromosome (as shown in the Punnett square above).
  - So, in theory, 50% of babies will be female and 50% will be male.

- It's the presence of a Y gene on the Y chromosome – the **sex-determining gene** – that determines whether the embryo is male or not (it triggers the development of testes in the embryo). In the absence of the Y chromosome, ovaries develop.
- **Higher** Because of the shape of the sex chromosomes, there are parts of the X chromosome that can have no matching alleles on the Y chromosome. If a defective gene is found on this part of the X chromosome, this can result in a sex-linked disorder.
- Sex-linked diseases, such as haemophilia (a blood-clotting disorder) and red-green colour blindness, are far more likely to be present in males.

Improve your grade

Genetic diagrams

Higher: A disease called cystic fibrosis is caused when two recessive alleles of a gene are present. The diagram shows the occurrence of cystic fibrosis in the family. What are the genotypes of:

a father A?

b daughter D?

Explain your answers.

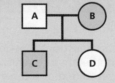

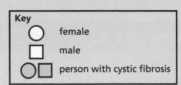

Key
- ○ female
- □ male
- ○□ person with cystic fibrosis

AO3 [4 marks]

Gene disorders

- Some disorders are caused by defective or faulty alleles.

- Huntington's disease is a dominant disorder, i.e. the presence of a *single* dominant allele will cause the disease. It occurs in middle age. Symptoms include tremors (uncontrollable shaking), memory loss, inability to concentrate and mood changes.

- *Both* alleles on the chromosome pair must be recessive for a person to get a recessive disorder, such as cystic fibrosis. Symptoms include the production of thick gluey mucus that affects the lungs and makes digesting food difficult, breathing problems and chest infections.

- For recessive single gene disorders, a person with a normal and a defective allele will be normal (because the normal gene is dominant), but they will be a **carrier**.

- If both parents are carriers they can give birth to a child with the disease.

Remember!
Only one defective allele in a pair is needed to cause the disease in a dominant disorder. Both alleles need to be defective for a recessive disorder to occur.

Genetic testing

- **Genetic screening** is used to check for a particular disorder, even when there is no history of it in the family. It is hoped this will minimise the damage such disorders can cause. For example, the heel prick 'blood spot test' is used on most newborn babies to diagnose rare genetic disorders.

- **Genetic testing** of individuals is carried out when a genetic disease, such as cystic fibrosis, runs in the family. This may allow people to get treatment for the disease or to plan for the future.

- Genetic testing sometimes raises ethical questions. For example, if a person has Huntington's disease should they tell their employer, insurance company or family?

- Genetic testing during pregnancy may involve **cell sampling** by:
  - amniocentesis (collecting cells from the developing **fetus** which are present in amniotic fluid)
  - chorionic villus sampling (testing a sample of cells taken from the placenta).

  Both tests carry risk – around 1% of babies are miscarried as a result of genetic testing.

- Production of **embryos** by *in vitro* fertilisation (IVF) allows doctors to check the genetic make-up of the embryos prior to implantation. This is known as 'embryo screening'.

- Embryo screening is used to investigate families with a known history of a disorder, such as cystic fibrosis. It allows doctors to remove any embryos suffering from a disorder and implant only normal embryos.

- Screening embryos prior to implantation and only using healthy embryos is called **pre-implantation genetic diagnosis (PGD)**.

- Procedures like PGD and embryo research are carefully monitored in the UK. Guidelines for clinics and research centres cover ethical and moral considerations for embryo use.

- Parents likely to pass on a genetic abnormality:
  - may decide not to have a family
  - may need to decide on whether to continue with the pregnancy or terminate it.

- Types of genetic tests may give false negative results and may *sometimes* give false positive results (where the test is positive but the person does not have the disorder).

Higher

⦿ Ideas about science

You should be able to:

- show awareness of, and discuss, the official regulation of scientific research and the application of scientific knowledge, e.g. the regulation of embryo testing and research

- say clearly what the ethical issue is, e.g. the use of information from genetic testing, and summarise different views

- identify and develop arguments that consider the best outcome for the greatest number of people, e.g. the mass screening of newborn babies.

⦿ Improve your grade

Gene disorders

Foundation: Write down one dominant genetic disorder and one recessive genetic disorder. For each disorder, list three symptoms the person will show.

AO1 [4 marks]

Cloning and stem cells

Cloning

- **Clones** are individuals with identical genes.
- In **asexual reproduction** only one parent is involved, so the offspring has identical DNA to the parent. **Bacteria**, along with some plants and simple animals (e.g. *Hydra*), reproduce asexually.
- Plants can reproduce asexually by:
  - using runners (for example strawberries) – shoots sent out that grow into identical plants
  - producing bulbs (for example daffodils).
- Identical twins are human clones produced when a fertilised egg splits, resulting in two genetically identical individuals.

- As clones have identical DNA, any differences between individuals in a clone and their parent must be a result of the environment and *not* genes.
- The advantages of producing clones / asexual reproduction are that:
  - successful characteristics are seen in offspring
  - asexual reproduction is useful where plants and animals live in isolation.
- The disadvantage of producing clones / asexual reproduction is that there is no genetic variation. This means that if conditions change or there is disease, the population could be wiped out.

- Clones have been produced by artificial animal cloning, for example Dolly the sheep and Snuppy the dog.
  - The nucleus from a body cell is extracted and inserted into an egg cell that has had its nucleus removed. This gives the egg cell a full set of genes without having been fertilised.
  - The embryo is implanted into a suitable surrogate mother.
- It is illegal to create clones of humans in many countries, including the UK.

Stem cells

- A human embryo develops from a single cell. This cell divides over and over again as the baby develops. Most of these cells become specialised to do different jobs (a process called **differentiation**).
  - After five days, the embryo is a ball of cells containing **embryonic stem cells**.
  - These cells are unspecialised. They divide and develop into the different types of cell in the human body.
- As adults, some **stem cells** – **adult stem cells** – remain in certain parts of our bodies.
- Adult stem cells can repair or replace certain cell types. For example, bone marrow cells are able to develop into different types of blood cells.

- Adult stem cells are used to treat various diseases, but have limited uses.
- Because embryonic stem cells can develop into other cell types they have huge potential. However, their use is controversial:
  - They are usually taken from unused embryos following fertility treatments.
  - Their use involves the destruction of the embryo.
- Recent research is focusing on reprogramming adult body cells into stem cells, and collecting cells from the umbilical cord blood when a baby is born.

- Stem cells could be used in:
  - the testing of new drugs
  - understanding how cells become specialised in the early stages of human development by the switching on and off of particular genes
  - renewing damaged or destroyed cells in spinal injuries, heart disease, Alzheimer's disease and Parkinson's disease.

EXAM TIP

Make sure you understand:
- the difference between embryonic and adult stem cells
- the difference in potential that these have in the treatment of disease.

Improve your grade

Stem cells

Higher: Discuss the future use of stem cells in medicine.

AO1, AO2 [4 marks]

B1 Summary

Genes carry instructions to control how an organism develops.

The instructions tell the cell how to make essential proteins.

Proteins are functional, e.g. enzymes, or structural, e.g. collagen.

How genes control our characteristics and development

Human characteristics are determined by:

- genes, e.g. dimples
- the environment, e.g. scars
- or by a combination of both, e.g. body weight.

Many characteristics are controlled by several genes working together, e.g. eye colour.

Genes control how an organism functions.

They are found in the nucleus of the cell. They are sections of the DNA molecules that make up chromosomes.

Sex determination:

- In human females the sex chromosomes are XX.
- In human males the sex chromosomes are XY.

The Y chromosome has the sex-determining gene, which causes testes to develop. Without this (i.e. in females), ovaries develop.

Sexual reproduction causes variation as the offspring has genes from both parents. Offspring are similar to parents because of genes inherited from them. Siblings differ because they inherit different combinations of genes.

Genetic diagrams – Punnett squares and family trees – are used to show genetic crosses.

Genetics and inheritance

Different forms of a gene are alleles. An individual:

- is homozygous when two alleles are the same for a characteristic
- is heterozygous when two alleles are different for a characteristic.

Alleles can be dominant or recessive. With one or two dominant alleles, the individual shows the characteristic. The recessive characteristic is seen only in individuals with two recessive alleles.

The genotype of an organism is its genetic make-up. The phenotype describes the organism's features.

Body cells have pairs of chromosomes; sex cells have one chromosome from each pair.

Chromosomes have the same type of genes in the same place on each chromosome of the pair.

Single gene disorders are caused by faulty alleles of a gene.

The faulty gene can be dominant or recessive.

For recessive single gene disorders, a person with a single recessive gene will not have the disorder but will be a carrier.

A Punnett square or family tree shows the risk of inheriting a disorder/being a carrier.

Implications of testing for genetic diseases:

- tests carry risk of miscarriage
- results not 100% reliable
- decision as to whether to have children or abort a pregnancy
- should an employer, the family or insurance company be told?

Genetic diseases

Genetic testing is carried out when genetic disease runs in the family.

Genetic screening is carried out on a large scale, e.g. in newborns, where there is no history of a disease.

Embryos produced by IVF can be screened before implantation (pre-implantation genetic diagnosis – PGD).

PGD and embryo research are carefully monitored in the UK.

Natural clones are individuals with identical genes, so any differences between individuals must be due to environmental factors.

Some organisms reproduce asexually to produce clones.

Identical twins are formed when cells of an embryo separate.

Clones

Artificial clones can be produced when a nucleus from a body cell is transferred to an unfertilised egg.

Embryonic stem cells can develop into any type of specialised cell. Adult stem cells can develop into fewer cell types.

Being unspecialised, stem cells have potential in the treatment of disease.

Microbes and disease

How microorganisms cause disease

- **Microorganisms** that cause disease and make us feel ill are called **pathogens**. Pathogens include bacteria and viruses.

- When microorganisms get into the body, they reproduce quickly and cause symptoms of disease.

	Diseases caused by bacteria	Diseases caused by viruses
Symptoms caused by:	release of poisons or **toxins** by the bacteria	damage to the cells as the viruses reproduce
Examples:	bacterial meningitis, tetanus, salmonella food poisoning, tuberculosis (TB)	influenza (flu), the common cold, measles and chickenpox

- Bacteria reproduce by dividing into two, which is a type of asexual reproduction called **binary fission**.

- Bacteria reproduce rapidly (this is called exponential growth) in the ideal conditions of the human body.

- Viruses need a 'host' cell to reproduce. They enter the host cell and 'hijack' the cell's mechanisms for making DNA and proteins, and make copies of themselves.

- The copies of the virus are released in very large numbers from the infected cell and go on to infect other cells and/or other people.

Remember!

Viruses always need a cell to live in. They can't live outside the body for very long.

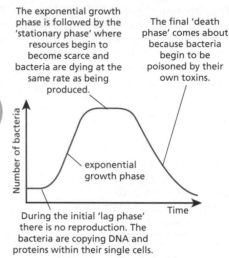

The exponential growth phase is followed by the 'stationary phase' where resources begin to become scarce and bacteria are dying at the same rate as being produced.

The final 'death phase' comes about because bacteria begin to be poisoned by their own toxins.

exponential growth phase

During the initial 'lag phase' there is no reproduction. The bacteria are copying DNA and proteins within their single cells.

The growth of bacteria in culture

EXAM TIP

It is likely that in the exam you will be asked to calculate the number of bacteria produced in a given time. Practise answering this type of question.

Our defence system

- The body's *external defences* include skin, saliva and tears, and acid in the stomach.

- The body's *internal defence* is our **immune system**.

- The immune system uses **white blood cells** to defend the body. White blood cells:
  - are made in the bone marrow
  - are found as several types: some engulf and digest invading microorganisms while others produce **antibodies**, which recognise and destroy invading microorganisms.

- Pathogens have chemicals on their surface that antibodies recognise as being foreign. These are called **antigens**.

- Different white blood cells carry different types of antibody on their surface. An antibody is specific to one antigen.

- The antibody corresponding to an antigen locks onto it. This sets off a series of events:
  1 The white blood cell divides to produce many copies.
  2 Each white blood cell produces many antibody molecules that lock on to the invading cells.

- Different types of antibody:
  - destroy the invading microorganism
  - enable white blood cells to recognise the microorganism as being foreign
  - cause the microorganism to clump together, meaning it is easier to engulf white blood cells.

- After the infection clears up, **memory cells** remain in the bloodstream. These can produce large numbers of antibodies very quickly if the microorganism enters the body again.

- At this point the person is said to be **immune** to that particular pathogen.

Improve your grade

Our defence system

Higher: A bacterium enters your blood stream. Describe the series of events leading to the bacterium being destroyed by your immune system. *AO1* [5 marks]

Vaccination

Vaccination programmes

- White blood cells make antibodies against chemicals on the surface of pathogens. These chemicals are called antigens.
- Memory cells left in the body can produce antibodies very quickly if they meet the living disease-causing microorganism.
- A **vaccine** contains a safe form of the microorganism that causes a disease, so that you don't become ill after receiving it.
- In the bloodstream, the immune system attacks the microorganism in the vaccine.

- Vaccination programmes protect children against diseases that are preventable.
- Babies and children undergo a course of **vaccinations** in their first year or so.
- Some pathogens do not change over time, so the same vaccines can be used against these, year after year.
- Other pathogens, such as the influenza (flu) virus, change rapidly, so new vaccines must be developed.

- An epidemic occurs if a disease spreads rapidly through a population, for example in a city or country.
- To avoid an epidemic, it is necessary to vaccinate a high percentage of the population – this leads to what is called 'herd immunity'.
- Widespread vaccination has eradicated one disease – smallpox – from the world. It has also reduced childhood diseases such as measles, mumps and rubella.
- The longer-term aim is to eradicate certain diseases altogether.

Making vaccination safe

- Scientists test new vaccines very carefully as they are developed to check for any **side effects**.
- Side effects can be more severe in some people than in others, because of genetic variation.
- No type of medical treatment can ever be *completely* risk-free (but people often think the risk is higher than it is).

- Vaccinations are *extremely* safe, and millions of people have benefited from them.
- *Occasionally*, a child may develop a minor adverse reaction (such as a rash or fever). In *very rare cases*, the reaction can be more serious. Any adverse reactions are recorded and followed up.
- Any *risks* of vaccination must be considered against the *benefits*. Any vaccine generating an unusual number of adverse reactions will be quickly withdrawn.

Other ways of reducing infection

- **Antimicrobials** are a group of substances that are used to kill microorganisms or inhibit (slow) their growth. They are effective against bacteria, viruses and fungi.
- **Antibiotics** are a type of antimicrobial which are effective against bacteria but not against viruses. They allow doctors to treat illnesses caused by bacteria, such as tuberculosis.
- Studies show that some microorganisms are developing **resistance** to these antimicrobials because of their very wide use.
  - Resistance to antimicrobials means that some strains of bacterial infections are now difficult to control.
  - Antimicrobial-resistant microbes can be a particular problem where antimicrobials are used frequently, such as in hospitals.

Improve your grade

Vaccination programmes

Higher: The graph shows the number of cases of measles, and deaths from measles, in England and Wales from 1940 to 2008.

Discuss what the data suggest about the effectiveness of the measles vaccines.

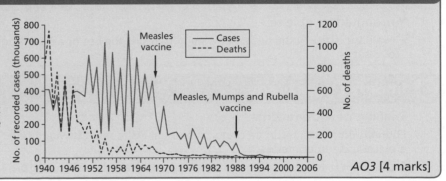

AO3 [4 marks]

Safe protection from disease

Antimicrobials

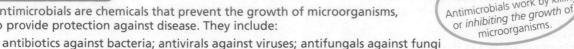
Remember!
Antimicrobials work by *killing* or *inhibiting the growth of* microorganisms.

- Antimicrobials are chemicals that prevent the growth of microorganisms, so provide protection against disease. They include:
  - antibiotics against bacteria; antivirals against viruses; antifungals against fungi
  - many cleaning products, along with antiseptics and disinfectants.

- Over a period of time, bacteria and fungi can develop resistance to antimicrobials. In a population of microorganisms, some may be resistant to the antimicrobial. These will survive the use of the antimicrobial and pass on their resistance. The resistance spreads through the population of microorganisms.

- Resistant microorganisms are sometimes called '**superbugs**'.

- The overuse of certain antibiotics has led to some microorganisms becoming resistant to them. For this reason a course of antibiotics should be:
  - prescribed *only* for more serious infections, when they are really needed
  - completed, so that the bacteria causing the infection are killed completely.

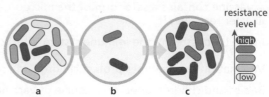

How the use of antimicrobials can lead to a more resistant population of microbes:
a initial population of microorganisms
b directly after treatment with antimicrobial
c final population of microorganisms

- Random changes in the genes – called **mutations** – give some bacteria resistance to antimicrobials.

- Because of the rapid reproductive rate of bacteria, antimicrobial-resistant genes spread through the population.

- Antibiotic resistance has led to some strains of bacteria that are very difficult to eradicate. These include MRSA, which is a problem in many hospitals.

(Higher)

Trialling new treatments

- New medicines, vaccines and other treatments are tested very carefully before being made available to the general public.

- Early stages of testing involve human cells grown in the laboratory, and animals.

- If the drug seems to be effective and safe, it is tested on humans in **clinical trials**. These are carried out on healthy volunteers (to check for safety) and on people with the illness (to test for safety and effectiveness).

- When drug trials are carried out, one group of people is given the new drug being trialled. The results are compared with those of a **control group**.

- One type of control group receives the existing treatment; another type of control group receives a **placebo** – a tablet or liquid made to look like the drug, but without the active ingredient.

- One ethical issue related to a drug trial is that it must not disadvantage the patient. If evidence from the trial suggests the new drug is effective, it is offered straightaway to patients receiving the placebo.

- In an 'open-label' trial, both researchers and patients know which drug the patient is receiving.

- In a 'blind' study, the patient doesn't know which drug they are receiving but the researcher does.

- In a 'double-blind' study, neither patient nor researcher knows which drug is being given.

- Some trials investigate the effects of the drug over a long period of time. This is because:
  - side effects may appear, or increase, over time
  - the drug may become less effective.

(Higher)

Ideas about science

You should be able to:
- evaluate critically the design of a study to test if the use of a new drug is effective in the treatment of disease.

Improve your grade

Trialling new treatments
Foundation: When testing a chemical on humans to see if it would be suitable as a new antibiotic, a placebo is sometimes used.
Explain why placebos are important, and how ethical issues with using placebos are overcome.   *AO1* [4 marks]

The heart

The heart and circulatory system

- The heart, blood vessels and blood make up the **circulatory system**.

- The blood carries nutrients and oxygen to the body's cells, and removes and carries waste products from the cells.

- The blood is pumped around the body in blood vessels by the heart.

- The heart is a 'double pump' – as one half is pumping oxygenated blood from the lungs to the body, the other half is pumping deoxygenated blood from the body to the lungs.

- At the lungs, deoxygenated blood absorbs oxygen and gets rid of carbon dioxide.

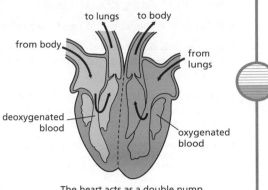

The heart acts as a double pump.

- The three types of blood vessels are **arteries**, **capillaries** and **veins**.

Blood vessel	Transport of blood	Structure
Arteries	*Away from the heart* under high pressure	Walls are very thick, elastic and muscular to withstand the pressure.
Capillaries	Link arteries and veins	Walls are one cell thick to allow the transfer of substances to and from cells.
Veins	Collect blood and *return it to the heart*	Walls contain elastic, muscular tissue, but are thinner than those of arteries. The blood is under low pressure and veins have valves to prevent the backflow of blood.

- Heart muscle has its own blood supply. The **coronary arteries** run over the surface of the heart. They provide the heart with the nutrients and oxygen it needs to contract, and remove waste products.

- The coronary arteries can become blocked by fatty deposits. This will prevent the heart from receiving the oxygen it needs, leading to a heart attack.

Reducing the risk of heart disease

- **Coronary heart disease (CHD)** is caused by the build-up of fatty substances in the arteries.

- The main lifestyle factors that increase a person's risk of CHD are: **1** Smoking cigarettes; **2** Poor diet (a diet high in **saturated fat** and salt); **3** Misuse of drugs (this includes excessive consumption of alcohol); **4** Stress.

- Data show there is a link, or **correlation**, between these lifestyle factors and heart disease.

- Regular exercise helps to prevent against CHD by strengthening heart muscle, providing a healthy body weight and reducing stress.

- A healthy diet, low in saturated fat, lowers blood **cholesterol** and reduces the risk of heart disease.

- Genetic factors contribute to a person's chances of having CHD – members of families that have a history of heart disease need to lower the risk factors.

- Researchers study the occurrence of heart disease (epidemiological studies) and genetic factors.

- Rates of CHD are higher in industrialised countries, such as the UK and USA, than less-industrialised nations, for example India and China.

Ideas about science

You should be able to:

- use the ideas of correlation and cause when discussing data on a lifestyle factor, e.g. obesity, and heart disease

- explain why an observed correlation between heart disease and a lifestyle factor, such as poor diet, does not necessarily mean that the factor causes the outcome.

Improve your grade

The heart and circulatory system

Foundation: The heart needs its own blood supply to live. Describe how the heart receives this and what happens in a person with coronary heart disease.   *AO1* [5 marks]

Cardiovascular fitness

Heart rate and blood pressure

- **Heart rate** is measured by recording **pulse rate**. This is the number of pulses as blood passes through an artery close to the skin.

- Pulse rate is measured in beats per minute (bpm).

- The **resting heart rate** is the heart rate when a person is relaxed. A resting heart rate of 70–100 bpm is normal for teenagers. A rate of 50–70 bpm in an adult is an indication of a good level of fitness.

- The misuse of drugs, such as nicotine, alcohol and Ecstasy (MDMA), has a negative effect on health, including the heart rate, and increases the risk of heart disease and heart attack.

- **Blood pressure** measurements record the pressure of blood on the walls of an artery.

- People with consistently high blood pressure have an increased risk of heart disease.

- High blood pressure damages the walls of the arteries and makes them more likely to develop fatty deposits and get narrower. It also puts a strain on the heart.

- Blood needs to be under pressure to reach every cell in the body.

- Blood pressure is an important **indicator** of health.
  - **High blood pressure** increases the chance of strokes and heart attack.
  - **Low blood pressure** can cause dizziness and fainting.

- Blood pressure is measured as millimetres of mercury (mm Hg) and given as two numbers, for example 110/80. The higher value is when the heart is contracting; the lower value is when the heart is relaxed.

Remember!
There is no such thing as a single 'normal' value for heart and pulse rates and blood pressure. Everyone is different and values given are averages or ranges across the population.

Epidemiological studies

- Studies of the occurrence of disease using large numbers of individuals are called **epidemiological studies**.

- Epidemiological studies have been carried out on the link between lifestyle factors and heart disease. Studies are carried out:
  - on samples of individuals who are matched on as many factors as possible and differ only in the factor being investigated, e.g. smokers and non-smokers; drug users and non-drug users
  - on individuals chosen at random
  - that investigate whether the genes carried by individuals affect their risk of suffering from particular health problems.

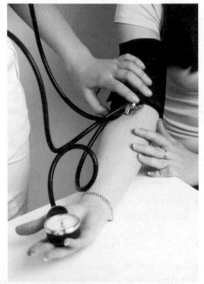

The blood pressure measurement is a standard check that doctors carry out to check the general health of a patient.

Ideas about science

You should be able to:

- discuss whether given data suggests that a given factor, such as smoking, does or does not increase the chance of a given outcome, such as heart disease

- evaluate critically the design of a study to test if a given factor increases the chance of heart disease, by commenting on the design and results of the test.

Improve your grade

Heart rate and blood pressure
Foundation: Explain the term 'blood pressure' and describe how blood pressure affects health.

AO1 [2 marks]

Keeping things constant

A constant internal environment

- The maintenance of a constant environment is called **homeostasis**.

- Temperature, pH and levels of sugar, water and salt must be kept at values within a very narrow range for the body to function.

- Homeostasis involves communication by the nervous and hormonal systems.

- Response to change is by these 'automatic' control systems throughout the body.

- The systems involved in homeostasis are in three parts:
  1 **Receptors** detect change in the environment.
  2 **Processing centres** receive information and determine how the body will respond.
  3 **Effectors** produce a response.

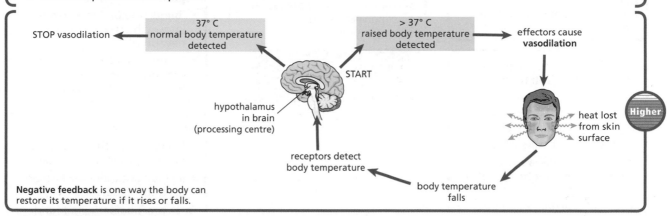

STOP vasodilation ← **37° C** normal body temperature detected ← **> 37° C** raised body temperature detected → effectors cause **vasodilation**

START

hypothalamus in brain (processing centre)

receptors detect body temperature

body temperature falls

heat lost from skin surface

Negative feedback is one way the body can restore its temperature if it rises or falls.

Water balance

- Water is *taken in* by drinking and eating and it's also *produced* by respiration.

- Water is lost in urine and faeces and when we sweat and breathe out.

- The water content of our bodies must be maintained to keep the body's cells bathed in **blood plasma**. So, the concentration of water in our cells must be kept constant.

- If the blood plasma is too concentrated, the cells will lose water.

- If the blood plasma is too dilute, the cells will absorb water and burst.

The kidneys

- Concentration of blood plasma is affected by external temperature, exercise level and intake of fluids and salt.

- The **kidneys** respond to changes in the blood plasma by changing the concentration of urine that is **excreted** from the body.

Remember!
You do not have to know how the kidneys work in detail. You just have to know that they play a vital role in balancing water, salt and other substances in the blood plasma.

- Some recreational drugs affect the water balance of the body:
  - Alcohol causes the kidneys to produce a large volume of dilute urine and the body becomes dehydrated.
  - Ecstasy (MDMA) causes the kidneys to produce very small volumes of concentrated urine. The body's cells will swell with water.

- The kidneys help to balance levels of water, **urea**, salts and other chemicals in the blood.

- **Anti-diuretic hormone (ADH)** is released by the pituitary gland in response to changes in the concentration of blood plasma. The secretion of ADH is controlled by negative feedback.

- ADH acts upon the kidneys to reduce the amount of water lost in the urine.

- Alcohol suppresses the release of ADH, so less water is reabsorbed by the kidneys.

- Ecstasy increases ADH production, so more water is reabsorbed by the kidneys.

EXAM TIP

Practise showing how ADH release is controlled by changes in the concentration of the blood plasma by drawing a flowchart.

Improve your grade

The kidneys

Higher: Explain the effect of Ecstasy on the water balance of a person's body.

AO1 [3 marks]

B2 Summary

Protection against infection

Organisms that cause infectious diseases include bacteria and viruses.

Damage to cells during an infection is because of toxins produced by the microorganisms.

Microorganisms reproduce very rapidly in the human body, to produce very large numbers.

Vaccines:
- contain a safe form of a disease-causing microorganism
- produce immunity (because memory cells remain after the vaccination)
- are very safe, but not risk-free (possible side effects; reactions vary because of genetic differences).

To prevent epidemics, a high percentage of the population must be vaccinated.

Drug and vaccine trials ensure safety and effectiveness. The trials first involve animals and human cells. Later human trials involve healthy volunteers and people with the illness. Control groups use an existing drug or a placebo. The use of placebos raises ethical issues.

Trials can be:
- open-label (new drug is known by researchers and patient)
- 'blind' (the patients do not know who is receiving the new drug)
- 'double-blind' (neither patient nor researcher knows who is being given the new drug).

The immune system has types of white blood cell that:
- destroy microorganisms by engulfing and digesting them
- produce antibodies against antigens on their surface.

Every antigen has a corresponding antibody that recognises it.

After an infection, memory cells remain in the body so that antibodies can be produced very quickly. The person now has immunity to the microorganism.

Antimicrobials:
- kill or inhibit the growth of bacteria, fungi and viruses
- include antibiotics, which are used to kill bacteria (only).

Over time, bacteria and fungi can develop resistance to antimicrobials.

To reduce this, only use antibiotics when necessary and complete the course.

Heart disease

The heart is part of the circulatory system. It is a double pump (left side to body, right side to lungs) and has its own blood supply.

The structure of the arteries, capillaries and veins is related to their functions.

Heart attacks are caused by fatty deposits blocking the blood supply to the heart.

Heart disease is caused by lifestyle factors (poor diet, stress, smoking, misuse of drugs), and/or genetic factors.

These factors are identified by large scale epidemiological and genetic studies.

Heart rate can be recorded by measuring the pulse rate.

Blood pressure is a measure of the pressure of the blood on an artery wall. It is measured as two numbers: the higher number is when the heart is contracting, the lower number when it is relaxed.

Values of 'normal' heart rate and blood pressure are given as a range, as they vary in individuals.

High blood pressure increases the risk of heart disease.

Water balance

Nervous and hormonal systems maintain a constant internal environment in the body (homeostasis). These control systems:
- are 'automatic'
- have receptors, processing centres and effectors.

Negative feedback is used to reverse any changes in the body's state.

Hormonal control of urine concentration is by ADH.

Alcohol reduces ADH secretion.

Ecstasy (MDMA) increases ADH secretion.

A balanced body water level maintains cell water concentration. This is vital for cell function.

Input of water is from drinks, food and respiration; losses are through sweating, breathing, faeces and urine.

The kidneys respond to water concentration in the plasma by producing dilute or concentrated urine.

Species adaptation, changes, chains of life

Species and adaptation

- A **species** is a group of organisms that can breed together to produce *fertile* offspring.

- Species are **adapted** to living in their environment. For example, a cactus is adapted to living in hot dry conditions by storing water in its stem; a camel stores fat in its hump.

- Adaptation of organisms to their environment means that they are able to *survive* to *reproduce* successfully.

- The organisms that live in a **habitat** are dependent on their environment and other species living there. They depend on other species for food and **compete** with each other for resources.

- Animals compete with each other for food, a mate, living space and territories. Plants compete for light, nutrients, water and space.

- The feeding relationships of organisms are shown in a **food web**.

- The feeding relationships of organisms in a habitat are often complex. They depend on each other, often in ways other than just providing food. This is called **interdependence**.

- Because of their interdependence, any change that affects one species in a food web is likely to affect all species in that food web.

Extinction

- A species can become **extinct** if it is unable to adapt rapidly to a change in the environment, e.g. climate change.

- Removal of habitats due to human activity threatens species, e.g. the Siberian tiger and mountain gorilla.

- The introduction of a new species can lead to extinction if the species is a competitor, predator or causes disease.

- The **extinction** of a species in a habitat will affect other organisms in the food web and may cause them to also become extinct.

Energy transfer

- Nearly all organisms on Earth are dependent on energy from the Sun.

- Plants absorb a small percentage of the energy from sunlight to produce their own food by **photosynthesis**. Plants store this energy in chemicals that make up the plants' cells and tissues.

- Other organisms get their energy by eating plants. Almost every food chain begins with a plant absorbing energy from the Sun.

- Energy is transferred from one organism to the next along a food chain.

- Only a small percentage of the energy transferred remains in each organism's body. In the transfer from a plant to an animal, some energy is lost because:
  - Some parts of the plant aren't eaten or can't be digested by the animal.
  - The animal uses some of the plant's energy for respiration. During respiration, some energy is lost as heat.
  - The waste products of the animal, for instance urine, contain some energy.

- Energy is *lost at each level of a food chain*. So the length of food chains is limited – they are rarely longer than four or five organisms.

- You can calculate the **efficiency** of energy transfer at any level using the equation:

$$\text{Percentage efficiency} = \frac{\text{energy in tissues}}{\text{energy in food eaten}} \times 100$$

- Energy transfer continues after an organism has died. Microorganisms such as bacteria and fungi feed on dead or decaying organisms. These are called **decomposers**.

- Partly decayed material is called detritus.

- **Detritivores**, such as earthworms and woodlice, feed on detritus and break it down further.

Improve your grade

Extinction

Foundation: The harlequin ladybird originated in Asia and arrived in Britain in 2004. It eats the same food as our native ladybird (aphids – greenfly and blackfly). It also **predates** on many insects (other ladybirds, and the eggs of butterflies and moths) and eats fruit.

Suggest why the spread of the harlequin ladybird might affect food webs. *AO2* [4 marks]

Nutrient cycles, environmental indicators

The carbon cycle

- **Carbon** is the key element of the chemicals that make up all living things. It is continually recycled through the **carbon cycle**.

- Carbon *enters* the carbon cycle as carbon dioxide from the air. Plants *fix* this carbon, so that it can be used and stored by organisms, by photosynthesis.

- Carbon is *returned* to the air in the following ways:
  - as a product of **respiration**, when plants and animals release energy from food
  - through the **decomposition** of dead organisms by soil microorganisms such as bacteria and fungi
  - by **combustion** of organic materials.

The nitrogen cycle

- Nitrogen is an essential component of living things. It is recycled through the **nitrogen cycle**.

- Plants take up nitrogen from the soil through their roots, in the form of nitrogen compounds including **nitrates**. These are converted into **proteins**.

- Protein is an important nutrient in animals' diets. It passes along food chains as animals eat plants and other animals.

- Nitrates are released back into the soil as animals excrete waste, and as plants and animals die and are decomposed by microorganisms.

> ### EXAM TIP
> Make sure that you can identify the stages of nitrogen fixation and denitrification in a diagram of the nitrogen cycle.

- Nitrogen *enters* the nitrogen cycle in two ways:
  1. Nitrogen molecules in the air are split by lightning. Nitrogen atoms then combine with oxygen in the air to form nitrates, which are washed into the soil by rain.
  2. **Nitrogen-fixing bacteria**, found in the soil and in the roots of leguminous plants such as beans and peas, convert nitrogen in the air into nitrates.

- Nitrogen *leaves* the nitrogen cycle when **denitrifying bacteria** convert nitrates in the soil into nitrogen gas. This process is called denitrification.

Higher

Indicators of environmental change

- Environmental change can be measured using:
  - **non-living indicators**, e.g. carbon dioxide levels, temperature and nitrate levels
  - **living indicators**, e.g. **phytoplankton** (microscopic aquatic plant-like organisms), **lichens** (dual organisms made up of a fungus and alga living together) and aquatic organisms such as mayfly larvae.

- Scientists monitor environmental change from a local through to a global scale.

- Observations of some living indicators can give us very precise information about levels of pollution and environmental change. This is called a biotic index. For example, mayfly larvae need high levels of oxygen in the water, so will indicate very low levels of pollution.

- Interpretation of data from non-living indicators and living indicators helps scientists to monitor environmental change and trends over a period of time.

- Measurements using non-living indicators, e.g. CO_2 in the air and water, are monitored continuously. We can also look at historical levels from CO_2 trapped in ice.

Improve your grade

Indicators of environmental change

Higher: Lichens are sensitive to sulfur dioxide in the air. The distribution of three different types of lichen was measured in a city centre, and at different distances from it. The results are shown opposite.

  a Suggest what the graph tells you about the resistance of the lichens to pollution.

  AO3 [1 mark]

  b Why did the scientists measure the frequency of the lichens in 50 quadrats?

  AO2 [2 marks]

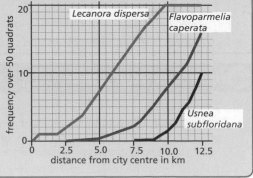

Variation and selection

Life on Earth

- Life on Earth began around 3500 million years ago. The first forms of life were very simple.
- Over millions of years these simple life forms gave rise to all the different species of organisms we see today, along with ones that are now extinct. This process of change is called **evolution**.
- The changes involved in evolution begin with variation between individuals. Variation has genetic and environmental causes.
- Evidence of how organisms changed over time is found in **fossils**. Fossils are the remains of organisms, or other traces of their lives such as footprints or eggs, that have turned into rock.
- Scientists can date fossils from the layer of rock they are found in.

Variation and evolution

- A **mutation** is a change in the genetic information in a cell. A mutation will result in a change in the characteristics of an organism.
- Mutations can occur as DNA is copied during the production of new cells.
- If a mutation occurs as sex cells are produced, the mutation is passed to the offspring.
- Most mutations are harmful, but sometimes new, useful characteristics are produced. Useful mutations will be passed on throughout the population.

- Evolution involves the development of new species. The theory of evolution provides the best explanation for the enormous number and variety of organisms on the planet today.
- In a population of organisms, the alleles of genes that occur in that population – the **gene pool** – will change because of mutations and other causes of genetic variation.
- It may be an advantage to have certain genes, rather than others, so that some become more common while others disappear. Over thousands or millions of years, with changes in the frequency of different genes, new species emerge.
- The theory of evolution is based on data and observations, both of organisms alive today and on the **fossil record**.

Variation and natural selection

- Because of genetic variation, some individuals will have characteristics that give them a better chance of survival than others.
- Individuals with advantageous genes will survive to reproduce and pass these to their offspring. This is **natural selection** – nature is selecting the most advantageous genes to be passed on.

- Selection does not just occur naturally – humans have produced crop plants, livestock and pets for thousands of years by a form of selection called **selective breeding**. This involves:
  - choosing the individuals with the characteristics that are closest to those required
  - breeding these (and preventing other individuals from breeding)
  - repeating the process over several generations.

- Natural selection is an important part of the evolutionary process. It results in an organism that is better able to survive in terms of:
  - reproduction, which will lead to an increase in the number of individuals displaying the characteristics in later generations.
  - competition with other animals, e.g. catching food, escaping predators, resistance to disease

> **Remember!**
> In selective breeding, humans are choosing the desirable characteristics. In natural selection it's nature that determines those individuals most able to survive.

Ideas about science

You should be able to:

- recognise data or observations (e.g. on the fossil record or the organisms on a volcanic island) that are accounted for by an explanation (such as evolution or natural selection)
- give good reasons for accepting or rejecting a proposed scientific explanation, such as evolution by natural selection.

Improve your grade

Variation and natural selection

Higher: Scientists have observed that the number of cases in which disease-causing bacteria are resistant to antibiotics has increased over the last 30 years. Explain why some people think that this is evidence of natural selection.
AO2 [4 marks]

Natural selection and evolution

- Over a long period of time, advantageous genes chosen by natural selection are likely to become the norm in the population. This is how evolutionary change takes place.

- However, natural selection can be seen in operation over *short* periods of time. One example is the development of bacteria that have become resistant to antibiotics.

- A number of factors influence the rate at which evolution takes place:
  - When the environment changes, only those organisms that are best adapted, or can re-adapt, will survive.
  - If organisms become **isolated**, for instance on an island, natural selection will act independently on the different populations. Over time the populations will become distinct and no longer able to reproduce with each other. They will be new species.

> **Remember!**
> It is natural selection, together with isolation or a changing environment, which provides the driving force for evolution

- In the search for evidence of evolution, scientists have investigated the relationships between organisms by:
  - examining the fossil record
  - observing similarities and differences in physical features, e.g. skeletons, flowers
  - analysing DNA sequences (more closely related organisms have more DNA sequences in common).

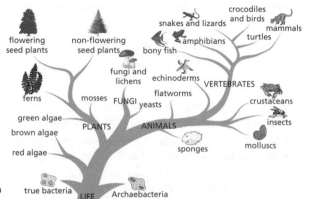

Relationships between organisms can be shown in a 'tree of life' that suggests where branches occurred.

Evidence for evolution

- Organisms are **classified** by putting them into groups. This helps us to understand the enormous diversity of organisms on Earth and how they evolved.

- Classification was based on grouping organisms with similar features. Today, classification is also based on analysing DNA sequences.

- A common ancestor is the most recent organism from which the species in a group descended.

- There is evidence to support the theory of evolution:
  - The simplest organisms are found in the earliest rocks. More complex ones appear in younger rocks.
  - More recent fossils have features that look like adaptations or developments of those of older organisms.
  - DNA analysis of today's organisms has confirmed predictions made from the fossil record, including when branches in the tree of life occurred.

- The theory of evolution was proposed by Charles Darwin and Alfred Russel Wallace in 1859.

- The theory was based on Darwin's observations on the Galapagos Islands. The islands are volcanic, so organisms on the islands must have arrived, at some point, from the mainland.

- Darwin observed that organisms, such as mockingbirds, were similar but had slight differences to those on the mainland. The mockingbirds were also different from one island to the next.

- Darwin's idea that species were not fixed, but could change over time, was the result of his observations and his creative thinking.

- By contrast, there was no evidence to support Larmarck's theory. Larmarck suggested that animals acquired characteristics during their lifetime that were then passed on to their offspring.

Improve your grade

Evidence for evolution

Higher: The following are sequences of very short sections of DNA from four species of primate – a human, Neanderthal man, a chimpanzee and a gorilla.

Human   CTGGGCGCGTGCGGTTGTCCTGGTCCTGCT   **Neanderthal** CCGGGCGCGAGCGGTTGTCCTGGTCCTGCA

Chimpanzee CCGGGCGCGTGCGGTTCACCAGGTCCTGCA   **Gorilla**   CAGGGCGCGGGGAGGTTTACCACATGCTTCA

Look at the sequences and suggest what the evolutionary relationships of the animals are, based on the data. Explain your reasoning. Explain the level of confidence you have in your conclusion, and suggest how this could be increased.   *AO2, AO3* [5 marks]

Biodiversity and sustainability

Conserving biodiversity

- **Biodiversity** is the variety of life on Earth and in different habitats, including:
  - the number of different species
  - the range of different types of organisms, for example plants, animals, microorganisms
  - the **genetic diversity** (variation) within each species.
- Habitats such as tropical rainforests have a very high biodiversity (they have many species).
- Many of these species could be valuable to us as food crops or medicines.

- Species are now becoming extinct more rapidly than at any other time, except for **mass extinction events** seen in the fossil record. It is thought that this is connected with human activity, as organisms are hunted and their habitats are destroyed. Climate change will accelerate this rate of extinction.

- To record and monitor species accurately, they need to be classified.
- Organisms are classified in the following way:
  - A kingdom is a large group with many organisms but fewer characteristics in common.
  - Moving down from the kingdom, the groups get smaller and have fewer organisms with more characteristics in common.
  - The level identifying the individual type of organism is the species.

> ### EXAM TIP
> You do not need to know the names of the groups between a kingdom and a species, but in the exam you may be asked to sort organisms into different groups based on the number of features they have in common.

Sustainability

- **Sustainability** is about meeting today's needs without stopping future generations from meeting theirs, e.g. farming the land in a way that enables future generations to also farm it.
- Sustainability means: limiting our impact on wildlife, habitats and the environment; actively supporting ecosystems and populations of living organisms.

- To ensure sustainability, we need to maintain biodiversity. The loss of a single species removes a food supply and can have a big impact on the whole ecosystem.

- **Intensive monoculture crop production** maximises crop yields but it is not **sustainable**. It reduces the biodiversity of the field by: growing just one crop species; removing hedgerows to create huge fields for planting; spraying crops with herbicides and pesticides.

Improving sustainability

- Disposable products, e.g. nappies, create large amounts of waste that are slow to decompose.

- We can improve sustainability in product manufacture, for example by:
  - using as little energy as possible and minimal packaging
  - using locally available materials and limiting transport of the product
  - creating as little pollution as possible.
- The **Life Cycle Assessment** tracks the environmental impact of a product from: sourcing of raw materials → manufacture → transport → use → disposal.

- **Biodegradable** packaging has become popular for some products such as plastic carrier bags.
- It is best to reduce all types of packaging, as even biodegradable materials break down very slowly in landfill sites, produce carbon dioxide, and require energy to produce and transport.

Ideas about science

You should be able to:
- explain the idea of sustainability, and apply it to specific situations
- use Life Cycle Assessment data to compare the sustainability of alternative products or processes.

Improve your grade

Conserving biodiversity

Foundation: When investigating species extinctions, explain why we can use real data of human populations but need a computer model for numbers of extinctions. *AO2* [4 marks]

B3 Summary

Systems in balance

All organisms are dependent on energy from the Sun.

Plants absorb this energy. A small percentage is converted to chemical energy by photosynthesis, and stored in the chemicals that make up plant cells.

Energy transfer between organisms:

- occurs when they're eaten
- occurs after death, when their bodies are fed on by decomposers and detritivores
- involves only a small proportion of the energy – the majority is lost as heat, waste products and uneaten parts.

The length of food chains is therefore limited.

Environmental change can be monitored using non-living indicators and living indicators.

Nitrogen is recycled in the nitrogen cycle:

- Certain microorganisms convert nitrogen in the air into nitrates (nitrogen fixation).
- Nitrates are converted by plants to protein, and transferred to animals that eat them.
- Nitrogen compounds are returned to the soil by excretion, death and decay of organisms.
- Nitrogen is returned to the air by microorganisms (denitrification).

A species is a group of organisms that can breed together to produce fertile offspring.

Adaptation of a species to its environment increases its chances of survival.

Organisms in a habitat are dependent on the environment and other species for survival, and compete for resources.

In a food web, organisms show interdependence – changes affecting one species will impact on others.

A species can become extinct if the environment changes and it cannot adapt.

The carbon cycle:

- recycles carbon through the environment
- involves the processes of combustion, respiration, photosynthesis and decomposition.

Microorganisms are involved in respiration and decomposition.

Evolution

All individuals of a species show variation.

Variation that is genetic is passed on to offspring.

Variation can arise from accidental changes in genes (mutations).

Mutations in sex cells can be passed to offspring.

Life began on Earth 3500 million years ago.

All living things on Earth evolved from simple living things.

Natural selection results from genetic variation and competition between organisms for reproduction and survival.

Organisms that are better able to survive than others will pass on their genes.

Natural selection is similar to selective breeding by humans.

A combination of mutations, natural selection, environmental change and isolation produces new species by evolution.

Darwin's theory of evolution was the product of many observations and creative thought. Alternative theories, such as Lamarck's, do not fit with our understanding of genetics.

Evidence for evolution comes from the fossil record and analysis of DNA sequences of organisms.

Biodiversity

Organisms are classified into groups using physical features and DNA.

- A kingdom is the largest group of organisms.
- The term species identifies the particular type of organism.

Classifying organisms:

- makes sense of the enormous range of organisms on Earth
- helps to show their evolutionary relationships.

Biodiversity refers to the variety among living things (the number of species; the range of types of organisms; the genetic variation in a species).

The rate of species extinction is increasing as a result of human activity.

Biodiversity is important for the future growth of crops and search for new medicines.

Sustainability is meeting the needs of the people today without damaging the Earth for future generations.

Biodiversity is essential for sustainability. In crop cultivation, large scale monoculture is not sustainable.

Improvements in sustainable production include the selection of packaging materials.

The changing air around us

The gases that make up the air

- Air (or **atmosphere**) is a **mixture** of **gases**:
  - it is mainly nitrogen, oxygen and argon
  - it contains small amounts of water vapour (H_2O) and **carbon dioxide** (CO_2).
- Clouds are water or ice and dust is a solid, so they are not parts of the air.

- Gases spread out to take up all the space available.

- Particles are very small, so in gases there is lots of empty space between gas **molecules**. This means gases can be squeezed into a smaller volume, like in a bike tyre.

- Dry air contains: 78% nitrogen (N_2); 21% oxygen (O_2). The remaining 1% is mainly argon (Ar) and very small amounts of other gases.

Most of a gas is empty space.

EXAM TIP
When drawing diagrams to show particles in air, remember that most of the available space will be empty.

Remember!
Two gases make up 99% of the air. All the rest fit into the last 1%.

- Oxygen gas reacts with most metals to make solid metal oxides.

- We can find the percentage of oxygen in air by passing air over heated copper.

The Earth's atmosphere

- Earth's atmosphere was probably formed about 4 billion years ago by gases given out by volcanoes.

- Volcanoes release huge amounts of carbon dioxide and water vapour. They also release **lava** and dust.

- Different processes have removed almost all of the carbon dioxide that was in the early atmosphere, leaving the air with the composition we have today:
  - Four billion years ago the Earth's atmosphere was very hot.
  - As the Earth cooled, oceans formed from the condensed water.
  - About 3 billion years ago simple **bacteria**-like creatures evolved to use **photosynthesis**.
  - This removed carbon dioxide from the air, and released oxygen, allowing animals to evolve.
  - Carbon dioxide was removed by plants and animals dying and becoming buried.
  - Over millions of years some of the buried material became **fossil fuels**.
  - Carbon dioxide **dissolved** in oceans reacts with salts to form insoluble calcium carbonate.
  - This forms **sediments** which become buried and cemented to form **sedimentary** rocks.

- Ideas about the composition of the Earth's atmosphere have changed over time:
  - Sixty years ago many scientists thought the early atmosphere was largely ammonia and **methane**.
  - Recent rock composition discoveries showed early ideas were not correct, and the early atmosphere was largely carbon dioxide.

Ideas about science

You should be able to:

- suggest reasons why collecting air quality data at the same location may give different values

- understand why taking several measurements of different air pollutants gives the best estimate of their true value.

Improve your grade

The Earth's atmosphere

Foundation: The early atmosphere was mainly composed of water and carbon dioxide.

Suggest how these were gradually removed.

AO1 [4 marks]

Humans, air quality and health

How has human activity changed air quality?

- Humans are changing the gases in the atmosphere by burning fuels. Fuels are used in factories, power stations, for transport and in homes.

- Gases we call **pollutants** are harmful to health. Examples are **carbon monoxide, nitrogen oxides** and **sulfur dioxide**.

- Pollutants are harmful to the environment and to the people and animals living there. For example, carbon monoxide reduces the amount of oxygen blood can carry.

- Air quality is 'good' if it has very few pollutants and 'poor' if there are lots of pollutants.

- Burning fuels releases carbon dioxide and solid **particulates** that float in the air, e.g. **carbon** (soot). Particulates are also released naturally as ash from volcanoes.

- In the last 50 years the amount of carbon dioxide in the air has increased by 25%.

- Carbon dioxide is linked to climate change, by acting as a 'greenhouse' trapping heat in the atmosphere.

- Human activity, like burning down forests to make more farmland, increases carbon dioxide and particulates.

- When air pollution levels are high, more deaths from asthma, heart disease and lung disease occur. There is a **correlation** between air quality and health.

- Air quality is more of a problem in large cities, such as Mexico City and Beijing. Some countries, e.g. the UK, have made laws to try to improve air quality.

- Sulfur dioxide and nitrogen dioxide are pollutants that make **acid rain**, which damages plants and animals.

- Sulfur dioxide and nitrogen oxides are asthma triggers for some people.

> ### EXAM TIP
> Make sure you know the names of the two gases responsible for acid rain. Do not get this mixed up with carbon dioxide, which is linked to global warming.

Measuring air quality

- Small amounts of carbon dioxide in the air are measured in parts per million (ppm). 1 ppm means that there is 1 gram of the pollutant substance in 1 million grams of air.

- The other pollutant gases – carbon monoxide, nitrogen oxides and sulfur dioxide – are measured in parts per billion (ppb).

- The amount of pollutant gases are measured in air quality monitoring stations throughout the UK. The data is transmitted automatically to a central computer for analysis.

> **Remember!**
> Some pollutants *directly* affect us, but some like acid rain harm the environment, so harm us *indirectly*.

Correlation and cause

- A correlation is a link between a factor and an outcome. Data is needed to show this.

- To establish a causal link, evidence needs to show that changing a particular factor is the only cause of a particular outcome. For example, when the levels of sulfur dioxide increase, statistics show that more people have asthma attacks. The data shows a correlation or link, but sulfur dioxide is not the only asthma trigger. Some asthmatics have attacks for completely different reasons.

Ideas about science

You should be able to:

- in a given context, identify the outcome, e.g. asthma cases, and the factors that may affect it, e.g. increase in air pollution

- in a given context, suggest how an outcome, e.g. car emissions, might alter when a factor is changed, e.g. stricter MOT testing

- identify, and suggest from everyday experience, examples of correlations between a factor and an outcome where the factor is (or is not) a plausible cause of the outcome.

Improve your grade

How has human activity changed air quality?

Foundation: How do humans affect air quality?   *AO1* [4 marks]

Burning fuels

What happens when fuels burn?

- Oxygen is needed for any fuel to burn and release energy.

- Fossil fuels such as petrol, diesel and fuel oil are mainly **hydrocarbons**. Hydrocarbons only contain carbon and hydrogen atoms.

- Coal is a fossil fuel mainly made of carbon atoms.

- When a hydrocarbon fuel burns: hydrocarbon fuel + oxygen ⟶ carbon dioxide + water (+ energy)
- **Oxidation** is when oxygen is added to a substance.
- **Reduction** is when oxygen is removed from a substance.
- **Combustion** (or burning) is an oxidation reaction.

EXAM TIP

If a question asks for a word equation, do not try to use symbols. Some words to use will probably be in the question.

- Gases in the atmosphere can be separated.

- Pure oxygen makes fuels burn more rapidly and at higher temperatures. One example is in an oxy-fuel welding torch, which can melt steel.

What happens to atoms in chemical reactions?

- **Atoms** do not change. In chemical reactions atoms get rearranged to make new substances.

- Atoms of non-metal elements join to form **molecules**.

- **Elements** are rearranged to make new **compounds** in chemical reactions.

- Atoms in **reactants** are rearranged into new **products** with different properties.

Remember!
The numbers and types of elements in the reactants are the same as in the products.

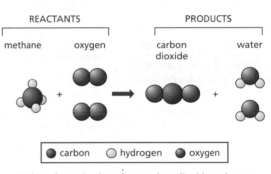

REACTANTS   PRODUCTS

methane   oxygen   carbon dioxide   water

● carbon   ○ hydrogen   ● oxygen

Methane burns in air to form carbon dioxide and water.

- Mass is conserved in a reaction because all the atoms in the reactants are just rearranged into the products. For example, all the carbon atoms in all the fossil fuels ever burnt are still present, but in different forms.

Burning sulfur

- Solid yellow sulfur burns and makes a colourless gas called sulfur dioxide.
- Sulfur is insoluble but sulfur dioxide dissolves in water make an acid solution.

- Fossils fuels contain small amounts of sulfur from the plants and animals that formed them.
- When fossil fuels burn the sulfur burns: sulfur + oxygen ⟶ sulfur dioxide ($S + O_2 \rightarrow SO_2$)
- Coal often contains the most sulfur, so burning coal can give off more sulfur dioxide than other fossil fuels.

- In the 1970s people noticed forests and aquatic life in ponds were dying.
- Scientists were able to use data to explain that sulfur dioxide caused acid rain.
- Acid rain lowers the pH when it falls on land or enclosed water, harming living things or eroding carbonate rock.
- Acid rain does not affect humans directly, so it is called an indirect pollutant.

Improve your grade

What happens when fuels burn?

Foundation: Methane (CH_4) reacts with oxygen to form carbon dioxide and water. Finish the diagram to show this reaction. Use ● to represent a carbon atom, ● to represent a hydrogen atom and ○ to represent an oxygen atom.

 methane + oxygen ⟶ carbon dioxide + water

AO2 [3 marks]

Pollution

How pollutants are formed

- Power stations and transport make most pollution because they burn most fuel. Electricity production and transport has increased over the last century.

Remember!
Fossil fuels all contain impurities, which when burnt can produce pollutants.

- Sulfur dioxide is made if the fuel contains sulfur.
- Carbon dioxide is always formed when fuels burns.
- If not enough air is available to burn the fuel:
  - poisonous carbon monoxide is made
  - bits of solid carbon (soot) called particulates are made, making surfaces they land on dirty.
- Car engines make nitrogen oxides when nitrogen and oxygen from the air react at high temperatures. This contributes to acid rain.
- Air quality measurements need repeating many times because results vary. On dry, hot, calm days, air pollutants can be trapped in cities.

EXAM TIP
Make sure you know the different between *complete* and *incomplete* combustion, and know the products formed.

- Carbon monoxide and particulate carbon are formed during incomplete combustion (insufficient oxygen).
- Only in the last 50 years have scientists discovered how different air pollutants form, and how they react with air to produce **smog**, acid rain and climate change.

- Nitrogen monoxide (NO) is formed in furnaces and engines at a temperature of about 1000°C.
- When nitrogen monoxide is released into the atmosphere it cools. It then reacts with more oxygen to form toxic nitrogen dioxide, a brown gas.
- Both NO and NO_2 pollutants can be in the air, so NO_x is used to represent both of them.
- NO_x damages buildings, contributes to acid rain, and can affect health.

Complex reactions

- Chemical formulae show how many atoms are joined together.
- You need to know the formula and be able to draw visual representations of the compounds shown opposite.

The formula of a compound shows the number of atoms of each element that are joined together in a molecule. The molecule diagrams show the arrangement of the atoms in the molecule.

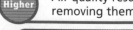

CO_2 CO H_2O

SO_2 NO NO_2

● carbon ○ hydrogen
◉ oxygen ○ sulfur ● nitrogen

Removing pollutants and monitoring air quality

- Atmospheric pollutants don't simply disappear – they have to go somewhere.
- Pollutants are removed from the air when:
  - particulate carbon settles on surfaces, making them dirty
  - sulfur and nitrogen oxides react with water and oxygen to produce a mixture of sulfuric acid and nitric acid in rain (acid rain)
  - carbon dioxide is used by plants for photosynthesis
  - carbon dioxide dissolves in rain water and in oceans.
- Climate scientists take the **mean** of many measurements for each pollutant. The mean is a good estimate of true value.
- The **range** is the difference between high and low results.

- Air quality results vary for several reasons. Potential **outliers** could actually be valid data, and removing them could lead to mistakes.

Ideas about science

You should be able to:

- from a set of repeated measurements of a quantity, use the mean as the best estimate of the true value
- explain why repeating measurements leads to a better estimate of the quantity
- from a set of repeated measurements of a quantity, make a sensible suggestion about the range within which the true value probably lies and explain this.

Improve your grade

How pollutants are formed

Foundation: Describe the damage acid rain causes, and explain why the UK is being blamed for acid rain damage in Europe.

AO1 [4 marks]

Improving power stations and transport

Improving power stations

- Reducing electricity reduces fossil fuel use in power stations.
- New electrical products use less electricity, but some is wasted if they are left on standby.

- Burning oil and gas makes less sulfur dioxide than burning coal.
- Sulfur can be removed from oil and gas before it is burnt, but it is harder to remove from coal.
- Power stations are developing ways of reducing pollution by cleaning waste gases.
- Power stations can remove solid particulates using electrostatic filters.
- Sulfur dioxide can be removed from waste gases by **flue gas desulfurisation**.

- Two 'wet scrubbing' methods used to remove sulfur dioxide from power station waste gases are:
  1 using an alkaline slurry of calcium oxide (lime) and water to make gypsum (calcium sulfate), which can be sold as plaster
  2 using sea water, a natural alkaline which absorbs sulfur dioxide.

Higher

Reducing CO$_2$ and replacing fossil fuels

- Burning less fossil fuel reduces the amount of carbon dioxide (CO$_2$) gas released.
- Ways to reduce our use of fossil fuels include: using alternative energy sources; improving building insulation; walking, cycling, using public transport.

- One alternative to fossil fuels is **biofuels**, which are made from plants. Examples are wood chips, palm oil and alcohol made from sugar.
- Biofuels are 'carbon neutral' – when they are burned they release the same amount of carbon dioxide that the plant originally took from the air to grow.
- Large areas of land are needed to grow biofuels. The land could be used for growing food.

- Gas produces less carbon dioxide than coal for the same amount of energy released.
- Fossil fuels are not renewable, so they are not a **sustainable** source of energy.

> **Remember!**
> To meet the same energy demand, we need to burn less fossil fuels and find alternatives. This will also reduce pollution.

Reducing air pollution from transport

- Air pollution from vehicles can be reduced by:
  - using cars less, especially for short journeys
  - using cleaner fuels and removing pollutants from exhausts
  - making public transport cheaper, more frequent and available in more places.

- Modern vehicles have more efficient engines that use less fuel.
- **Catalytic convertors** contain a platinum catalyst that allows pollutant gases to react with each other: carbon monoxide + nitrogen monoxide → nitrogen + carbon dioxide. In this reaction:
  - carbon monoxide gains oxygen so it is **oxidised**
  - nitrogen monoxide loses oxygen to its is **reduced**.
- Low sulfur fuels are needed as sulfur damages the catalyst. Using low sulfur fuels also reduces sulfur dioxide emissions.
- Legal limits for exhaust emissions are enforced by strict MOT tests.

- Electric cars do not give out pollutant gases when being used, but the electricity produced by fossil fuel power stations used for charging does.
- Research continues into improving batteries and improving charging times. Few charging points are available at present.

EXAM TIP
In questions that start with 'Suggest', try to include a wide range of answers.

Higher

Improve your grade

Reducing air pollution from transport

Foundation: Suggest ways of reducing air pollution from cars.

AO1 [4 marks]

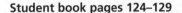

C1 Summary

99% of the Earth's atmosphere is mainly nitrogen and oxygen. The remaining 1% is mainly argon, with tiny amounts of carbon dioxide and other gases.

The Earth's atmosphere

Pollutants are released by natural processes such as volcanoes and by human activity such as burning fuels.

Earth's atmosphere was probably formed by volcanic activity.

Carbon dioxide dissolved in the sea and became trapped in sedimentary rocks and fossil fuels.

Changes in the atmosphere

As the Earth's first atmosphere cooled, water vapour condensed to form oceans.

Plants evolved. Photosynthesis removed carbon dioxide and released oxygen.

Elements are made of atoms. Non-metal atoms combine to make molecules.

Air pollution

UK air pollution levels are regularly monitored.

Chemical reactions involve arranging atoms into new substances which have different properties.

Coal is carbon.
In a plentiful air supply it burns to form carbon dioxide.
In limited air it burns to form poisonous carbon monoxide. Some does not burn and makes soot.

Carbon monoxide, sulfur dioxide, nitrogen dioxides and particulate carbon (soot) are all air pollutants.

Nitrogen oxides are made when nitrogen and oxygen from the air react in hot engines and furnaces.

Making air pollution

Small amounts of sulfur are in all fossil fuels.
When the fuel burns, the sulfur forms sulfur dioxide.
Sulfur dioxide dissolves in water forming acid rain.

Power stations and transport make the most pollution as they burn most fuels.

New electrical products are designed to save energy, reducing electricity demand.

Improving power stations and transport

We can reduce pollution by not using cars as much, using cleaner fuels, and having cars serviced regularly.

Catalytic convertors remove nitrogen oxides and carbon monoxide from exhaust fumes.
Carbon monoxide is oxidised into carbon dioxide.
Nitrogen oxides are reduced into nitrogen.

Power stations can reduce pollution by flue gas desulfurisation.

Biofuels are made from plants and are classed as 'carbon neutral'. They can be used in cars and by power stations.

Using and choosing materials

- Each material has properties that make it suitable for the job it is doing.
  - Rubber is used for car tyres because it hard and **elastic**.
  - **Fibres** are used to weave cloth into clothes.
  - **Plastics** keep their shape when moulded into objects like washing-up bowls.
- Different plastics have different properties. Manufacturers make products by choosing plastics that give the best properties.

> **Remember!**
> Materials have a range of properties, and all need considering before choosing one for a particular job.

- Properties describe how a material behaves.
  - **Melting point** is the temperature at which a solid turns into a liquid.
  - **Tensile strength** is the force needed to break a material when it is being stretched.
  - **Compressive strength** is the force needed to crush a material when it is being squeezed.
  - **Stiffness** is the force needed to bend a material.
  - **Hardness** is how well a material stands up to wear. Hardness can be a compared by scratching two materials together.
  - **Density** is the mass of a given volume of the material. It compares how heavy something is for its size.

- Some properties depend on the size and shape of the material being tested.
- Density is mass per unit volume. The units are g/cm^3 or kg/m^3.
- The effectiveness and durability of a product depend on the materials used to make it:
  - Some materials can be drawn into thin filaments with greater tensile strength. They can be spun into fibres and woven into cloth.
  - Ropes are made by winding fibres together. The more that are wound, the greater the strength.
- Rubber is an elastic material that bounces back when a force is removed. Different types of rubber have different compressive strengths and hardnesses.

> ### EXAM TIP
> Questions about properties often give data in tables. Take time to understand what the data is showing before answering the questions.

Errors and variation in measurement

- A single result may vary, so repeats are needed.
- A result which is very different might be an **outlier** – an incorrect result.
- Calculating the **mean** (average) is a good way to estimate the **true value**.

- Many measurements need to be taken to find the true value.
- The **range** is the smallest to the largest result, excluding outliers.
- We can never be sure if a set of measurements gives the true value.

- **Errors** in measurement produce variations in data.
- Outliers can only be discarded if an error occurred in the measurement.

Higher

Ideas about science

You should be able to:
- suggest reasons why several measurements of the same quality may give different values
- when asked to evaluate data, make reference to its repeatability and/or **reproducibility**
- estimate the true value.

Improve your grade

Comparing materials and measuring properties

Foundation: Sub aqua divers should always leave a marker buoy at the dive site to warn other boats to stay away, reducing the risk of divers being hit when they surface.

Describe the main properties that the marker buoy would need.   AO1 [4 marks]

Natural and synthetic materials

Natural and synthetic materials

- All the materials we use are chemicals or mixtures of chemicals.
  - **Metals** are chemicals which are shiny, **malleable** and electrical conductors.
  - **Ceramics** include clay, glass and cement. They are hard and strong.
  - **Polymers** are large molecules used to make rubbers, plastics and fibres.
  - Concrete is a mixture of sand and cement.
  - Bronze is a mixture of copper and tin.

A nylon fibre is synthesised by reacting chemicals in two solutions together. The solutions do not mix and the nylon is formed at the interface between them.

- **Natural materials** from living things which need little processing are cotton and paper from plants and silk and wool from animals.
- Other natural raw materials which are extracted from the Earth's crust are limestone, iron ore and **crude oil**.
- **Synthetic materials** are manufactured by chemical reactions using raw materials.
- Synthetic materials are alternatives to natural materials from living things.

- Synthetic materials have replaced natural materials because:
  - some natural materials are in short supply
  - they can be designed to give particular properties
  - they are often cheaper and can be made in the quantity needed.

Remember!
Synthetic materials are made by chemical reactions, and can be designed to do particular jobs.

Crude oil and using hydrocarbons

- Crude oil (petroleum) is a mixture of thousands of **hydrocarbons**. Hydrocarbons are compounds of just carbon and hydrogen atoms.
  - Most hydrocarbons from crude oil are used as fuels.
  - When fuels burn in oxygen, carbon dioxide and water are made (see diagrams below and on page 25).

EXAM TIP

When representing a chemical reaction, count the atoms to make sure the reactants and products contain the same number and same types.

- Burning a fuel like methane is a chemical reaction, so atoms are rearranged into new products.
- The number of atoms of each element in the reactants must be same in the products.

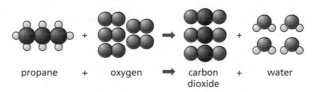

| propane | + | oxygen | → | carbon dioxide | + | water |

Propane burns in air to produce carbon dioxide and water.

- Crude oil consists mainly of a mixture of hydrocarbons, which are chain molecules of varying length up to 100 carbon atoms long.
- As crude oil is a mixture, its composition varies from place to place.

name	methane	ethane	propane
formula	CH_4	C_2H_6	C_3H_8
atomic model			

Some of the alkanes.

- Nearly 90% of crude oil is used as fuels.
- Around 3% of crude oil, mainly smaller hydrocarbon molecules, is used to synthesise other chemicals. Examples of synthesised chemicals from oil are ethanol and plastics.

Improve your grade

Natural and synthetic materials

Higher: What are synthetic materials, and what advantages do they have over natural materials?

AO1 [4 marks]

Separating and using crude oil

Separating out the substances in crude oil

- Crude oil is separated by **fractional distillation**:
  - The oil is heated up which turns it all into gases.
  - The distillation tower gets cooler as it gets higher.
  - Gas molecules **condense** into liquids when they cool.
  - Liquids with similar boiling points collect together. We call these **fractions**.

- Hydrocarbons in each fraction have boiling points within a range of temperatures.

- Molecule chain lengths are similar sizes within each fraction.

- The smaller the molecule chain length, the lower the boiling point.

- The smaller the molecule chain length, the smaller the forces between molecules.

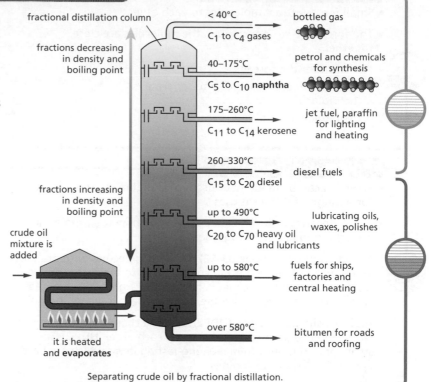

fractional distillation column

fractions decreasing in density and boiling point

fractions increasing in density and boiling point

crude oil mixture is added

it is heated and **evaporates**

Temperature	Fraction	Use
< 40°C	C_1 to C_4 gases	bottled gas
40–175°C	C_5 to C_{10} naphtha	petrol and chemicals for synthesis
175–260°C	C_{11} to C_{14} kerosene	jet fuel, paraffin for lighting and heating
260–330°C	C_{15} to C_{20} diesel	diesel fuels
up to 490°C	C_{20} to C_{70} heavy oil and lubricants	lubricating oils, waxes, polishes
up to 580°C		fuels for ships, factories and central heating
over 580°C		bitumen for roads and roofing

Separating crude oil by fractional distillation.

Investigating boiling points

- Attractive forces exist between molecules in crude oil, holding them together.

- As the hydrocarbon chain length increases, the force between these molecules increases.

- Larger molecules need more energy to break them out of a liquid to form a gas, so have higher boiling points.

Remember!
The lower the boiling point, the smaller the molecule chain length, and the higher it will rise during fractional distillation.

Making polymers

- A **polymer** is a large molecule made by joining many smaller molecules called **monomers**. A polymer can have a chain of anything from hundreds to millions of carbon atoms.

- A polymer is made by a process called **polymerisation**.

- Polymers with better properties mean some older materials have been replaced. Examples are plastic buckets and carbon fibre tennis rackets.

- Ethene is a monomer used to make polyethene.

- Different monomers produce different polymers.

- PET (polyethylenetetraphthalate) is a polymer used to make drinks bottles. PET is clear, strong, has a low density and does not shatter. This makes it a superior material to glass.

- Polymer chains can be altered by replacing hydrogen atoms with other atoms or groups of atoms.

- Each new polymer has its own set of properties and uses.

- Materials such as Kevlar have advantages over alternatives, but can also have disadvantages.

- Material choice will depend on comparing properties for different jobs, with cost being a factor.

Improve your grade

Investigating boiling points

Higher: Explain why during the distillation of crude oil, small hydrocarbon molecules rise to the top of the tower.

AO1 [4 marks]

Polymers: properties and improvements

Attractive molecules

- Small forces attract molecules to each other.
- The forces are strongest when the molecules are close together.
- The stronger the force:
  - the more energy is needed to separate the molecules
  - the higher the melting point.

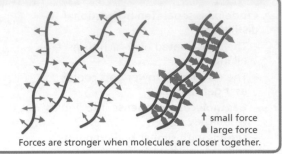

↑ small force
⬆ large force

Forces are stronger when molecules are closer together.

Polymer differences

- Polymers can be made with properties that make them suitable for a range of different uses.
- The properties of polymers depend upon how their molecules are arranged and held together.
- *Low density polyethene* (LDPE) has long molecules with branches. The branches keep molecule chains apart, so the forces between different molecules are weak. Items made from LDPE, e.g. plastic carrier bags, are weak, flexible, soft and have low melting points.
- *High density polyethene* (HDPE) has long chains but no branches, so the molecules are aligned close to each other. HDPE is much stronger and is used to make long-lasting items which are hard and stiff, such as water pipes.

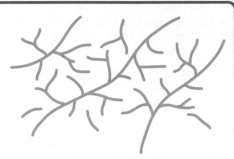

LDPE has branches between the molecular chains, which reduce the attractive forces between them.

- HDPE has a high degree of crystallinity. This means there are lots of areas with regular patterns in the way the molecules line up.
- High **crystalline** polymers are strong with high melting points, but can be brittle.

Improving polymers

- Making the molecule chain longer makes it stronger.
- Longer chains need more force to separate them.
- Longer chains have higher melting points than short chains.

- **Plasticisers** are used to make a polymer softer. They are small molecules inserted into polymer chains to keep them apart, weakening the forces between them.
- Plasticised **PVC** is still hardwearing and waterproof, but it is also flexible, making it a suitable material for rain coats.
- **Thermoplastics** soften when heated and can be moulded into shape.
- **Thermosetting** plastics do not soften when heated. They contain **cross-links** which lock the molecules together so they cannot melt.

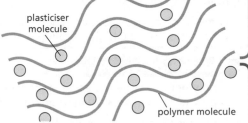

plasticiser molecule

polymer molecule

The plasticiser molecules make the material less rigid.

Remember!

All plastics can be classified as either thermosetting (remains rigid when set) or thermoplastics (melt easily when heated).

- Crystallinity can be increased by removing branches on the main polymer chain and making the chains as flat as possible. This is so that the molecule chains can line up neatly.
- Drawing polymers through a tiny hole when heated makes the molecule chains line up, increasing crystallinity and forming a higher tensile strength fibre.
- Materials which have been treated in this way include bullet-proof vests and sail material (Kevlar).

Improve your grade

Improving polymers

Higher: A company is making rotor blades for a radio-controlled toy helicopter.

A polymer needs to be made stronger but more flexible. What could be done, and how will it change the properties?

AO2 [3 marks]

Nanotechnology and nanoparticles

Small natural nanoparticles

- The width of a human hair is about 0.1 millimetres.
- Microscopes are used to view small objects like human cells.
- Molecules and atoms are thousands of times smaller still.
- **Nanoparticles** are materials containing up to a thousand atoms.
- Nanoparticles:
  - occur naturally, such as salt in seaspray
  - occur by accident, such as solid particulates made when fuels burn
  - can be designed in laboratories.

Nanotechnology

- **Nanotechnology** is the use and control of very small structures. The size of these structures is measured in **nanometres (nm)**. A nanometre is one millionth of a millimetre.
- An atom is about one-tenth of a nanometre in diameter.
- Nanoparticles can be built up from individual atoms. These structures are about the same size as some molecules.
- 'Buckyballs' (see opposite) are very strong carbon spheres made of 60 carbon atoms.
- Carbon **nanotubes** are being designed in laboratories.
- Some nanoparticles are effective **catalysts** as they have a large surface area. Increasing surface area provides more sites for reactions to take place.
- Surface area increases when a lump of solid is cut up into bits.

A model of a natural nanoparticle of carbon called a 'buckyball'.

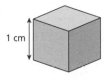

volume 1 cm³
surface area 6 cm²

1 cm

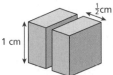

volume 1 cm³
surface area 8 cm²

½ cm

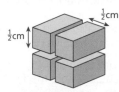

½ cm

volume 1 cm³
surface area 10 cm²

1 cm³ of gold could make 1×10^{21} nanoparticles, each with a volume of 1 nm³. If each nanoparticle was a cube it would have a total surface area of 6×10^{21} nm², which is 6×10^{6} cm². This is one million times the surface area of the original piece of gold.

- Nanotechnology builds structures from 10 atoms across (1 nm) up to a thousand atoms across (100 nm).
- While the diameter of nanotubes is measured in nanometres, they can be millimetres long.
- Nanoparticles have very large surface areas. Because of this, they show different properties to larger particles of the same material.

Remember!

A nanometre is one-thousandth of a micrometre, or one-millionth of a millimetre.

1 000 millimetres = 1 metre so, 1 mm = 1×10^{-3} m

1 000 000 micrometres = 1 metre so, 1 µm = 1×10^{-6} m

1 000 000 000 nanometres = 1 metre so, 1 nm = 1×10^{-9} m

These nanostructures made from carbon atoms are called nanotubes. They may be as small as 3 nm in diameter.

Improve your grade

Nanotechnology

Foundation: Explain what nanoparticles are, and suggest why some act as catalysts. *AO1* [4 marks]

Using nanoparticles

- Silver nanoparticles are very good at killing bacteria. They can be: added to fibres and woven into socks; put into wound dressings; put into plastic and made into food containers.

- Titanium oxide nanoparticles are put into sunscreen. They make the sunscreen **transparent** (no white residue) and absorb UV light.

- Nanoparticles can be mixed with other materials like metals, ceramics and plastics. These combined materials are called **composites**.

- Composite materials are stronger and harder wearing. Adding nanoparticles to:
  - plastic sports equipment makes it stronger
  - tennis balls make them stay bouncy for longer
  - rubber used in tyres make them harder wearing.

- Nanotechnology is the science of making and using nanoparticles.

- Graphite forms in strong sheets that separate easily. Individual graphite sheets one-atom thick are known as graphene sheets.

- Graphene sheets can be rolled into carbon nanotubes. These are super-strength materials.

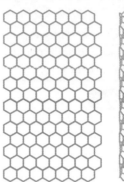 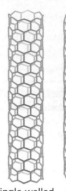 

single layer of graphene   single-walled nanotube   multi-walled nanotube

Modern composite technology using graphene nanotubes results in super-strength materials.

Are nanoparticles safe?

- Silver nanoparticles can be washed out of clothes containing them and get into sewage works.

- Sewage works use bacteria to clean water. Silver nanoparticles could kill these useful bacteria.

- If silver nanoparticles are released into the environment they could kill lots of useful microorganisms.

- Nanoparticles are also used in cosmetics and sunscreens. The nanoparticles are added to materials that have already been used and tested.

- Nanoparticles are small enough to pass through skin into blood, and into body organs. The possible medical effects of this are not yet known.

- While a lot of research is taking place into the use of nanoparticles, little has been carried out into their possible harmful effects.

- One fear is that nanoparticles in the air might be breathed in and cause lung or brain damage.

Remember!
Nanoparticles may have risks as well as benefits and both need to be considered.

- Some people think that because nanoparticles occur naturally, such as in soot and volcanic dust, they pose no danger.

- Others disagree because new nanoparticles with new properties have been manufactured.

- No one knows if nanoparticles used in solids, like windows and paintwork, can escape into the air.

- Some people want proof that new nanotechnologies will not create health and environmental **risks**.

- Risk is defined as the change of an event occurring, and the consequences if it did.

Ideas about science

You should be able to:
- explain why it is impossible for anything to be completely safe
- identify examples of risks which arise from nanotechnology
- suggest ways of reducing a given risk.

EXAM TIP
If asked to identify risks and benefits, use different examples for each one.

Improve your grade

Are nanoparticles safe?

Higher: Fresh Crop is a company selling mixed salads. They are considering adding silver nanoparticles to their food packaging to help prevent bacterial decay.

Explain why some people believe this may have risks.   *AO2* [3 marks]

C2 Summary

Measuring the properties of materials

Materials like plastics, fibres and rubber have many different properties. All these need to be considered when choosing them for a job.

Some properties are melting point, strength, stiffness, hardness and density.

Materials vary slightly, so they need to be measured many times to establish the range, eliminate outliers, find the mean, and estimate the true value.

Natural and synthetic materials

Synthetic polymers are man-made materials from the Earth's crust. Examples are:
- plastics from crude oil
- aluminium metal from bauxite ore.

Natural polymers come from living things. Examples are:
- cotton from plants
- silk from animals
- limestone and iron ore from the Earth's crust.

Synthetic materials are alternatives to using natural materials. They include:
- plastic washing-up bowls instead of metal or ceramic bowls
- neoprene wetsuits instead of rubber ones
- Kevlar body armour instead of metal plates.

Separating crude oil

Crude oil is made of hydrocarbon chains of varying lengths.

Crude oil is separated by fractional distillation into fractions.

Larger hydrocarbons have more attractive forces between the molecules, so more energy is needed to separate them, resulting in higher boiling points.

Different fractions are used mainly as fuels and lubricants, with a small amount used for raw materials, to make materials like plastics.

Each fraction contains a small range of molecule sizes of similar boiling points.

Plastics

The strength, stiffness and hardness of a plastic links to the amount of energy needed to separate the molecule chains.

Small molecules called monomers can join together to make polymers.

Polymer strength depends on chain length, cross-linking, plasticisers and increased crystallinity.

Nanotechnology

Nanotechnology involves very small structures up to 100 nm in size.
They occur naturally, by accident, and by design.

Nanoparticles have different properties compared to larger particles of the same material. One key factor is their larger surface area compared to their volume.

Carbon nanoparticles can be used to strengthen sports equipment and body armour.
Silver nanoparticles give fibres antibacterial properties.

The health effects of using nanoparticles are unknown as they have not been around very long.

Moving continents and useful rocks

Looking at rocks

- **Geologists** study rocks to see how the Earth's surface has changed. They look at how rocks form, how they change, and when changes happened.
- Geological changes happen by slow movements of **tectonic plates**. Plates can move by sliding past each other, colliding or pulling apart.

- Plate collisions build mountain ranges, which erode over time.
- Geologists can explain most of the past history of the Earth by processes they can observe today.

Making Britain

- Over millions of years, Britain has moved across the Earth's surface.

- 600 million years ago, England and Wales were separated from Scotland by ocean, and both were near the South Pole.
- Gradually, different continents drifted and crashed together to form a **supercontinent**, Pangea.
- Britain is made from rocks from different ancient continents.
- Originally, Britain was nearer the equator with a warmer climate.
- Different climates existed in Britain, from tropical swamps to ice ages.

> **Remember!**
> The UK has many different rocks and raw materials due to continental movement and climate changes.

Stories in magnetism

- As volcanic **lava** solidifies, **igneous rock**s are formed.
- Magnetic materials in the lava line up along the Earth's **magnetic field**.
- The Earth's magnetic field changes over time.
- Geologists can date rocks and track the slow movement of continents using changes in magnetic patterns, linked to radioactive decay.
- This evidence supports **plate tectonic theory**.

Limestone, coal and salt

- Rocks are raw materials found buried in the Earth's crust. Coal, **salt** and limestone are three important raw materials.
- 200 years ago the industrial revolution started in north-west England. Chemical industries built up near to raw materials and transport links.
- There was coal in south Lancashire, salt in Cheshire and limestone in the Peak District. The port of Liverpool and the canal system provided good transport links.

- Limestone formed while Britain was covered by sea:
  - Shellfish died forming **sediments** on the sea bed.
  - Sediments compacted and hardened to form limestone, a **sedimentary rock**.
  - Tectonic plate movements pushed the rock to the surface.
  - Gradually the rocks above were **eroded** away until the limestone was exposed.
- Coal formed in wet swampy conditions when plants like trees and ferns died and became buried. This excluded oxygen, slowing down decay.
- Salt formed while Cheshire was covered by a shallow sea:
  - Rivers brought **dissolved** salts into the sea.
  - Climate warming **evaporated** the water, leaving salt that mixed with sand blown in by the wind.
  - Rock salt formed and was buried by other sediments.

> **EXAM TIP**
> Make sure you know how limestone, coal and salt are formed.

- Geologists have found evidence for limestone, coal and salt formation.
- Coal contains fossils of the plants that formed it.
- Limestone contains bits of shell fragments from sea creatures.
- Rock salt contains different-shaped water-eroded grains and wind-eroded grains.
- Ripple marks in rocks indicate water flow from rivers or waves in the sea.

Improve your grade

Limestone, coal and salt

Foundation: Coal, limestone and salt are major raw materials for industry. Choose one and state how it is formed.

AO1 [4 marks]

Salt

Extracting and using salt

- Salt is used in: the food industry; as a source of chemicals; to treat icy roads in winter.
- Salt can be obtained from: collecting and evaporating sea water; mining underground deposits of rock salt.

- Salt is sodium chloride (NaCl) and has many industrial uses.
- Rock salt is spread on icy roads because:
  - the rock is insoluble but the sand in the rock salt gives grip
  - it shows up so people know when roads have been gritted
  - the salt in solution lowers the freezing point, preventing ice forming as easily.
- Only one rock salt mine exists in Britain (in Cheshire). It mines a million tonnes a year.
- If more salt is needed it is usually imported.
- Salt extraction from sea water is only economical in hot climates.
- Purer salt can be obtained by solution mining, which is mainly automatic.

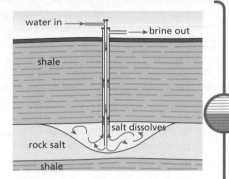

Solution mining – water is pumped at high pressure into the rock salt, the salt dissolves and the salt solution is pushed to the surface.

- Mining rock salt and solution mining can cause subsidence. About half the rock salt cannot be mined, as it is left in place for support.
- Mining can allow water in mines, which may let salt leach out into water supplies, contaminating them.
- Evaporating salt from sea water takes up large areas and spreads salt into the local environment, damaging habitats.

Remember!
Salt is a valuable raw material, but its extraction can have an environmental impact.

The risks of eating salt in foods

- Salt is used in food both as flavouring and as a preservative.
- A higher salt level prevents bacteria growth.
- Too much salt is bad for your health.

- Many people are worried about salt intake, which can cause **high blood pressure**, heart failure and strokes.
- This means salt is classified as a **hazard**.
- A **risk** is the chance of getting ill, and the consequences if you did.
- Risk can be estimated by measuring salt intake.
- Food labels show the amount of salt contained in the product.
- Knowing the risk allows you to make decisions.

Food labels give the amount of salt in the food and the percentage of the recommended daily allowance.

- The government Department of Health (DH) and the Department of the Environment, Food and Rural Affairs (Defra) are responsible for carrying out risk assessment for chemicals in food and advising the public about how food affects health.

EXAM TIP
Make sure you know how to determine risk.

Ideas about science

You should be able to:
- identify, and suggest, examples of unintended impacts of human activity on the environment, such as those resulting from salt extraction
- discuss a given risk, taking account of the chance of it occurring, and the consequences if it did
- discuss the public regulation of risk, e.g. salt levels in food, and explain why it may in some cases be controversial, e.g. whether the daily guideline allowance for salt should be the same for everyone.

Improve your grade

Extracting and using salt

Higher: Describe how salt is obtained by solution mining, and suggest why the chemical industry prefers this method of extraction.   *AO1* [4 marks]

Reacting and making alkalis

About alkalis

- **Alkalis** make **indicators** change colour. Litmus turns blue in alkalis and red in **acids**.
- Alkalis neutralise acids to make **salts**. This is called **neutralisation**.
- The word equation for neutralisation is: acid + alkali ⟶ salt + water

Using alkalis

- Alkalis are used for: dying cloth; neutralising acid soil; making soap; making glass.
- Stale urine and ash from burnt wood were used in the past as sources of alkalis.
- Due to increased industrialisation, by the 1900s demand for alkalis outstripped the supply.

- In the past, one major use of soap was for cleaning wool. Soap was made by mixing the ashes from burnt wood (called potash) with animal fat and boiling it.
- In coastal areas, seaweed or seaweed ash (called soda) could be used to neutralise acidic soils.
- The first alkali to be manufactured was lime (calcium oxide). This was done by heating limestone (calcium carbonate) in a lime kiln, using coal as fuel.
- Lime is used for: neutralising acidic soils; making glass when heated with sand; removing impurities when iron is made.

- Before modern dyes, clothes were coloured using dyes from plants and animals.
- Alum is a mordant that 'sticks' dye to a fabric. It was purified by reacting it with ammonia contained in stale urine.

Making alkalis

- In 1787 the Frenchman Nicholas Leblanc discovered how to manufacture an alkali.

- The Leblanc process made sodium carbonate by reacting salt and limestone, heated with coal.
- It gave off large amounts of hydrogen chloride (an acidic, harmful gas). It also produced heaps of solid waste, called galligu, that slowly released hydrogen sulfide, a foul-smelling, toxic gas.
- Later, a process was invented to change the harmful hydrogen chloride into useful substances:
  – chlorine used to bleach textiles prior to dying
  – hydrochloric acid, which is a starting material for making other chemicals.

- Chlorine can be made by reacting hydrochloric acid and manganese dioxide.
- **Oxidation** converts hydrogen chloride to chlorine.
- **Compounds** have different properties from those of the **elements** they contain.

Remember!
Pollution problems can sometimes be solved by turning waste into useful chemicals.

Patterns of reactions

- An alkali is a solution with a **pH** greater than 7. It turns pH indicator blue or violet.
- Alkalis are **soluble** metal hydroxides and soluble metal carbonates. Some examples are:

Soluble hydroxides	sodium hydroxide – NaOH potassium hydroxide – KOH
	calcium hydroxide – $Ca(OH)_2$
Soluble carbonates	sodium carbonate – Na_2CO_3 potassium carbonate – K_2CO_3

- Most metal hydroxides and metal carbonates are insoluble. They are not alkalis but are called **bases**. Bases react with acids in a similar way to alkalis, but do not affect indicators.
- The general pattern of these reactions is: hydroxide + acid ⟶ salt + water

  carbonate + acid ⟶ salt + water + carbon dioxide gas
- For example: sodium hydroxide + sulfuric acid ⟶ sodium sulfate + water
- The salts produced by different acids are:

Acid	hydrochloric – HCl	sulfuric – H_2SO_4	nitric – HNO_3
Salt	chloride – Cl	sulfate – SO_4	nitrate – NO_3

Higher

Improve your grade

About alkalis

Foundation: Use the tables above to write a word equation for making potassium nitrate.   *AO2* [4 marks]

Uses of chlorine and its electrolysis

The benefits and risks of adding chlorine to drinking water

- In the 19th century, many people died from drinking dirty water.
- Chlorine is now added to water supplies to kill microorganisms.

- The introduction of chlorination made a major contribution to public health.
- Chlorination killed water-borne microorganisms that cause diseases like cholera and typhoid.
- A **correlation** exists between the start of water chlorination in the USA and a fall in the number of deaths from typhoid.

- Chlorine is a toxic gas and can affect human health if too much is present in water.
- Some people disapprove of adding chlorine to water supplies.
- People using mains water supplies have no choice about chlorination.
- Chlorine can react with organic materials in water supplies, forming toxic or carcinogenic compounds called disinfectant by-products (DBPs).
- In the UK the government has decided that the risk from DBPs is very small, so the benefits of disinfecting water outweigh the risks.

> **Remember!**
> Chlorination is used throughout UK as the benefits of killing dangerous microorganisms outweigh the possible risk of cancer.

Electrolysis of brine

- **Electrolysis** breaks up compounds using an electric current.
- The electrolysis of brine (sodium chloride solution) makes: chlorine gas; hydrogen gas; sodium hydroxide solution.
- All three products have uses, so there is no waste.

> **EXAM TIP**
>
> Learn the products of brine electrolysis and a use for each one.

- Electrolysis causes a chemical change, making new products.
- The **anode** is the positive **electrode** and the **cathode** is the negative electrode.
- Large amounts of electricity are needed for electrolysis, so it is expensive.
- The **membrane cell** method is one way to electrolyse brine continuously.
- During brine electrolysis, chlorine forms at the anode and hydrogen at the cathode.
- Industrial uses of these products are:
  - chlorine for making **plastics** like **PVC**, in medicines and crop protection
  - hydrogen for making margarine, as rocket fuel, in fuel cells in vehicles
  - sodium hydroxide for paper recycling, industrial cleaners and refining aluminium.

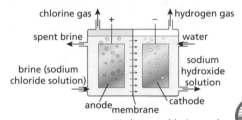

Hydrogen, chlorine and sodium hydroxide are the products of electrolysing sodium chloride solution in the membrane cell.

- Brine electrolysis is one of the most widely used industrial processes.
- While the products have many uses, they can have an environmental impact.
  - Chlorine products, e.g. from fridges and aerosols, are linked to ozone depletion and have been banned.
  - Chlorine used in paper bleaching releases dangerous dioxins, increasing the risk of cancer.
  - The mercury diaphragm method of electrolysis, which is still used, releases mercury waste. This can enter the food chain and is a cumulative poison.
  - Plastics made using chlorine are non-biodegradable.

Ideas about science

You should be able to:

- offer reasons for people's willingness (or reluctance) to accept the risk of a given activity
- explain why it is impossible for anything to be completely risk-free.

Improve your grade

The benefits and risks of adding chlorine to drinking water

Foundation: Worldwide, over 100 000 people die from a disease called cholera each year, through drinking dirty water. In Britain, cholera is rare.

Explain these statements, and suggest advice for people who drink water directly from rivers.

AO2 [4 marks]

Industrial chemicals and LCA

Are chemicals safe?

- Chemicals contain elements. Elements cannot be destroyed, so they remain in the environment forever.
- A risk assessment is used to find out how dangerous substances are.

- Chemicals in the form of solids, liquids and gases can spread out into the environment.
- Some toxic chemicals persist in the environment, can be carried over large distances, and may accumulate in food chains, ending up in human tissues.
- To decide the level of risk of a particular chemical we need to know:
  - how much of it is needed to cause harm
  - how much will be used
  - the chance of it escaping into the environment
  - who or what it may affect.

- Thirty years ago, European laws made risk assessment compulsory for new chemicals.
- Many substances we have used for years have not been tested or there is insufficient data about them for risks to be judged.
- Many people perceive the risk is greater for newer chemicals with less familiar names, while the actual risk could be less.

Should we worry about PVC?

- PVC is a plastic containing carbon, hydrogen and chlorine.
- Small molecules called **plasticisers** are added to PVC to make it softer.
- Plasticised PVC is used to cover electrical wires, for clothing and for seat covers.

- Plasticiser molecules can leach out of PVC into the surroundings, where they may have harmful effects.
- Although chemicals used for plasticisers have passed safety tests, they may have long-term effects on fish, and large amounts have been shown to harm animals.
- As a precaution, plasticised PVC children's toys have been banned in the USA and Europe.

- If PVC is burnt, it gives off toxic gases including dioxins. If eaten, these chemicals build up in fat and are thought to cause cancer.
- Plasticisers are relatively new so long-term studies are not possible. Because of this many people dispute the risks.

Life Cycle Assessment

- A **Life Cycle Assessment** (LCA) measures the energy used to make, use and dispose of a substance, and its environmental impact.

- There are four main stages of an LCA (see diagram), which is sometimes called a 'cradle, use, grave' assessment.
- At each stage of an LCA we need to consider:
  - How much natural resources are required?
  - How much energy is needed or produced?
  - How much water and air is used?
  - How is the environment affected?
- When an LCA has been completed, different products can be compared fairly.

1 Preparing the chemicals from raw materials found in plants, animals, rocks, the oceans or the air

2 Making the product from the chemicals – including transporting the chemicals and the finished product

3 Using the product

4 Disposing of the product and the materials in it when it is of no more use

Stages in the life cycle of a product.

- To produce a fair and **accurate** LCA, a lot of data is required.
- Some aspects are hard to measure, e.g. the lifetime of a car and its disposal method can vary.

Remember!
A Life Cycle Assessment shows the total energy and environmental impact of a product, from getting the raw materials to making it to final disposal (cradle to grave).

Improve your grade

Should we worry about PVC?

Higher: Cling-film is used to wrap food. It may be made from thin sheets of PVC, which contain plasticiser molecules called phthalates to increase flexibility. These molecules have passed safety tests. Despite this, some people are still worried about their safety.

Suggest reasons for some people's reluctance to accept the risk. *AO2* [4 marks]

C3 Summary

Rocks and continents

Geologists look for fossils, shell fragments, grain shape and ripple marks for evidence of conditions when sedimentary rocks form.

The Earth's crust is made very slowly by moving tectonic plates.

Mountain building, erosion, sedimentation, dissolving and evaporation have led to the formation of raw materials.

The climate in Britain has varied from tropical swamps to ice ages.

Britain is made from rocks that originally came from ancient continents.

Salt and alkalis

Rock salt is obtained by mining. Salt is obtained by evaporating sea water and solution mining.

Salt is used as a preservative or flavouring in food, but too much can cause health problems.

All methods of obtaining salt can have environmental impacts.

Government agencies assess food safety and give advice about healthy eating.

Reacting and making alkalis

Alkalis neutralise acid soils, and are used in dying clothes, and making soap and glass.

Early alkali manufacture produced toxic by-products.

In the past, ash from burnt wood and seaweed were used as alkalis.

Soluble hydroxides and carbonates are called alkalis. Insoluble ones are bases.

Uses of chlorine and its electrolysis

Adding chlorine to water supplies killed microbes, and this greatly improved public health.

Electrolysis is used to separate sodium chloride solution into hydrogen, chlorine and sodium hydroxide, all of which have many uses.

Some people disagree with adding chlorine to drinking water as products might cause cancer, but the benefits outweigh the risk.

Perceived risk and calculated risk are different.

Electrolysis requires large amounts of electricity, and can have an environmental impact.

Industrial chemicals and Life Cycle Assessment

Small molecules called plasticisers make PVC more flexible.

Life Cycle Assessment measures the energy and environmental impact of a product from its cradle to grave.

Plasticisers can leach out and may have harmful effects.

Our solar system and the stars

The solar system

- The centre of our **solar system** is a star called the Sun.

- The eight **planets** in our solar system are spherical and have nearly circular **orbits** around the Sun. The four planets closest to the Sun are solid rock; the four outer planets are gas giants.

- Each planet may have a number of **moons** (balls of rock) in near-circular orbit around it.

- **Asteroids** are irregular lumps of rock, mostly in near-circular orbit between Mars and Jupiter.

- **Comets** are small objects made of rock and ice, with very elongated orbits around the Sun.

- **Dwarf planets**, such as Pluto, are small spherical lumps of rock in orbit around the Sun.

- Nearly all (99%) of the solar system's mass is in the Sun.

- Jupiter is the heaviest planet, followed by Saturn, Uranus, Neptune, Earth, Venus, Mars and Mercury.

- Neptune has the largest orbit, followed by Uranus, Saturn, Jupiter, Mars, Earth, Venus and Mercury.

The Universe: distances and sizes

- The Sun is one of thousands of millions of stars in the **Milky Way**.

- The Milky Way is one of thousands of millions of **galaxies** which make up the **Universe**.

diameter of the Milky Way
(100000 light-years)
25000 times larger

diameter of solar system
(0.006 light-years)
200 times larger

Earth's orbit
(300000000km)
200 times larger

distance to the
next nearest star
(4.2 light-years)
700 times larger

distance to the
nearest galaxy
(2300000 light-years)
23 times larger

Sun's diameter
(1400000km)
100 times larger

Earth's diameter
(13000km)

1 light-year = 95000000000000km

Comparing sizes and distances in the Universe.

- Distances to objects outside the solar system are measured in **light-years**. A light-year is the distance that light travels in a year. One light year is 9.5 million million kilometres.

- Light travels through space (a vacuum) at a speed of 3.0×10^5 km/s (300 000 km/s).

- Light takes a very long time to reach us from the stars. We can only observe what stars were like in the past, when the light left them.

EXAM TIP

It is easier to deal with large numbers if they are in standard form. So use 3.0×10^5 km/s instead of 300 000 km/s when you do calculations with the speed of light.

Finding out about the distance to stars

- All the evidence we have about distant stars and galaxies comes from the **radiation** which astronomers can detect.

- Two stars which have the same **real brightness** (appear as bright as each other) can have different **relative brightness**. The star which is further away has a smaller relative brightness.

- If you know the distance to one of the stars, the difference in their relative brightness can be used to calculate the distance to the other one.

- There are uncertainties with this method of measuring the distance to stars:
  - It is based on the assumption that similar types of stars have the same real brightness.
  - It is based on estimating the distance to one of the stars.
  - Many things can make it difficult to make precise observations of stars at night. These include dust, rain, clouds and **light pollution** from streets and buildings.

- Telescopes in space take measurements without distortions from the Earth's atmosphere.

- As the Earth orbits the Sun, nearby stars move slightly against the fixed background of distant stars. This is called the **parallax** effect. Although small, it can be used to find the distance of a star.

- Only stars nearby have a parallax effect which is large enough to be measured.

Improve your grade

The solar system

Foundation: Describe the motion of moons and planets in the solar system.

AO1 [4 marks]

The fate of the stars

Fusion of elements in stars

- The Sun's energy comes from hydrogen. Hydrogen nuclei are jammed together so hard that they combine in pairs to form the element helium. This process releases loads of energy and is called **nuclear fusion**.
- Fusion in stars forces hydrogen nuclei together to make other elements as well. These other elements spread through space when a star explodes at the end of its life.
- All chemical elements with atoms heavier than helium were made in stars.

- Nuclear fusion is only possible when there are very high densities and temperatures.
- At high enough densities, nuclear fusion can make heavier elements up to iron.
- Heavy stars end their lives as a **supernova**. This is a massive explosion where all the different chemical elements, including those heavier than iron, are made.

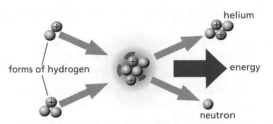

forms of hydrogen

helium

energy

neutron

The fusion reaction that takes place in the Sun.

- The solar system was made from a collapsing cloud of dust and gas about 5000 million years ago.
- Apart from the hydrogen, all of the material in that cloud came from the explosions of large stars. Evidence for this comes from elements in the Sun other than hydrogen and helium.

Remember!

Nuclear fusion is when nuclei are crushed together to form new elements, releasing energy. Nuclear fission breaks up heavy nuclei into lighter ones, releasing energy.

The expanding Universe

- Most of the galaxies appear to be moving away from us.
- This motion of the galaxies increases the wavelength of the light we receive from them.

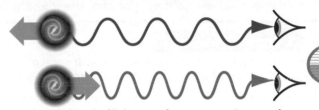

The wavelength of light waves from a star moving away from or towards you changes.

- The increase of wavelength from a galaxy moving away is called **redshift**.
- Apart from a few close galaxies, the amount of redshift increases with distance from Earth. In general, the further away from Earth a galaxy is, the faster it is moving away from us.

Higher

- The redshift in the light coming from distant galaxies provides evidence that all parts of the Universe are expanding, with galaxies moving apart from each other as time goes on.

The Big Bang

- The Universe started expanding rapidly from a single point about 14 000 million years ago.
- The Sun was created about 5000 million years ago.
- The Earth was created about 4500 million years ago.

- Scientists believe that the Universe began with a 'Big Bang'.
- The detection of cosmic background radiation provides evidence to support the **Big Bang theory**.

The future

- The ultimate fate of the Universe depends on how it continues to expand. If there is enough mass in the Universe, gravity will slow down the expansion and make it collapse again.
- The fate of the Universe is difficult to predict because:
  - we can only measure the mass of those parts of the Universe which emit radiation
  - precise measurements of the speed and distance of galaxies is difficult because their radiation travels such a long way to get to us.

Improve your grade

Fusion of elements in stars

Higher: Explain how most of the material in and around you was created by stars. *AO1* [4 marks]

Earth's changing surface

The changing Earth: erosion and sedimentation

- The surface of the Earth is always changing.
- Material **erodes** slowly from mountains and becomes **sediments** which make rocks.
- **Volcanoes** erupt quickly, spewing out **lava** to make new mountains or a crater.
- Sometimes plants and animals are buried by sediments or lava to become fossils.
- Slow movements of the crust can make fold rocks and new mountains.
- **Geologists** study rocks. Their findings provide evidence of how the Earth has changed.

- Rocks are eroded by moving water, glaciers, wind and rockfalls. Mountains are made smaller and smoother by erosion. Valleys are made deeper by erosion of riverbed rocks.
- Eroded rock fragments are transported by the wind, water and ice, broken up further, and deposited on riverbeds and in the sea. This is called **sedimentation**.
- Over millions of years the sediments are crushed together to form layers of **sedimentary rocks**.
- The build-up of sedimentary rock layers eventually makes seas shallower.

> **Remember!**
> The processes that made and destroyed rocks in the past are still going on today.

- The age of the Earth can be estimated from, and must be greater than, the age of its oldest rocks, which are 4000 million years old.
- If no new rocks had been created, erosion for that length of time would have worn all of the continents down to sea level.
- Breaks in the Earth's **crust** allow molten rock to escape from volcanoes and create new mountains.
- Collisions between different parts of the crust also push rocks up to make new mountains.

Continental drift

- Alfred Wegener's 1915 theory of **continental drift** says that millions of years ago there was a single land mass on Earth. Since then it has split into several continents which have drifted apart.
- Wegener's theory was based on the following evidence:
  - the way continents fit together so well
  - similar fossils and rocks are found on continents now separated by oceans.
- Collisions between moving continents also explains the folding of rocks into mountains.

- Geologists are scientists who study the Earth. They did not accept Wegener's theory because:
  - they already had other, simpler theories which explained some of his observations
  - nobody could explain or measure the movement of the continents
  - Wegener was not a trained geologist and his theory was very different from the others.

- Continents move because they sit on the mantle, whose rocks move slowly by convection as they carry heat away from the Earth's hot core.
- The seafloor between continents moving apart can increase by a few centimetres each year. This is called **seafloor spreading**.
- **Oceanic ridges** form on the expanding seafloor where liquid rock from the mantle fills the gap.

Higher

- The solidifying rock in oceanic ridges is magnetised by the Earth's field.
- The Earth's magnetic field changes direction over millions of years.
- Each time the Earth's field reverses, so does the magnetisation of the oceanic ridges. So the seafloor has strips of reversed magnetism parallel to the gap where new rock is created.

Ideas about science

You should be able to:

- suggest plausible reasons why scientists disagreed with Wegener's theory
- understand that data from experiments are only valid if they can be obtained by other people.

Improve your grade

Continental drift

Foundation: Explain why Wegener's theory of continental drift was not accepted when it was first published.

AO1 [4 marks]

Tectonic plates and seismic waves

Tectonic plates

- The Earth's crust has solid **tectonic plates** which float on semi-solid rocks.
- Tectonic plates meet at a **plate boundary**. Here, earthquakes, volcanoes and mountains are found.

- Volcanoes occur when liquid **magma** is forced through cracks where tectonic plates are moving apart.
- Volcanic mountains form when one tectonic plate is forced under another heading towards it.
- **Fold mountains** form when two tectonic plates meet head-on.
- Earthquakes are releases of energy from tectonic plates sliding past each other.
- The **rock cycle** can be explained by the movement of tectonic plates.

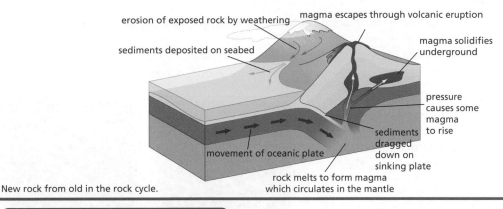

New rock from old in the rock cycle.

Higher

The structure of the Earth

- The core of the Earth is mostly liquid iron.
- Semi-liquid rock in the **mantle** floats on the core.
- The outer core is a layer of liquid nickel and iron about 2200 km thick.
- The inner core is solid nickel and iron about 1250 km thick.
- A thin layer of solid rock in the crust floats on the mantle.

The structure of the Earth.

Seismic waves

- Two types of **seismic waves** are generated when tectonic plates suddenly move.
  - **P-waves** move quickly through solid crust and liquid core.
  - **S-waves** only move slowly through the solid crust.
- Seismometers on the Earth's surface record these waves after an earthquake.

- We can work out the structure of the Earth by measuring the time of arrival of seismic waves across the Earth from an earthquake.
- Seismic waves speed up and change direction when they enter denser regions of the Earth's core.
- P-waves are **longitudinal**, which means the particles vibrate along the direction of motion of the wave.
- S-waves are **transverse**, so the particles vibrate at right angles to the direction of motion of the wave.
- The core of the Earth must be liquid because only P-waves pass through it.

EXAM TIP

Make sure that you can clearly describe the difference between a longitudinal wave and a transverse one.

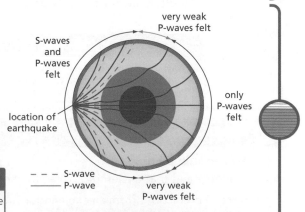

The timing and strength of detected seismic waves gives information about the Earth's interior.

Improve your grade

Tectonic plates

Higher: Explain why there are volcanoes at plate boundaries.

AO1 [4 marks]

Waves and their properties

Finding out about waves

- A **wave** transfers energy away from a vibrating source. The wave creates a series of disturbances as it moves, vibrating material that it passes through.

- There is no overall transfer of matter in the direction of motion of the wave, just energy.

- The **amplitude** of a wave is the maximum height of the disturbance from the undisturbed position.

- The **wavelength** is the distance from one maximum disturbance to the next.

- The speed of a wave is how fast each maximum disturbance moves away from the source:

$$\text{wave speed (m/s)} = \frac{\text{distance travelled (m)}}{\text{time taken (s)}}$$

- The **frequency** of a wave is the number of vibrations of the source in one second.

The height of the wave above the surface is the amplitude

The distance between ripples is the wavelength

undisturbed water level

Measuring the amplitude and wavelength of a wave.

- An **oscilloscope** is a machine that displays waves on a screen. A grid on the screen lets you compare the wavelength and amplitude of waves:
  - a sound is louder if it has a larger amplitude
  - a sound is higher pitched if it has a shorter wavelength.

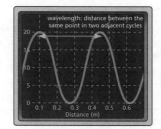
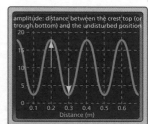

Two different sound waves on the same oscilloscope grid. How are they different?

- The scale on an oscilloscope is used to measure wavelengths in metres or to measure amplitude.

- The wavelength and amplitude of the left-hand wave can be found like this:
  - Each horizontal square is 0.1 m. The wavelength is 3.5 squares. So the wavelength is 0.35 m (0.1 m × 3.5 squares).
  - Each vertical square is 5 units. The amplitude is 2 squares. So the amplitude is 10 units (5 units × 2 squares).

Calculating wave frequency and speed

- The unit of frequency is **hertz** (Hz). 1 Hz means 1 vibration per second.

- Waves obey this **wave equation**:

  wave speed (m/s) = frequency (Hz) × wavelength (m).

- The higher the frequency, the shorter the wavelength. The wavelength is always **inversely proportional** to the frequency.

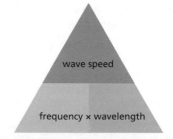

wave speed

frequency × wavelength

Cover up what you want to find out. Use the quantities you know to work out the answer.

EXAM TIP

Always write down the quantities you know with their units, as well as the one you are asked to find. This will help you to select the correct equation.

Ideas about science

You should be able to:

- understand that a scientific explanation can successfully predict the results of experiments

- make predictions from an explanation and compare them with data from experiments to test the explanation.

Improve your grade

Finding out about waves

Higher: Sound in steel has a speed of 2 km/s. What is the wavelength of a sound wave in steel which has a frequency of 80 000 Hz?

AO2 [2 marks]

P1 Summary

Scientific explanations

Scientists use journals and conferences to inform about their experiments and explanations.

A scientist's judgement about an explanation may be affected by their personal background.

Scientists think up explanations for data from experiments.

The explanations are tested by comparing their predictions with data from new experiments.

An explanation is only accepted when it can account for all the data from experiments.

Scientists repeat experiments done by others to check their data.

The solar system

The solar system was formed from gas and dust in space about 5000 million years ago.

The distance of a star can be estimated from its brightness and parallax.

The source of the Sun's energy is fusion of hydrogen nuclei, making it hot enough to emit light, which moves through space at 300 000 km/s.

The star at the centre of the solar system is called the Sun.

Earth is one of eight planets which orbit around the Sun.

Moons orbit some of the planets.

Comets, asteroids and dwarf planets orbit the Sun.

Galaxies

The Sun is one of thousands of millions of stars in the Milky Way galaxy.

The Universe is made up of thousands of millions of galaxies.

Redshift and distance data suggest that the Universe expanded from a single point about 14 000 million years ago.

The redshift of light from galaxies can be used to measure how fast they appear to be moving away.

The expansion of space means that the greater the distance to a galaxy, the faster it appears to be moving away.

The changing Earth

The oldest rocks on Earth are about 4000 million years old.

Mountain building, earthquakes, seafloor spreading and volcanoes happen at the edges of tectonic plates.

The Earth's solid crust floats on a soft mantle and is continually worn down by erosion.

Convection currents in the mantle split the crust into tectonic plates and move them.

The theory of tectonic plates explains Wegener's idea of continental drift.

Seismic waves

Earthquakes produce waves which carry energy as they travel through the Earth:
- P-waves are longitudinal waves which can move through the liquid core.
- S-waves are transverse waves which can only move through the solid mantle and crust.

Distance moved by a wave = wave speed × time of travel

Waves have an amplitude, wavelength, frequency and speed.

Wave speed = frequency × wavelength

Waves which ionise

Electromagnetic radiation

- You see things only when your eyes (detectors) absorb **electromagnetic radiation** called light.
- Some of the things you see are light sources. They emit light, often by glowing hot.
- Everything else that you see **reflects** light. Black objects don't reflect much light, they mostly **absorb** it. Shiny objects reflect most of the light that falls on them.

- Light is one part of the **electromagnetic spectrum**.
- All waves in the electromagnetic spectrum **transmit** through a **vacuum** at 300 000 km/s.
- The energy of a wave in the spectrum increases with increasing frequency. Waves with higher frequencies carry more energy.

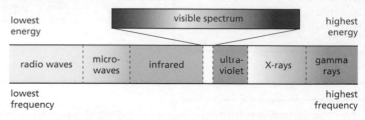

The electromagnetic spectrum. Within visible light, red light has the lowest frequency and violet light has the highest frequency.

- The figure shows how electromagnetic waves are grouped in ranges of frequency.

Light photons

- Electromagnetic waves transfer energy in packets called **photons**.
- The energy in a photon depends only on the frequency of the wave. Increasing the frequency of an electromagnetic wave increases the energy of its photons.

> **EXAM TIP**
>
> Although electromagnetic waves transfer energy in packets (like particles do), they also have wave properties such as wavelength, amplitude and frequency.

Radiation intensity

- Solar cells produce electricity. They work by absorbing electromagnetic radiation from the Sun. The solar cell transfers this energy to electrical energy.
- The energy absorbed in each second from an electromagnetic wave depends on its **intensity**. This depends on:
  - the number of photons per second (intensity increases with the number of photons)
  - the energy transferred by each photon (intensity increases with energy).

- The energy of a wave is spread over an increasing area as it moves away from its source.
- This means that the intensity of the wave decreases with increasing distance from its source.

- The intensity of an electromagnetic wave is the energy transferred to each square metre of absorbing surface in each second. The units of intensity are therefore $J / m^2 / s$.
- **Higher:** The intensity of a wave in a vacuum is inversely proportional to the square of its distance from its source.
- If the wave is partially absorbed by the medium it is passing through, the intensity drops even more rapidly than an inverse square law.

Ionisation

- **Atoms** and **molecules** have no overall electric charge. **Electrons** are negatively charged. **Ions** have either positive or negative charge.
- Atoms or molecules are **ionised** when they lose electrons.

- The photons of an **ionising radiation** have enough energy to ionise atoms or molecules.
- The only ionising radiations in the electromagnetic spectrum are high energy ultraviolet, **X-rays** and **gamma rays**.

- Ionisation of a molecule can start off a chemical reaction involving that molecule.

Improve your grade

Ionisation

Higher: Bacteria are single-cell organisms that can pollute drinking water. Explain why exposing the water to ultraviolet light removes the bacteria, but exposing it to visible light does not affect the bacteria.

AO2 [4 marks]

Radiation and life

Effects of ionising radiation

- **Radioactive** materials emit gamma rays. Gamma rays pass easily through the human body.
- X-rays pass through muscle but are absorbed by bone.
- Cells are ionised and damaged when they absorb gamma rays or X-rays. The damaged cells can either die or develop into cancer.

- Physical barriers absorb some ionising radiation.
- X-rays are absorbed by dense materials, so X-rays are used to make shadow pictures of people's bones or their luggage.
- Hospital workers such as **radiographers** are protected from ionising radiation by lead and concrete screens.

Microwaves

- Things which absorb radiation heat up.
- Cells which absorb **radiation** are damaged if they get too hot.
- You can increase the thermal energy transferred from radiation by increasing its exposure time and intensity.

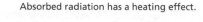

microwaves produced in microwave oven

- Water molecules are good at transferring energy in **microwaves** to thermal energy.
- As the water absorbs the microwaves they vibrate more and share this energy with other molecules around them.
- Cells contain water so heat up when they absorb microwaves.

cold water   hot water

Absorbed radiation has a heating effect.

- Microwave ovens can only be used to heat up food which contains water. Microwaves reflect off the metal in the walls and door or are absorbed by them instead of escaping. This protects people from the radiation.

- Mobile phone networks use low intensity microwaves. The heating effect of these microwaves is very small, but some people are concerned about the health risks from the radiation.
- The risk of cell damage from mobile phones is measured by comparing cancer **rates** in large groups of people who do and do not use mobile phones.
- There is no evidence of harm from mobile phone use.

Ozone

- Sunlight contains ultraviolet radiation. Ultraviolet radiation causes sunburn and skin cancer.
- Sunscreens and clothing protect people by absorbing ultraviolet radiation emitted by the Sun.
- A layer of **ozone** at the top of the atmosphere absorbs ultraviolet radiation from the Sun. This protects living things on Earth from some of the harmful effects of ultraviolet radiation.

- Increased exposure to ultraviolet radiation increases the risk of developing skin cancer.
- If the ozone layer thins by 1%, the risk of skin cancer increases by about 4%.

- Ultraviolet radiation from the Sun causes chemical changes to molecules in the ozone layer. **Higher**

Ideas about science

You should be able to:
- understand the possibility of the risk of harm from science-based technologies
- argue that the consequences of a risk, such as skin cancer, need to be considered as well as the chances of it happening
- understand why scientists use large matched samples to investigate correlations between an outcome and a factor
- understand why people underestimate the risk of activities they enjoy, such as sunbathing, and overestimate the risk from unfamiliar or invisible sources, such as ultraviolet radiation.

Improve your grade

Microwaves

Higher: Explain the risks of cooking food with microwaves.

AO2 [3 marks]

Climate and carbon control

The greenhouse effect

- Radiation from the Sun contains a range of frequencies. Only some of those frequencies pass through the **atmosphere** of the Earth.
- The Earth warms up when it absorbs radiation from the Sun.
- The infrared radiation emitted by the Earth makes it cool down. Radiation from the Earth may pass into space, reflect off clouds or radiate back from gases which absorb it.
- When the Earth's radiation is absorbed or reflected back, this keeps the Earth warmer. We call this the **greenhouse effect**.

Electromagnetic radiation frequencies

- Everything emits some electromagnetic radiation with a range of frequencies.
- The **principal frequency** of that radiation is the one with greatest intensity.
- The principal frequency increases with increasing temperature.
- Radiation from the hot Sun has a higher frequency than radiation from the cool Earth.

Carbon dioxide in the atmosphere

- **Carbon** is found in all living things. It is constantly being recycled through the carbon cycle.
- **Carbon dioxide** in the atmosphere is found in very small amounts. It is absorbed by plants through **photosynthesis** and released by living organisms as they rot (decompose) and **respire**.
- Carbon dioxide is one of the main **greenhouse gases** found in the Earth's atmosphere.
- The level of carbon dioxide in the atmosphere has been steady for thousands of years, because the rates of absorption and release of carbon dioxide have been the same.
- In the last 200 years the level of carbon dioxide has steadily increased because:
    - the burning of **fossil fuels** as an energy source has increased the rate of release
    - cutting down forests (**deforestation**) to clear land has decreased the rate of absorption.

> **Remember!**
> Fossil fuels contain carbon removed from the atmosphere millions of years ago by organisms which were buried before they could decompose.

- **Higher** • The recent increase in the temperature of the Earth **correlates** with the rise in carbon dioxide levels in the Earth's atmosphere. Many scientists believe that this correlation is caused by the carbon dioxide because it is a greenhouse gas.

Global warming

- The greenhouse effect is slowly increasing the average temperature worldwide. This is called **global warming**.
- Global warming will have many effects, including: changes to the crops which can grow in a region; flooding of low-lying land due to rises of sea level as the sea expands and glaciers melt.
- Various gases in the atmosphere are responsible for the greenhouse effect:
    - Water vapour has the most effect, because there is so much of it.
    - The small amount of carbon dioxide has an effect.
    - **Methane** is a strong absorber of infrared radiation, but there is very little of it.
- **Higher** • Scientists use computer models to predict the effects of global warming. The models are tested by seeing if they can use past **data** to predict today's climate.
- Models suggest that global warming will result in more extreme weather events because of:
    - increased water vapour in the hotter atmosphere
    - increased **convection** in the atmosphere increasing wind speed.

Ideas about science

You should be able to:
- understand why correlation, for example between raised carbon dioxide levels and global warming, is not accepted as cause and effect until an explanation has been found.

Improve your grade

Global warming

Foundation: Explain how increasing carbon dioxide in the atmosphere results in global warming.

AO2 [4 marks]

Digital communication

Electromagnetic waves for communication

- Some electromagnetic waves can carry information from one place to another. That information includes text, voice, music and pictures.

- Radio waves and microwaves are not absorbed by air. This means that they can carry radio and TV broadcasts through the atmosphere.

- Infrared and light are not easily absorbed by the glass of **optical fibres**. This means that they can be used for long-distance telephone and internet communication.

- Radio waves use a **carrier wave** to transfer information. The information is used to change the amplitude or frequency of the carrier, in a process called **modulation**.

- A radio receiver demodulates the carrier wave to recover the information.

- An **analogue signal** varies continuously with any value. Sound is an example of an analogue signal.

- Modulation which varies the amplitude of a radio wave continuously makes analogue signals.

Digital signals

- A **digital signal** has only a few values. The digital code for sending sound or pictures has only two values, called 1 and 0.

- By contrast, analogue signals are those which vary continuously.

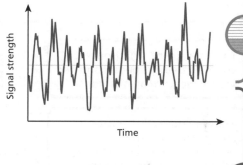

- Analogue signals, such as sound, can be sent by radio wave in a digital code:
  - The value of the **signal** is measured.
  - The value is coded as a string of 1s and 0s.
  - The 1s and 0s make a digital signal by pulsing the carrier wave on and off (where 0 = no pulse and 1 = pulse).
  - The process is repeated many times a second.

- Radio receivers use the strings of 1s and 0s in the digital signal to recover (**decode**) the original analogue signal and produce a copy of it.

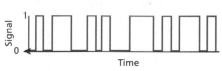

Analogue and digital signals.

- Radio waves can be affected by unwanted information called **noise**. Noise changes the amplitude of the carrier continuously.

- Noise always reduces the quality of the signal recovered by an **analogue** receiver, because the original signal cannot be separated from the noise.

- The wave **pulses** which carry information can usually be separated from the continuously varying noise in a digital receiver, so the quality of the signal is not affected.

Storing digital information

- Digital information is stored as **bytes**.

- A byte is a string of eight **binary digits**. Each binary digit can be either 1 or 0.

- Digital information builds up images from many small dots called **pixels**. The colour and brightness of each pixel is set by binary digits.

- Increasing the number of binary digits for a picture increases the sharpness of the image.

- Digital information builds up sound from a rapid series of values called samples. The value of each sample is set by binary digits.

- Increasing the number of binary digits for a sound increases its quality.

- An advantage of using digital signals is that they are easily stored in electronic memories so can be processed by computers.

Improve your grade

Analogue and digital

Foundation: Describe the difference between analogue and digital signals used for radio broadcasts.

AO1 [4 marks]

Cause and effect

If changes to a factor always result in changes to an outcome, then they are correlated.

Scientists will only accept that a factor causes an outcome if they can explain it with theories.

Scientists establish correlation by comparing the outcome of large samples whose only difference is the factor.

Risk

People's perceptions about risk may be unrealistic: they fear the unfamiliar, but will minimise risks from things they enjoy doing.

We have to make decisions about risk all the time – some risks are greater than others.

Risk can be quantified by measuring the chance of it happening in a large sample over a period of time.

Both the chance of a risk happening and its consequences are considered when making decisions about risk.

Electromagnetic radiation

The intensity of radiation is the energy it delivers per second per square metre of an absorber.

Water in our cells transfers energy from microwave photons to heat. The heating effect depends on both the intensity of the radiation and the length of exposure.

The risk of harm from microwave radiation can be reduced by surrounding sources with metal reflectors.

Low intensity microwaves from mobile phones may be a health risk.

Ultraviolet, X-ray and gamma photons have enough energy to ionise atoms which absorb them.

Cells exposed to ionising radiation can die or become cancerous.

Dense materials absorb X-rays and gamma rays. Sun screens, clothing and the ozone layer absorb ultraviolet radiation.

Electromagnetic radiation carries photons of energy at 300 000 km/s through empty space.

The seven types of electromagnetic radiation in order of increasing photon energy are: radio waves, microwaves, infrared, visible, ultraviolet, X-rays, gamma rays.

The energy of a photon increases with increasing frequency or decreasing wavelength.

Global warming

The Sun transfers energy through the Earth's atmosphere as short wavelength radiation (light). The light is absorbed by the Earth's surface, heating it up.

Burning fossil fuels has increased CO_2 in the atmosphere, causing climate change.

The warmed Earth emits long wavelength radiation (infrared), some of which is absorbed by greenhouse gases (carbon dioxide, water vapour and methane) in the atmosphere, heating it up.

Communication

Radio waves and microwaves carry TV and radio broadcasts because they pass through the atmosphere.

Sound and images can be transmitted as digital signals by switching waves on and off.

The quality of a received sound or image increases with the number of bytes of information transmitted.

Light and infrared carry information long distances along optical fibres.

Although signals become weaker and noisier as they travel, the pattern of 1s and 0s can often be restored at the receiver.

A signal is an electromagnetic wave which has been changed by some information.

An analogue signal is the result of a continuous change.

A digital signal is the result of switching the wave between two values called 1 and 0.

Energy sources and power

Energy sources

- A **primary energy source** is used in the form that it is found. Primary energy sources include:
  - **fossil fuels** such as coal, oil and gas
  - **nuclear fuel**, e.g. uranium and plutonium
  - **biofuels** from plants and animals
  - wind, waves and sunlight.
- Primary energy sources can transfer their energy to **secondary energy sources**, e.g. electricity.

- Many different primary energy sources are used to make electricity.
- Most of our electricity is made from burning fossil fuels. Fossil fuels are a non-renewable energy source (one day there will be none left).

- The burning of fossil fuels is increasing the amount of carbon dioxide, a **greenhouse gas**, in the atmosphere.
- The greenhouse effect is causing **global warming**, an increase in the temperature of the Earth's surface.
- Global warming is likely to result in climate change, leading to floods and storms.

Power

- **Power** is the amount of energy transferred in one second, i.e. the rate of energy transfer.
- A power of one **watt (W)** transfers one **joule** of energy in one second. This is shown by the equation:
  Energy transferred (in J) = power (in W) $\times$ time (in s)
- A joule is a very small amount of energy, so domestic electricity measures energy transfer in **kilowatt-hours (kWh)**. There are 1000 watts in 1 **kilowatt** (kW).
- A kilowatt-hour (kWh) is the energy transferred by a power of 1 kW in an hour. This is shown by the equation:
  Energy transferred (in kWh) = power (in kW) $\times$ time (in hours)

- The flow of electricity in a circuit is called **current**. It is measured in **amperes**.
- The **voltage** of a power supply is measured in volts (V).
- Electric current in an appliance transfers energy to it from the power supply. The rate at which this is done is found using the equation:
  Electrical power (in W) = current (in A) $\times$ voltage (V)

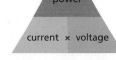

Cover up what you want to find out.

- The power of an electrical device is the rate at which it transfers energy from the power supply. This is usually given as a **rating**.
- For an electrical device of a given power rating, e.g. 1000 W, and an electricity supply of a given voltage, e.g. 230 V, the current the device needs to work is found using the equation:

  Appliance current (in A) = $\dfrac{\text{appliance power (in W)}}{\text{supply voltage (in V)}}$

Ideas about science

You should be able to:

- understand that scientists may identify unintended impacts of human activity, such as climate change

- explain how scientists may be able to devise ways of reducing this impact, for example by using renewable fuel sources rather than fossil fuels.

Improve your grade

Power

Foundation: A kettle comes with this warning: *This kettle must be used with a 230 V, 50 Hz supply. The current in the leads will be 8.7 A.*

How much energy in kilowatt-hours is transferred from the supply when the kettle is used for 15 minutes?   *AO2, AO3* [3 marks]

Efficient electricity

Buying electricity

- Current in a circuit results in both useful and unwanted energy transfers.
  - Current in a **component** transfers electrical energy into other useful forms.
  - Current in the connecting wires wastefully transfers electrical energy to heat.
- An electricity meter records the amount of electrical energy transferred into a house.
- Electricity is metered in kWh or Units to keep the numbers small and manageable.
- Electricity supplied to a component (in units) = power (kW) × time (in hours)
- Cost of electricity = number of units supplied × cost per unit

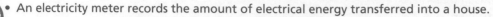

Remember!
Always convert power in W to kW before calculating the number of units used.

Energy diagrams

- Information about energy use can be displayed as:
  - a pie chart to compare different uses of electricity
  - a bar chart to show the amount of electricity used by different components
  - a line graph to show how the number of units consumed changes with time.
- **Sankey diagrams** show the energy transfers in a component. The sum of the energy transfers out of a component equals the input energy. This shows that energy is conserved.

- In the Sankey diagram for a component:
  - Energy flows in from the left. The amount of **energy input** is shown at the left of the arrow tail and is proportional to the thickness of the arrow tail.
  - The sum of the widths of the new arrows at a split is equal to the width of the arrow before the split.
  - Each **energy output** should be shown on the right with an arrow head. Useful energy transfers flow to the right. Wasteful energy transfers flow down.

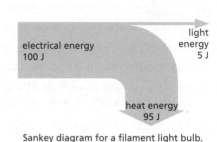

electrical energy
100 J

light energy
5 J

heat energy
95 J

Sankey diagram for a filament light bulb.

Efficiency

- The **efficiency** of a component tells you the proportion of electricity that it transfers into a useful form. It is found using the equation:

$$\text{Efficiency} \times \frac{\text{energy usefully transferred}}{\text{total energy supplied}}$$

- Values for efficiency can range from 1.00 (100%) to 0.00 (0%)
- Components with a high efficiency, e.g. 95%, don't waste much electricity.
- A component with a low efficiency, e.g. 20%, wastes a lot of electricity.

- As a person, you can use less electricity by:
  - using high efficiency A-rated appliances
  - turning off components when they aren't needed
  - not boiling more water than is necessary and cooking food in a microwave oven.
- As a nation, we can use less energy by:
  - using more efficient cars
  - living in houses with better insulation
  - building more efficient power stations and improving the output of old power stations.

- Global demand for energy and other resources will rise in the future as the population increases and the quality of life for many people improves.
- All human activity has an impact on the environment. This impact can be reduced by:
  - recycling resources such as metals, glass and plastics
  - generating electricity from renewable sources of energy, e.g. water, wind and solar power.

Improve your grade

Efficiency

Foundation: Explain why governments have passed legislation which forces people to use energy-efficient lamps in their homes.

AO2 [4 marks]

Generating electricity

Generators

- Moving a magnet near a circuit causes an electric current to flow in the circuit.
- The current flows only when the **magnetic field** is changing – that is, when the magnet is moving.
- Power stations use this idea to produce mains electricity by **generators**.
- A generator contains an **electromagnet** near a coil of wire. There is a voltage across the coil when the electromagnet spins.

Remember!
The magnet has to be spinning near the wire coil for any electricity to be generated – just leaving it there doesn't work.

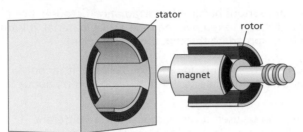

Parts of a generator.

- Power stations use primary fuels, such as fossil, nuclear and biofuels, to boil water into steam.
- The steam passes though a **turbine** (a set of blades on an axle), making its shaft spin round. The turbine shaft spins the magnet inside the generator.
- Increasing the current drawn from the coil requires an increase in the rate of transfer of energy from the primary fuel, i.e. an increase in the amount of primary fuel used each second.
- Primary energy sources used to turn turbines directly include wind and water.
- New developments in electricity generation must conform to government regulations.

How power stations work

- The turbine spins the shaft of the generator to make electricity.
- It is set spinning by steam, hot exhaust gas, wind and water.
- Thermal power stations use coal, oil, gas and nuclear power to spin the turbine from high pressure steam.
  - Coal-fired power stations do this by burning coal to transfer energy into water.
  - Gas-fired power stations burn natural gas to make hot gas for a turbine; another turbine is spun by steam from water heated by the hot gas.
- Hydroelectric power stations use a jet of high pressure water at the base of a dam to spin a turbine.
- Wind-driven power stations use convection currents in the atmosphere caused by the heating effect of the Sun on the land.

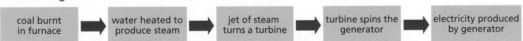

Processes in a coal-fired power station. Oil-fired power stations are very similar – oil is burned instead of coal to produce heat. They are called 'thermal' power stations.

- Nuclear power stations make high pressure steam by:
  - putting fuel rods close to each other in the **reactor** so that they heat up
  - taking away the heat with high pressure water circulating around the rods
  - using the high pressure water to boil low pressure water in a boiler.

Ideas about science

You should be able to:

- understand why the development and application of many areas of science are controlled by governments.

Improve your grade

How power stations work

Foundation: Describe how a power station transfers energy in coal to electricity.　　　　　*AO1* [5 marks]

Waste from power stations

- Waste from nuclear power stations is **radioactive** and a serious health risk.
- Nuclear waste must be carefully stored until it becomes safe.

- Nuclear waste emits **ionising radiation**, which affects body cells.
- An object is only **irradiated** when it is placed in the path of the radiation.
- An object is **contaminated** when it gets mixed up with radioactive material.
- Contamination can be a more serious hazard than irradiation because: it can result in a longer exposure to radiation; it is difficult to remove the radioactive material; it is difficult to stop the radioactive material from spreading through the environment.

- We often overestimate the risk from ionising radiation because: it cannot be seen or felt; its effects take a long time to develop; people worry about unfamiliar technology.
- Statistics about death rates can be used to: compare risks from different technologies; decide which technologies need to be controlled; decide which risks are too small to worry about.

Higher

> ### EXAM TIP
> When you compare the risks of different technologies, don't forget the consequences of failure. The greater the consequences are, the smaller the acceptable risk becomes.

Renewable energy sources

- A renewable energy source can be used over and over again.
- Renewable sources that can spin turbines to make electricity are: **1 hydroelectric** schemes; **2 wind turbines; 3 wave technology**.

Advantages of hydroelectric schemes	Disadvantages of hydroelectric schemes
• They can provide large amounts of electricity. • They can be turned on and off quickly. • They can pump water back behind the dam to store energy.	• They flood large areas of land. • Rotting plants in the water produce methane gas. • They cost a lot to build
Advantages of wind turbines	*Disadvantages of wind turbines*
• They are inexpensive to make. • They need very little maintenance.	• They need to be put in windy places. • They can only generate electricity when there is enough wind.

- Wave technology is still being developed.

- The environmental impact of renewable energy sources can be reduced by careful planning, e.g. sitting wind turbines offshore, using hydroelectric dams to control rivers.
- The financial impact of building these sources can be offset by the benefits of their use.

The National Grid

- Electricity is convenient because it can transfer energy over long distances for many uses.
- The **National Grid** is a network of cables which carries electricity throughout the UK.

- Power cables warm up because they wastefully transfer electrical current to thermal energy.
- Increasing the voltage of a power cable reduces the current. The National Grid carries electricity at a very high voltage to reduce wasteful energy transfers in the cables.
- Substations connected to the National Grid reduce the voltage to 230 V for our homes.

- There are wasteful transfers of electricity as it is generated and transmitted to the consumer.
- Wasteful energy transfers in transmission are much smaller than those in the power station.
- The efficiency of electricity generation (including the efficiency of the National Grid) is the proportion of the energy in the original fuel that is finally delivered to customers as electricity. In the Sankey diagram opposite it is 33 kJ / 100 kJ = 0.33 (or 33%).

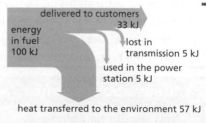

delivered to customers 33 kJ
energy in fuel 100 kJ
lost in transmission 5 kJ
used in the power station 5 kJ
heat transferred to the environment 57 kJ

Where energy is lost in transmission.

Improve your grade

Renewable energy sources

Foundation: The majority of the electricity in the UK is generated from fossil fuels. Explain the advantages and disadvantages of using wind and hydroelectric technology instead. *AO1* [4 marks]

Electricity choices

Choosing the best energy source

- The energy sources that can produce large amounts of electricity are: fossil fuels; nuclear power; hydroelectricity; biofuels.

- The energy sources that rely on the right sort of weather are: wind; waves; solar.

- Energy sources that do not produce greenhouse gases: nuclear power; wind; waves; solar; geothermal.

- Some energy sources have environmental impacts:
  - Fossil fuels produce greenhouse gases; extracting them is dangerous and a pollution risk.
  - Nuclear power creates radioactive waste.
  - Wind farms can cause noise and visual pollution.
  - Hydroelectric and tidal dams flood large areas.

- Some energy sources will eventually run out: fossil fuels; nuclear power.

- Some energy sources are free: wind; hydroelectric; tidal; solar; geothermal.

- When choosing an energy source you should consider:
  - its impact on the environment
  - the cost of building and running the power station
  - how much carbon dioxide and other waste it produces
  - the reliability of the energy source
  - the cost of using the energy source
  - the efficiency of the transfer of energy to electricity.

- The power output of a power station is measured in millions of watts or megawatts (MW).

- A power station which uses nuclear or fossil fuel has a steady output of about 1000 MW and a lifetime of about 40 years.

- Wind farms have a power output of 300 MW, although this varies with the weather, and a lifetime of about 20 years.

- Hydroelectric power stations can have power outputs of about 10 000 MW with lifetimes of about 80 years.

Dealing with future energy demand

- In response to global warming, many countries have agreed to limit their production of carbon dioxide. They can do this by using less energy for transport, heating and electricity.

- Vehicles, factories and power stations emit less carbon dioxide if they become more efficient.

- Switching off appliances at home when they aren't needed reduces energy use.

- Global demand for energy is likely to rise in the future because:
  - there will be more people, especially in developing countries
  - many of them want the high-energy lifestyle of industrialised countries.

- To reduce global energy demand, people in industrialised countries therefore have to use less.

- Industrialised countries already use cleaner, more efficient technology in their workplaces.

- Many goods imported by industrialised countries are made in the developing world with polluting and inefficient technology.

- A secure energy supply in the future requires us to: generate enough electricity to avoid power cuts; have constant access to energy sources; replace old power stations with more efficient ones; use a mix of renewable energy sources; use more renewable energy sources as fossil fuels run out.

Higher

◉ Ideas about science

You should be able to:

- understand that some technologies can impact adversely on the environment and some people's lives, for example by contributing to global warming

- explain that the benefits of using a technology, such as nuclear power stations, should be weighed against the risks, e.g. contamination.

◉ Improve your grade

Dealing with future energy demand

Foundation: It has been suggested that every person on Earth should only be allowed to produce a certain amount of carbon dioxide each year. Explain what impact this could have on your lifestyle. *AO2* [4 marks]

P3 Summary

Making decisions

People may disagree about which technologies should be allowed. Some decisions may be based on the best outcome for the greatest number of people or a moral sense of right and wrong.

The benefits of science-based technologies must be weighed against their costs.

Human activity may have unintended impact on the environment. Scientists may be able to suggest ways of reducing this impact.

The development of science-based technology is often subject to government regulation.

Electricity

Many power stations use a primary energy source to boil water into high pressure steam.

The steam passes through turbines, forcing them to spin round. The turbines spin the magnets inside generators so that a voltage appears across the coil inside.

Increasing the rate of energy transfer from the primary source increases the current in the generator.

Electricity can transfer energy easily from one place to another along the National Grid.

Domestic electricity consumption is measured in units of kilowatt-hours (kWh).

Electric current transfers energy from a power supply to an electrical appliance. Energy transferred = power × time

The power of an appliance is the energy transfer per second. Power = voltage × current

Electricity is a secondary energy source made from primary energy sources such as fossil fuels, uranium, wind, water and sunshine.

The efficiency of a power station is the ratio of its useful output to the total energy input.

Nuclear power

Nuclear power stations produce radioactive waste which emits harmful ionising radiation.

Nuclear power stations are expensive to build but their fuel does not emit carbon dioxide.

Exposure of people to ionising radiation can result in cancer.

Exposure by contamination is often more dangerous because the radioactive material has been incorporated into the body so cannot be removed.

Exposure by irradiation can be limited by shielding and keeping away from the source of radiation.

Renewables

Solar cells produce electricity directly from sunlight without producing carbon dioxide.

Solar power requires a lot of land and, like wind power, is very weather dependent.

Hydroelectricity has a large environmental impact as well as being expensive to build.

Geothermal power can only happen in places where rocks are hot near the Earth's surface.

Wind turbines are spun directly by the wind. They do not produce carbon dioxide but have a visual impact.

Carbon fuels

Biofuels, e.g. wood, plant oils, are renewable and result in no net change of carbon dioxide in the atmosphere.

Some fossil fuels will last longer than others, but they will run out eventually.

Coal, oil and gas produce carbon dioxide when they are burnt to make electricity.

Fossil fuel power stations can easily make large amounts of electricity but are expensive to build.

The chemical reactions of living things

Photosynthesis and respiration

- **Photosynthesis** is the process by which *green plants* make their own food. During the process, light energy is converted to chemical energy. The end product of photosynthesis is **glucose**.

- Photosynthesis plays a vital role in making energy available to organisms through food chains.

- **Respiration** is the process by which *all organisms* release energy from food.

- At night, plants respire only and give out **carbon dioxide**. During the day, plants photosynthesise as well as respire. There will be a net output of oxygen.

- Organisms use energy for many activities, including movement and producing large **molecules** required for growth.

> **Remember!**
> All organisms respire; only plants photosynthesise.

Enzymes

- **Enzymes** are chemicals that speed up the rate of chemical reactions.

- Enzymes are **proteins**. They consist of long chains of **amino acids** joined together.

- Cells assemble enzymes using the instructions provided in genes.

- The chemicals that enzymes work on are called **substrates**.

- The chemicals produced in the reaction are called **products**.

- The order and types of amino acid in an enzyme give it a complicated three-dimensional shape. This shape is essential for the enzyme to work.

- Part of the enzyme called the **active site** has a special shape that the substrate fits neatly into, like a key in a lock.

- Substrate molecules locked into the active site take part in the chemical reaction, and product molecule(s) are released. This is the lock and key model of enzyme action.

- Enzymes need a specific **pH** and temperature to work at their optimum. They stop working if the pH is inappropriate or the temperature is too high.

- The body temperature of mammals and birds is around 37 °C. This is the optimum temperature of *most* enzymes.

> **Remember!**
> Enzymes are *destroyed* by high temperatures. They are just chemicals, so they can't be 'killed'!

2 Substrate molecule fits in active site in enzyme.

3 Reaction occurs and products made; the enzyme speeds up this reaction.

1 Substrate molecules move towards active site in enzyme.

4 Product molecules do not fit in the enzymes as well so are released. The enzyme can then be used again with new substrate molecules.

The lock and key model of enzyme action.

- As the temperature is increased, enzyme activity increases because the reaction rate increases.

- High temperatures change the shape of the active site of the enzyme. At a point where the change in shape is permanent, the enzyme has been **denatured**.

- pH also affects enzyme activity by changing the shape of the active site. The change in shape can be temporary or permanent (it is denatured).

EXAM TIP

In the exam it is likely that you will be asked to interpret information on the activity of enzymes from data or graphs. Practise answering this type of question.

Higher

Ideas about science

You should be able to:

- in experiments on enzymes:
  - when planning an investigation, identify the effect of factors on the outcome, and control factors that might affect the outcome, other than the one being investigated
  - calculate the mean of a set of repeated measurements
  - when asked to evaluate data, make reference to its repeatability
  - use data (rather than opinion) to justify an explanation.

Improve your grade

Enzymes

Higher: Explain what happens to enzyme activity as the temperature is increased.   *AO1* [5 marks]

How do plants make food?

Glucose: making it and using it

- Photosynthesis is a series of chemical reactions that use energy from sunlight to build large food molecules in plant cells and some microorganisms such as **phytoplankton**.

- It is summarised in the word equation: carbon dioxide + water $\xrightarrow{\text{light energy}}$ glucose + oxygen

- Light energy from the Sun is required to drive the reaction between carbon dioxide and water to build up glucose.

- Sunlight is absorbed by the green chemical **chlorophyll** which, along with the enzymes for photosynthesis, is found in structures in the cell called **chloroplasts**.

- Chloroplasts are found in parts of the plant exposed to sunlight.

- The product of photosynthesis, glucose, is:
  - converted to chemicals needed for the plant's growth, e.g. **cellulose**, protein, chlorophyll
  - converted into starch for storage
  - used in respiration to release energy.

- Glucose from photosynthesis and **nitrates** taken up by the plant roots are used to synthesise amino acids, which are assembled into proteins.

- Oxygen is produced as a waste product of photosynthesis.

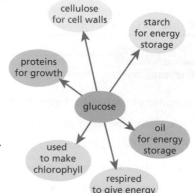

Possible fates of glucose made by photosynthesis.

Higher

- The word and symbol equations for photosynthesis are:

carbon dioxide + water $\xrightarrow{\text{light energy}}$ glucose + oxygen

$6CO_2 + 6H_2O \xrightarrow{\text{light energy}} C_6H_{12}O_6 + 6O_2$

> **Remember!**
> Amino acids are used to make proteins in plant, animal and microbial cells.

The cell as a factory

- The basic unit of life is the cell.

- The cell is surrounded by the **cell membrane**, which controls what enters and leaves cells.

- The jelly-like **cytoplasm** is where most of the chemical reactions in the cell occur.

- Animal cells and microorganisms such as yeast have a nucleus, cytoplasm and a cell membrane.

- The diagram opposite shows the structure of a plant cell.

- The cell membrane allows gases and water to pass in and out of the cell freely, but is a barrier to other chemicals.

- In plant cells, yeast cells and the cells of **bacteria**, the cell membrane is surrounded by the cell wall. It lets water and other chemicals pass through freely. In plants the cell wall is made from cellulose.

- The nucleus contains **DNA**, which stores the **genetic code**. The genetic code carries information the cell uses to make enzymes and other proteins. (Bacteria have no nucleus. Instead, they have a circular DNA molecule in their cytoplasm.)

- The cytoplasm contains **mitochondria**. Animal and plant cells, and microorganisms such as yeast, have mitochondria. These contain enzymes required for the release of energy by **aerobic respiration**.

Plant cell

Generalised plant cell. It includes all the features found in plant cells and is not a specific type.

	Animals	Plants	Microorganisms
Nucleus	✓	✓	Bacteria have cells with no nucleus; their circular DNA is in the cytoplasm. Yeast cells have a nucleus
Cytoplasm	✓	✓	✓
Cell membrane	✓	✓	✓
Mitochondria	✓	✓	Yeast; not bacteria (here, enzymes for respiration are associated with cell membrane)
Cell wall	✗	✓	✓

> **Remember!**
> The cell wall is found in plant cells, but not in animal cells. Microorganisms such as bacteria and yeast also have a cell wall, but it is not usually made of cellulose.

Improve your grade

Glucose: making it and using it

Foundation: Explain how the products of photosynthesis are used by the plant.   *AO1, AO2* [5 marks]

Providing the conditions for photosynthesis

Moving chemicals in and out of plants by diffusion

- **Diffusion** is the movement of chemicals from a high to low concentration.
- In photosynthesis:
  - Water is taken up by plant roots.
  - Carbon dioxide enters plant leaves by diffusion.
  - Oxygen leaves plant leaves by diffusion.
- Diffusion is:
  - passive (it just happens, and does not require energy)
  - the movement of molecules from an area where they are in high concentration, to an area where they are in lower concentration.

Remember!
Diffusion happens because of the random movement of molecules. This movement does not stop when the concentrations of molecules in two areas are equal; there is just no *overall* movement.

Moving chemicals in and out of plants by osmosis

- **Osmosis** is a special kind of diffusion involving water.
- It happens when chemicals are separated by a **partially permeable membrane**, e.g. the cell membrane. The cell membrane lets water pass through, but keeps other chemicals in or out.
- Osmosis is the overall movement of water from an area of high concentration (water, or a dilute solution) to an area of low concentration.
- The movement of water into plant roots from the soil, and across the roots, occurs by osmosis.

EXAM TIP
You may be asked to plot data showing the movement of water into and out of plant tissue and explain this movement, draw conclusions about the concentration of cell sap, and evaluate the data.

Active transport

- Minerals taken up by plant roots are used to make chemicals essential to cells, e.g. nitrogen taken up as nitrates is used to make proteins.
- Nitrates are normally in a higher concentration in plant cells than the soil, so root cells cannot take up nitrates by diffusion.
- Root cells use **active transport** to take up nitrates.
- Active transport uses energy from respiration to transport chemicals across cell membranes.

Higher

Factors that limit the rate of photosynthesis

- The rate of photosynthesis can be limited by temperature, carbon dioxide and light intensity.
- Carbon dioxide is present in the air at only 0.04%, so commercial plant growers often increase levels in their greenhouses.
- If a factor such as light intensity or carbon dioxide concentration is increased, the rate of photosynthesis increases, and then levels off.
- At the point where the graph levels off, something is limiting the rate. This is a **limiting factor**.
- A rise in temperature increases the rate of photosynthesis, up to a certain point. The rate then decreases, because of the effect of temperature on enzyme activity.

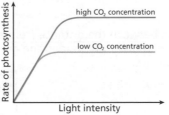

Increasing the concentration of carbon dioxide increases the rate of photosynthesis until another factor becomes limiting.

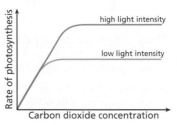

Here, light is limiting, so the rate can be increased by increasing the light intensity.

Higher

Ideas about science

You should be able to:

- in an investigation on osmosis: outline how a proposed scientific explanation might be tested; identify and treat **outliers** in a set of data
- in an investigation on the rate of photosynthesis: identify the outcome and factors that may affect it; suggest how an outcome might alter when a factor is changed.

Improve your grade

Moving chemicals in and out of plants by osmosis

Higher: Explain how plant roots take up water, and how this water moves across a plant root.

AO2 [5 marks]

Fieldwork to investigate plant growth

Investigating the effects of light on plant growth

- When investigating the effects of light intensity on plant growth, whether at different locations within an area or in different areas, ecologists need to:
  - Use an **identification key**, such as the one shown opposite, to identify the plants they find.

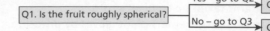

  - Measure how common the plants are.
  - Use a light meter to measure the light intensity.

- Ecologists investigate the **abundance** and distribution of plants:
  - at different locations in an area, and make a comparison
  - in different areas, and make a comparison.

- It is usually not appropriate to count all the plants in the area, so ecologists take appropriate samples.

- A sample is usually taken using a metal or wooden frame of known area called a **quadrat**.

- The ecologist makes a decision about the size of quadrat to use based on the size of the organism, and the size of area to be sampled.

- A quadrat is put on the ground:
  - a number of times, e.g. 10, so that an average is taken, and the number per m² can be calculated
  - at random in a location (by throwing, or preferably, picking coordinates at random from a grid system), *if there is no obvious change in the distribution of plants in the area.*

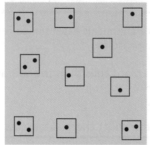

Quadrats used to survey an area.

> **Remember!**
> By using information of plant distribution from quadrats, ecologists can relate the plants' distribution to the availability of light and other factors.

- Alternatively, *if there is an obvious change in the plants in the area*, use the quadrat at measured distances along a line called a **transect**.

- Trees, bushes and hedges, or a group of the same species of plant, can cast shade on an area and therefore affect the growth of plants around them.

- There may therefore be a **correlation** between the distance from a tree, bush and hedge and the growth of other plants, and the growth of plants grown at high and low densities.

- However, other factors may be involved, such as competition for nutrients.

> ### EXAM TIP
> You may be asked to analyse, interpret and evaluate data on the effect of light, as well as other factors, on the growth of plants in an area.

Ideas about science

You should be able to:
- for data of plant distribution:
  - identify where a correlation exists
  - use the ideas of correlation and cause when discussing data and show awareness that a correlation does not necessarily indicate a causal link
  - evaluate critically the design of a field study, e.g. commenting on sample size and how well the samples are matched.

Improve your grade

Investigating the effects of light on plant growth

Foundation: Describe and explain how an ecologist would compare how a plant is distributed in two meadows.

AO1, AO2 [5 marks]

How do living things obtain energy?

Aerobic respiration

- Life processes depend on energy released from food by **respiration**. Respiration is a series of chemical reactions in cells that release energy by breaking down large food molecules.

- All organisms respire, including animals, plants and microorganisms.

- Respiration occurs in every cell in the body.

- The chemical used for respiration is glucose (chemicals in our food are converted into glucose, or we eat glucose itself).

- Respiration using oxygen is called **aerobic respiration**. The word equation for this is:
  glucose + oxygen $\longrightarrow$ carbon dioxide + water (+ energy released)

- Aerobic respiration takes place in animal and plant cells, and some microorganisms.

- Organisms require the energy released by respiration for the synthesis of large molecules and movement.

- Plants use the food produced by photosynthesis for respiration and **active transport**. Excess glucose is stored as starch.

- Respiration takes place as a series of enzyme-controlled reactions, with energy being released in stages.

Remember!
Don't forget that plants carry out respiration as well as photosynthesis. It's just that during bright daylight, photosynthesis occurs much faster than respiration, so the plant gives out oxygen and not carbon dioxide.

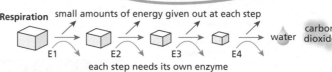

Respiration — small amounts of energy given out at each step
E1 E2 E3 E4 — water — carbon dioxide
each step needs its own enzyme

- The word and symbol equations for aerobic respiration are:
  glucose + oxygen $\longrightarrow$ carbon dioxide + water (+ energy released)
  $C_6H_{12}O_6 + 6O_2 \longrightarrow 6CO_2 + 6H_2O$ (+ energy released)

Higher

Anaerobic respiration

- Some organisms can still obtain energy when oxygen is in low concentration or absent, by **anaerobic respiration**.

- Anaerobic respiration occurs in human muscle cells during vigorous exercise, plant roots in waterlogged soil, and bacteria in deep puncture wounds.

- Aerobic respiration releases much more energy per molecule of glucose than anaerobic energy.

- The word equation for anaerobic respiration in animal cells and some bacteria is:
  glucose $\longrightarrow$ lactic acid + ENERGY

Fermentation

- **Fermentation** is a type of anaerobic respiration used by some microorganisms.

- Some fermentation products are useful to us, e.g. the ethanol produced by yeast (a fungus).

- Some microorganisms, including yeast, and some plants, produce alcohol (ethanol) by fermentation. The equation is: glucose $\longrightarrow$ ethanol + carbon dioxide + ENERGY

- This is important in the production of alcoholic drinks (beer, wine, etc.) and bread.

- Bubbles of carbon dioxide make bread rise, and alcoholic drinks sparkle. In bread, alcohol is evaporated off as it's cooked.

- Anaerobic respiration is also important in biogas generation.

Improve your grade

Fermentation

Foundation: Explain what the graph opposite tells us about changes in concentration of sugar and ethanol as yeast grows.

- sugar concentration
- ethanol concentration

AO1, AO2, AO3 [5 marks]

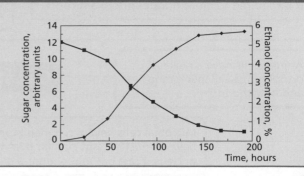

B4 Summary

All living cells require energy from respiration.

Photosynthesis in plants makes food and energy available to food chains.

Enzymes are proteins that speed up chemical reactions. They are assembled in the cytoplasm from the instructions carried by genes.

Each enzyme works best at an optimum temperature and pH.

The rate of an enzyme-controlled reaction increases as temperature increases.

High temperatures and extremes of pH prevent enzyme action by changing the shape of the active site (denaturing the enzyme).

The chemical(s) that an enzyme works on is the substrate. This must be the correct shape to fit to the active site of the enzyme – the lock and key mechanism.

Chemical reactions in cells

Photosynthesis uses energy from sunlight to build large food molecules.

Light is absorbed by the green chemical, chlorophyll.

Chloroplasts contain chlorophyll and the enzymes needed for photosynthesis.

The equation for photosynthesis is:

$$\text{carbon dioxide} + \text{water} \xrightarrow{\text{light energy}} \text{glucose} + \text{oxygen}$$

$$6CO_2 + 6H_2O \xrightarrow{\text{light energy}} C_6H_{12}O_6 + 6O_2$$

The glucose produced is converted to other chemicals that the plant needs, e.g. starch, cellulose and protein.

Fieldwork techniques are used to investigate the effect of light on plants in the wild. Identification keys show the types of plants affected. Plant distribution is shown using quadrats placed at random or along a transect.

Photosynthesis

The rate of photosynthesis is:
- affected by temperature, carbon dioxide and light intensity
- limited if any one of these is in short supply.

Respiration releases the energy from food used to drive chemical reactions in cells.

Aerobic respiration uses oxygen, and is shown by the equation:

$$\text{glucose} + \text{oxygen} \rightarrow \text{carbon dioxide} + \text{water } (+ \text{ energy released})$$

$$C_6H_{12}O_6 + 6O_2 \rightarrow 6CO_2 + 6H_2O \; (+ \text{ energy released})$$

It is carried out by animal cells, plant cells and some microorganisms.

A series of enzyme-controlled reactions release energy in stages.

Respiration

Anaerobic respiration takes place when oxygen is absent (or in very low concentration).

The equation in animal cells and some bacteria is: glucose $\rightarrow$ lactic acid (+ energy released)

The equation in plant cells and some microorganisms including yeast is:
glucose $\rightarrow$ ethanol + carbon dioxide (+ energy released)

It is important economically in the production of alcoholic drinks, bread, yogurt and biogas.

The cell membrane regulates what enters and leaves cells. It allows gases and water to move freely but is a barrier to other substances.

Diffusion and osmosis

O_2 and CO_2 move into and out of leaves by diffusion – the passive movement of molecules from an area of high concentration to an area of low concentration.

Active transport of molecules uses energy to transport molecules across the cell membrane.

It is required to move nitrates into plant roots.

Osmosis (a special form of diffusion) is the overall movement of water, through a partially permeable membrane, from a dilute to a more concentrated solution.

Animal cells have a nucleus, cytoplasm, a cell membrane and mitochondria.

Cell structure

Plant cells have a nucleus, cytoplasm, a cell membrane, mitochondria and a cell wall.

Microbial cells have a cell membrane and a cell wall. Cells of bacteria have no nucleus, but circular DNA in their cytoplasm. Yeast cells have a nucleus. Yeast has mitochondria but bacteria do not.

How organisms develop

Cell specialisation in animals

- In organisms that are **multicellular**, cells are **specialised** to do different jobs.
- Cells of the same type are grouped into **tissues**, e.g. muscle cells → muscular tissue.
- Different tissues are grouped together, and work together, in **organs**. For example, the heart has muscular tissue, epithelial (lining and covering) tissue, blood and nervous tissue.
- Organs work together as body systems, e.g. the circulatory system.

- Organisms begin life as a **zygote** – a fertilised egg.
- The zygote divides by **mitosis** – into 2, 4, 8, etc. – to form an **embryo**.
- In humans, up to and including the eight-cell stage, the cells are *identical*. These cells are **embryonic stem cells** – they will produce *any* cell type in the body.
- After the eight-cell stage, cells become specialised (this is called **differentiation**), and different tissues form.
- As adults, stem cells remain in certain parts of the body. **Adult stem cells** can differentiate into a limited number of cell types, e.g. bone marrow cells into different types of blood cell.

- In specialised cells, only the **genes** are needed to enable the cell to function, as that type of cell is switched on. In embryonic stem cells, any gene can be switched on.

> **Remember!**
> All body cells have the same genes. In most cells, only those required for the cells to function are switched on.

Cell specialisation in plants

- Specialised plant cells form tissues such as the **xylem**, which transports water and mineral salts, and **phloem**, which transports the products of photosynthesis.
- Tissues are organised into organs, e.g. stems, leaves, roots and flowers.
- Cells in regions called **meristems** are unspecialised.

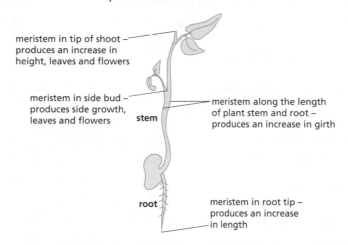

meristem in tip of shoot – produces an increase in height, leaves and flowers

meristem in side bud – produces side growth, leaves and flowers

stem

meristem along the length of plant stem and root – produces an increase in girth

root

meristem in root tip – produces an increase in length

- When meristem cells divide into two, the new cell produced can differentiate into different cell types (the other stays as a meristematic cell).
- In plants, the only cells that divide are in meristems.
- Meristems produce growth in height and width (by division of meristem cells, followed by enlargement of one of the **daughter cells**).

> **Remember!**
> At a certain point, animals stop growing. Only certain cells are capable of dividing to produce new cells. Unlike animals, plants keep growing for their whole lifetime.

Improve your grade

Cell specialisation in animals

Foundation: Compare how cells become specialised in animals and plants.

AO2 [5 marks]

Plant development

Plant clones

- New plants can be grown by placing the cut end of a shoot in water or soil.
- Roots grow at the base of the stem, while the shoot continues to grow.
- Plants grown in this way include garden plants and houseplants.

- Pieces of plants, e.g. plant stems, that have meristems and are used to produce **clones**, are called cuttings.
- Cuttings:
  - can be used to produce new plants with the same desirable features as the parent
  - produce clones that are genetically identical to the parent plant.
- Root growth in cuttings is promoted by plant **hormones** (using hormone rooting powder).
- Another method of cloning is called **tissue culture** – a small piece of tissue, or a few cells are placed on agar jelly containing nutrients and plant hormones. Each will grow into a small plant or plantlet.

- Plant hormones called **auxins** are included in the agar for tissue culture and in hormone rooting powder.
- Auxins increase cell division and cell enlargement, promoting growth of the plant tissue.

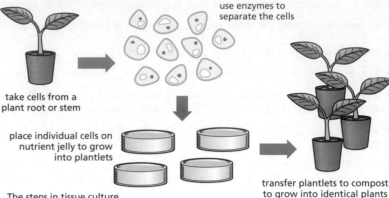

take cells from a plant root or stem

use enzymes to separate the cells

place individual cells on nutrient jelly to grow into plantlets

transfer plantlets to compost to grow into identical plants

The steps in tissue culture.

> **Remember!**
> The presence of a meristem (as sources of unspecialised cells) is necessary to clone a plant from cuttings.

Plant growth and development

- Plant growth and development is affected by the environment.
- Plants' response to the direction of light is called **phototropism**.
- Plants grow *towards* the light, so they are *positively* phototropic.
- Light is essential for photosynthesis, so by growing towards a source of light, plants increase their chances of survival.

- The plant hormone auxin is produced in the growing tip of plant shoots. It moves down the shoot and produces growth below the tip.
- If a plant is illuminated from one side:
  - the auxin produced in the tip is distributed towards the shaded side
  - the auxin produces growth on the shaded side
  - the shoot grows towards the light.

> **Remember!**
> Light affects how the auxin is *distributed* in a shoot – it is moved away from the light. This doesn't affect the overall amount of auxin within the shoot.

① Auxin made in shoot tip
Auxin is mainly found on the shaded side
Not much auxin is found on the light side
light
Cells on the shaded side elongate more
Cells on the sunny side elongate less

② The shoot grows towards the light

The role of auxin in phototropism.

Ideas about science

You should be able to:

- in experiments on phototropism and auxins:
  - in an account of scientific work, identify statements which report data and statements of explanatory ideas (hypotheses, explanations, theories)
  - identify where creative thinking is involved in the development of an explanation
  - recognise data or observations that are accounted for by, or conflict with, an explanation
  - draw valid conclusions about the implications of given data for a given scientific explanation.

Improve your grade

Plant clones

Foundation: Explain how and why plant breeders who have produced a new variety of plant take many cuttings from it.

AO1, AO2 [5 marks]

Cell division

Mitosis

- **Mitosis** is the type of cell division that takes place when an organism grows, and cells divide to repair tissues.

- Mitosis results in the production of *two* **daughter cells** that are genetically identical, i.e. have the same number of **chromosomes** as the parent cell.

This shows four of the cell's chromosomes (two pairs)

Each chromosome makes an identical copy of itself

Chromosomes line up along the centre of the cell

Four chromosomes go into each daughter cell. The cell divides in two

nucleus

chromosomes

Each daughter cell is an exact copy of the original cell

- Before mitosis, the **DNA** in each chromosome is copied. Each chromosome is now a double chromosome, with two DNA molecules.

Note that mitosis involves all the chromosomes in the nucleus.

- During mitosis, each double chromosome separates, so that two nuclei and two cells are produced.

- The events between and leading up to cell division, and cell division itself, are called the cell cycle. The main processes of the cell cycle are:

  1 *Cell growth:* the cell increases in size; numbers of organelles increase; the DNA in each chromosome is copied.

  2 *Mitosis:* two daughter cells, each identical to the parent cell and containing an identical set of chromosomes, are produced as the strands of each double chromosome separate and two nuclei are formed.

Meiosis

- **Meiosis** is the type of cell division used to produces **gametes** (**sex cells** – **eggs** and **sperm** in animals; eggs and pollen grains in flowering plants).

- In humans, gametes contain half the number of chromosomes (23) as body cells (which contain 46, or 23 pairs, of chromosomes), i.e. only one chromosome from each pair.

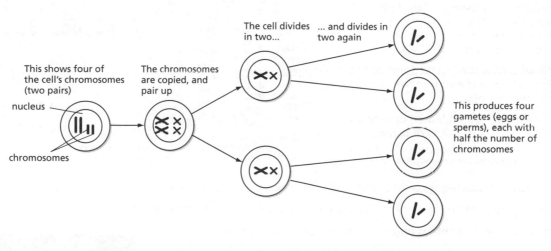

This shows four of the cell's chromosomes (two pairs)

nucleus

chromosomes

The chromosomes are copied, and pair up

The cell divides in two...

... and divides in two again

This produces four gametes (eggs or sperms), each with half the number of chromosomes

Meiosis ensures that sex cells have 23 chromosomes and not 23 pairs.

- At fertilisation, gametes (sperm and eggs) join to form a zygote with 46 chromosomes.

- It's important that gametes contain 23 chromosomes (1 chromosome from each pair), otherwise the zygote would end up with 92 chromosomes.

- The zygote contains a set of chromosomes from each parent.

Remember!
Mitosis produces two daughter cells with an identical number of chromosomes as the parent cell. Meiosis produces four daughter cells with half the number of chromosomes.

Improve your grade

Mitosis and meiosis

Higher: Explain why gametes (sex cells) are produced by meiosis and not by mitosis.   *AO1, AO2* [4 marks]

Chromosomes, genes, DNA and proteins

Chromosomes, genes and DNA

- Chromosomes:
  - are thread-like structures found in the nucleus
  - are made from a DNA molecule
  - can be grouped into pairs (humans have 23 pairs).

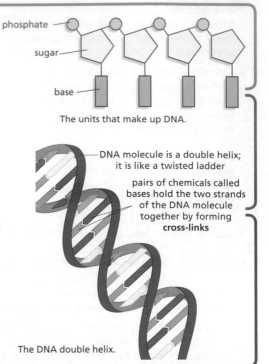

The units that make up DNA.

- The DNA molecule is a **double helix**.

- The DNA molecule is two strands facing each other.

- The strands of DNA are made up of units linked by chemicals called **bases**.

- There are four bases – A, T, C and G, which always pair up in the same way – A with T, and C with G.

- The order of bases in a gene makes up the **genetic code**. This is the code that gives instructions for the assembly of a protein (the **amino acids** that are in the protein, and the order in which they're arranged).

DNA molecule is a double helix; it is like a twisted ladder

pairs of chemicals called bases hold the two strands of the DNA molecule together by forming **cross-links**

The DNA double helix.

Remember!
The DNA molecule is like a twisted ladder. The rungs of the ladder are bases.

Protein synthesis

- In plant and animal cells, the genetic code that carries the instructions for protein synthesis is on the DNA *in the nucleus.*

- Protein synthesis occurs *in the* **cytoplasm**.

- Genes do not leave the nucleus, so in order to carry the genetic code to the cytoplasm, **messenger RNA (mRNA)** is produced in the nucleus, using DNA as the template (during which time the two strands of the DNA separate).

- mRNA carries the instructions for the assembly of proteins (protein synthesis) into the cytoplasm.

- Proteins are assembled on organelles in the cytoplasm called **ribosomes**.

Proteins are assembled from amino acids according to instructions provided by a gene.

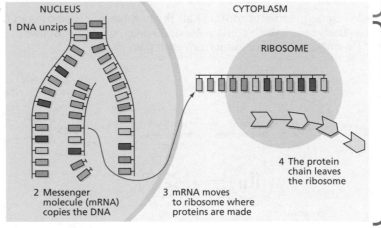

NUCLEUS
1 DNA unzips

CYTOPLASM
RIBOSOME

2 Messenger molecule (mRNA) copies the DNA

3 mRNA moves to ribosome where proteins are made

4 The protein chain leaves the ribosome

- The number and sequence of amino acids determines the type of protein and its properties.

- The sequence of amino acids in the protein is determined by the genetic code.

- The bases work in threes (base triplets) to code for an amino acid.

- mRNA is a copy of the base sequence of the DNA that makes up a gene.

- The mRNA leaves the nucleus and attaches to a ribosome.

- Transfer RNAs (tRNA) ferry amino acids to the ribosome, where they are bonded together.

EXAM TIP

You need to be able to explain this process in terms of how the order of bases in a gene provides the code for making a particular protein.

Ideas about science

You should be able to:
- identify how creative thinking was involved in developing the idea of, and explaining, the genetic code.

Improve your grade

Protein synthesis

Higher: Describe the process by which proteins are produced.

AO1 [5 marks]

Cell specialisation

Switching genes on and off

- The cell only produces the proteins it needs to carry out its function.

- The **genes** to make these proteins are **switched on**; the others are **switched off**.

- Up to the eight-cell stage of the embryo, the cells (embryonic stem cells) are identical.

- The cells produced by the division of embryonic stem cells undergo differentiation to produce specialised cells.

- Specialised cells begin to make specific proteins. They usually change shape and structure, e.g. muscle cells produce the proteins that enable them to contract.

- In embryonic stem cells, any gene can be switched on, so they can produce any type of cell.

- Embryonic stem cells (and adult stem cells) therefore have the potential to replace cells needed to replace damaged tissues.

Stem cell research and therapy

- Stem cells are used to produce new cells to replace damaged or diseased cells.

- Adult stem cells are found at various locations in the body, e.g. the bone marrow.

- These cells can be used to produce a limited number of cell types, e.g. bone marrow cells will differentiate to produce types of blood cells.

> **EXAM TIP**
>
> Make sure that you understand the difference between embryonic and adult stem cells.

- Using embryonic stem cells raises ethical issues, because in removing cells, the embryo is destroyed.

- According to some, embryos have a right to life from when they're conceived.

- Embryonic stem cells are usually removed from surplus embryos from in vitro fertilisation (IVF).

- The creation of embryos produced with the intention of destroying them would be even more controversial.

- Work with stem cells is therefore the subject of government regulation.

- **Therapeutic cloning** overcomes some ethical issues of using embryonic stem cells. It involves:
  - replacing the nucleus of an egg by the nucleus of a body cell
  - stimulating the egg cell to divide to produce an 'embryo'.

- The technique does not require fertilisation, and the cells will be genetically identical to the patient's (so will not be rejected by the immune system). The 'embryo' produced is still destroyed after stem cells are extracted.

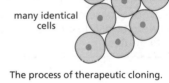

The process of therapeutic cloning.

- Using chemical treatment, scientists have managed to transform mammalian body cells into stem cells.

- Using this technique, inactive genes in the nuclei of body cells have been reactivated. The hope is that the transformed cells will be able to form cells of all cell types.

Ideas about science

You should be able to:

- identify ethical issues that arise when using stem cells in therapy

- develop arguments based on decisions relating to the use of embryonic stem cells in medical research, focused on how the outcomes might affect people

- consider that, for some people, certain actions (such as the use of embryonic stem cells) are considered wrong whatever the consequences.

Improve your grade

Stem cell research and therapy

Higher: Scientists have reprogrammed skin cells to function as stem cells. Explain, in principle, how this technique is carried out, and why this might be preferable to using embryonic stem cells.   *AO2* [5 marks]

B5 Summary

Plant cuttings can be produced from sections of plant stems that include parts of meristems.

Plant cuttings can develop into new plants that are clones of the parent.

Rooting of plant cuttings can be promoted by plant hormones called auxins.

Plant cuttings are used to grow plants with identical, desirable features.

A fertilised egg cell (zygote) divides by mitosis to form an embryo.

At the eight-cell stage, all the cells are embryonic stem cells. Embryonic stem cells can produce any type of cell.

After the eight-cell stage, most of the embryo's cells become specialised.

In the adult, some (adult) stem cells remain; these can develop into certain cell types.

How organisms develop

In multicellular organisms, cells become specialised.

Groups of specialised cells are grouped into tissues.

Groups of tissues form organs.

Growth and development of plants is affected by the environment.

A plant's response to light – it grows towards it – is called phototropism. Phototropism increases a plant's chances of survival.

When a plant is exposed to light from one direction, auxin is redistributed towards the shaded side, where it causes growth.

In plants, only cells in meristems can divide by mitosis.

New cells produced by meristems are unspecialised and can develop into any cell type.

Specialised cells are grouped into tissues, e.g. transporting tissues (xylem and phloem).

Tissues are grouped into organs, e.g. stems, roots, flowers, leaves.

Cell division

When a cell divides by mitosis, two *daughter* cells are produced.

Each daughter cell is genetically identical to the parent and to each other.

Meiosis is the type of cell division that produces gametes (sex cells). In meiosis, cells with half the number of chromosomes are produced. Meiosis is essential to produce sex cells.

The zygote contains a set of chromosomes from each parent, so if sex cells were to divide by mitosis, at fertilisation, the chromosome number would double.

The cell cycle involves:
- *Cell growth:* when numbers of organelles (structures in cells having a function) are increased and the DNA molecule of each chromosome is copied, producing 'double chromosomes' with two strands of DNA.
- *Cell division:* the two stands of DNA in each of the chromosomes separate, two nuclei form and the cytoplasm divides.

Gene function

DNA is a double helix, like a twisted ladder. The rungs of the ladder are bases. There are four bases in DNA. The bases always pair in the same way – A with T; C with G.

The order of bases in a gene is the instruction for the production of a protein. It defines the order in which amino acids are assembled into a protein.

The genetic code is part of the DNA in the nucleus.

A copy of the gene is made (messenger RNA – mRNA) using DNA as the template.

mRNA leaves the nucleus, attaches to a ribosome, and proteins are assembled on it in the cytoplasm.

All the cells in an organism have the same genes.

Only those genes required for a type of cell to function are switched on.

Other genes are switched off, so the cell only synthesises the proteins it needs.

In embryonic stem cells, any gene can be switched on, to produce any type of specialised cell.

Embryonic and adult stem cells have the potential to produce cells to replace damaged tissues.

Use of embryonic stem cells is subject to government regulation because of ethical issues.

In the carefully controlled conditions of mammalian cloning, it is possible to switch on genes to produce the cell types required.

The nervous system

Sending messages

- **Multicellular** organisms need communication systems, so that the body works as a whole and not as individual cells or organs.

- The two communication systems are the **nervous system** and the hormonal system.

- The nervous system:
  - sends messages using nerve cells or **neurons**, which produce a quick, short response; the nerve message, or nerve **impulse**, is *electrical*
  - has specialised organs called the brain and spinal cord.

- The hormonal system produces chemical messages in the form of **hormones**. The system is slower than the nervous system, but the response is longer-lasting.

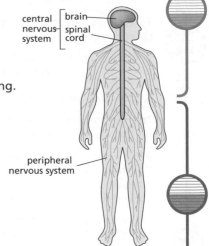

- In humans and other vertebrates, the **central nervous system (CNS)** consists of the brain and spinal cord.

- In the mammalian nervous system, the CNS is connected to the **peripheral nervous system (PNS)**. This is the neurons, which connect the CNS to the whole body.

- There are different types of neurons:
  - **Sensory neurons** connect **receptors**, e.g. in the eyes, ears and skin, which detect changes in the environment (called **stimuli**), with the central nervous system.
  - **Motor neurons** connect the central nervous system to **effectors**, e.g. muscles, which produce a **response**.

The central and peripheral nervous systems.

- Hormones are chemicals that are produced by **glands**. They are transported in the blood. This means that all organs of the body are exposed to them, but they affect only their 'target' cells.

- In the hormonal system, responses are slower and longer-lasting. For example:
  - **Insulin** is produced by the **pancreas**. It acts on the liver, muscles and body cells to take up **glucose** from the blood.
  - **Oestrogen** is produced by the **ovaries**. It is a sex hormone that controls the development of the adult female body at puberty, and the menstrual cycle.

> **Remember!**
> The nervous system uses electrical messages; the hormonal system uses chemical messages.

Neurons

- Neurons are cells **specialised** for carrying nerve impulses, so they are often very long.

- Neurons consist of the **cell body**, which contains the nucleus of the cell, and a long **axon**. Branches on the cell body called **dendrites** receive inputs from other cells (receptors and **nerves**) and conduct impulses towards the cell body.

- Axons carry impulses away from the cell body (to other nerves and muscles).

cell body
nucleus
dendrites
direction of impulse
fatty sheath (insulation)
axon

A motor neuron.

- The axon is a long extension of the **cytoplasm** in a neuron that communicates with the CNS or effector. Some neurons are therefore the longest cells in the body.

- Some axons are covered with an insulating fatty sheath called the **myelin** sheath.

> **Remember!**
> Neurons communicate with other neurons, but they do not physically touch each other.

- The speed of the nerve impulse is affected by:
  - temperature (the speed is increased; it's always faster in warm-blooded animals than cold-blooded animals)
  - the diameter of the axon (the wider the axon, the quicker the response)
  - the myelin sheath (as well as insulating the neuron from neighbouring cells, the presence of the myelin sheath speeds up the nerve impulse – it is able to 'jump' from gap to gap along the sheath, making it travel much more quickly).

Improve your grade

Neurons

Higher: Explain how nerve cells (neurons) are adapted to transmitting nerve impulses. *AO2* [5 marks]

Linking nerves together

Synapses

- Some neurons send messages to other neurons. There is a small gap called a **synapse** between one neuron and the next, through which the message has to be transmitted.
- As the nerve impulse reaches the end of the nerve, it is changed to a chemical signal, which crosses the synapse and sets up an electrical impulse in the next neuron.

- Sometimes a neuron has many synapses so that it can communicate information with all these neurons.
- There is no physical connection between neurons. The presence of a synapse means that a nerve is able to communicate better with several neurons that may go to different locations.

Higher

- As the nerve impulse reaches the end of the first neuron, a chemical **transmitter substance** is released.
- The transmitter diffuses across the synapse and binds with **receptor molecules** on the membrane on the next neuron. This initiates a nerve impulse in the next neuron.
- After an impulse has been transmitted across, the chemical transmitter is removed from the synapse (it is taken back up by the neuron or broken down by an enzyme).
- There are many different types of transmitter molecules. These work on different nervous pathways, e.g. serotonin is a transmitter that is important in brain function.
- Some transmitters work by inhibiting the next nerve instead of exciting it. Others work on muscles instead of nerves.
- Different transmitters have different receptor molecules.

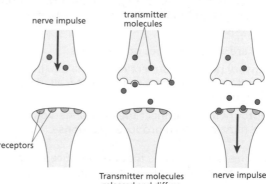

Transmitter molecules released and diffuse across synapse

Transmitter molecules bind to receptors. Nerve impulse is propagated to second neuron

The transmission of a nerve impulse across a synapse.

Nervous co-ordination

- The nervous system responds to changes in the environment called **stimuli** (singular **stimulus**).
- Stimuli are detected by special cells called **receptors**, e.g. light receptors, temperature receptors.
- Sometimes the receptors are grouped together or form part of organs, e.g. the eye and ear.

- A response to a specific stimulus may be required. The CNS co-ordinates the response.
- The response is made by an **effector**. Effectors include **glands** and muscles.
- Glands make and release chemicals such as enzymes and hormones, e.g. the hormone insulin is released after a meal when blood sugar rises.
- Muscles are used for movement. Their contraction helps the body to move away from dangerous stimuli and towards pleasant ones. Muscles are also used for movement we're not conscious of, e.g. our heartbeat.

> **EXAM TIP**
> You need to be clear about the definitions of receptor cells and effector cells.

Ideas about science

You should be able to:

- identify ethical issues when carrying out investigations on how neurotransmitters work in humans and other mammals
- consider that investigation of these could benefit people with deficiencies in neurotransmitters (acetylcholine in Alzheimer's disease; dopamine in Parkinson's disease), so that research might be justified whatever the consequences.

Improve your grade

Synapses

Higher: Describe how a nerve impulse is transmitted from a sensory nerve to a nerve close to it in a spinal cord.

AO1, AO2 [5 marks]

Reflexes and behaviour

Reflexes

- A **reflex** is a simple response to a stimulus, e.g. removing your foot automatically if you step on a sharp object or a hot one.

- The pathway of a reflex action through the nervous system is called the **reflex arc**.

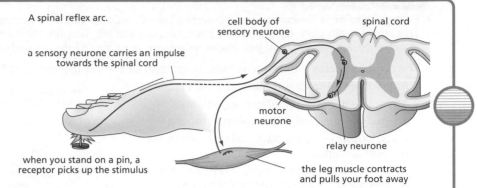

A spinal reflex arc.

a sensory neurone carries an impulse towards the spinal cord

cell body of sensory neurone

spinal cord

motor neurone

relay neurone

when you stand on a pin, a receptor picks up the stimulus

the leg muscle contracts and pulls your foot away

- The pathway is:

stimulus → receptor → sensory neuron → relay neuron in CNS → motor neuron → effector (muscle)

- Simple reflexes are automatic; they require no learning and happen from birth.

- Reflexes enable us to respond quickly to stimuli that could harm us. Simple reflexes in humans include:
  - dropping a hot object, the knee-jerk reflex, the pupil reflex (when a bright light is shone into your eye, the pupil gets smaller)
  - when a baby is born, newborn reflexes, e.g. stepping (taking steps when a baby is held under the arms, with head supported), grasping (when a finger is placed into the baby's palm), sucking (when a nipple or finger is placed in a baby's mouth).

- Responses are rapid because they use fixed pathways that do not involve the brain making a *conscious* decision (**relay neurons** are in the CNS, but the brain is not involved at all if the stimulus is below the neck).

- Relay neurons in the CNS connect with other neurons that run to the brain, so we:

  1 know what's happened after the reflex action has occurred

  2 can override an action, e.g. keeping hold of a hot object when you'd rather not drop it (the brain then sends a message to motor neurons, which changes the response of the muscles in your hand, so you keep hold of the object).

Higher

Instinctive and learned behaviour

- Animals have certain **behaviours** that help them to survive in their environment. Behaviour can be **instinctive** or **learned**.

- Instinctive behaviours are controlled by reflex responses, e.g. woodlice moving away from the light. Simple animals have simple nervous systems, so they can't *learn* behaviours.

- A reflex response to a stimulus can be learned by introducing an *unrelated* stimulus in association with the first. This is called **conditioning**.

- Two examples of conditioning are:
  - *Ivan Pavlov's work with dogs*. Dogs produce saliva (salivate) in response to the smell, sight and taste of food. Pavlov rang a bell immediately before giving dogs food. This process was repeated. Soon, the dogs would produce saliva at the sound of the bell, even if not given food. They had learned to associate the sound of the bell with food.
  - *John B. Watson's study with eight-month-old Albert*. Albert liked, and showed no fear of a white lab rat. Albert was then shown the rat while Watson made a loud noise which made Albert cry. Later, when Albert was shown the rat, he showed signs of distress even when there was no loud noise.

- With Pavlov's dogs, the bell is called the **secondary stimulus**; the **primary stimulus** is the food. Here, the secondary stimulus has no direct connection with the primary stimulus.

- **Conditioned reflexes** are a simple form of learning that help us, and other animals, to survive. For example, association of a plant's bright colours with the fact that the plant is poisonous helps an animal that's likely to eat the plant to avoid it and survive (if it survives its first encounter).

EXAM TIP

You need to be able to describe and explain two examples of conditioning, including Pavlov's dogs.

Higher

Improve your grade

Instinctive and learned behaviour

Higher: A bird eats a poisonous, brightly coloured caterpillar. It is sick, but survives. In future, it avoids eating this type of caterpillar. Explain how this is an example of a conditioned reflex and how it might help the bird's survival.

AO2 [5 marks]

The brain and learning

Brain structure

- Humans and other mammals have complex brains made up of billions of neurons. This larger brain gave early humans a better chance of survival; it enables learning by experience, including social behaviour, where we are able to interact with others.

- The **cerebral cortex** – the thin, folded, outer layer of the brain – is involved with:
  - *Intelligence* – how we think and solve problems.
  - *Memory* – how we remember experiences.
  - *Language* – how we communicate verbally.
  - *Consciousness* – being aware of ourselves and our surroundings.

  A larger number of folds in the cerebral cortex increases our ability to process information.

The cerebral cortex (the wrinkled surface layer of the brain), which is responsible for conscious thought and actions

The cerebellum, which controls movement and posture

The medulla, which controls breathing and heart rate

The main areas of the human brain.

- **Neuroscientists** map the regions of the brain using invasive and non-invasive methods.

- Invasive methods include:
  - studying how a person is affected when a certain part of the brain is damaged
  - during brain surgery, using electrodes to stimulate parts of the brain electrically, and seeing how patients are affected, including reporting memories and sensations.

- Non-invasive methods include producing images and mapping activity with scanning techniques, e.g. magnetic resonance imaging (MRI). These are useful in:
  - comparing non-diseased brains with the brains of people with brain disease, e.g. Alzheimer's disease
  - looking at activity in the brain when it's stimulated (by music, language or images).

Learning

- Transmitting impulses in the brain leads to links forming between the neurons. This is called a **neuron pathway**.

- If an experience is repeated, more and more impulses follow the same pathway. The pathway is strengthened.

- Neuron pathways are also strengthened by strong stimuli using colour, light, smell and sound.

- Learning happens in the brain as neuron pathways develop in the brain.

- Repeating actions strengthens neuron pathways; we get better at certain skills the more we practise.

- Learning results from experience where:
  - new neuron pathways form (and other pathways may be lost)
  - certain pathways in the brain become more likely to transmit impulses than others.

- Neuron pathways are formed more easily in children than adults.

- With billions of neurons in our brains, the potential number of neuron pathways is huge. This means we can adapt to new situations and respond to new stimuli.

- Children are born with certain instinctive responses to stimuli, e.g. the rooting reflex, where they turn their face towards a stimulus to aid breast feeding, but soon develop learned behaviours.

Higher
- Children not presented with new, appropriate stimuli, or those isolated during development, may not progress in their learning.

- Evidence suggests that children can only acquire certain skills at a particular age. **Feral** children (children who have lived away from human contact since a very early age) develop only limited language skills when returned to civilisation.

Remember!
It's the interaction between humans and their environment that enables neuron pathways to develop.

Ideas about science

You should be able to:

- identify that some forms of scientific research into the development of learning in humans and other mammals have ethical implications

- consider arguments and actions in ethical issues concerning techniques used to map the human brain.

Improve your grade

Brain structure

Foundation: Describe how scientists have mapped the areas of the brain to see how it works.   *AO1* [3 marks]

Memory and drugs

Memory

- **Memory** is the storage and then retrieval (bringing back) of information.
- There are two types of memory:
  - **Short-term memory** involves information from our most recent experiences, which is only stored for a brief period of time.
  - **Long-term memory** involves information from our earliest experiences onwards that can be stored for a long period of time
- You are more likely to remember information if:
  - There is a *pattern* to it. To remember information with no obvious link, you could try to put a pattern to it.
  - You use *repetition* (repeating things), especially over an extended period of time. You could read or rehearse something several times. Evidence suggests that the time intervals between the repeats is important.
  - There is a strong stimulus associated with it. Strong colours, bright light, strong smells or loud sounds associated with information can help us to remember it.

> **EXAM TIP**
>
> You need to understand the terms storage, retrieval, repetition and forgetting when referring to memory models.

- Scientists use models to try to explain how we store and retrieve information.
- The **multi-store model** splits memory into sensory memory, short-term memory and long-term memory, and shows how these work together.
- If information arrives in a memory **store** that is not passed on or retrieved, the information is lost, i.e. forgotten.

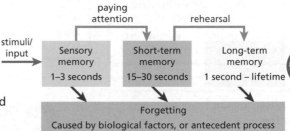

The multi-store memory model.

- Models are limited in explaining how memory works. This is because:
  - Memory is more complicated than shown in the model.
  - No models have an exact explanation of how long-term memory works.
  - The multi-store model is too linear, and doesn't provide sub-divisions of short-term and long-term memory.
  - The model does not differentiate between different types of stimulus and the difference in performance of individuals.

Drugs and the nervous system

- Many drugs and **toxins** work by affecting the transmission of nerve impulses across synapses, stopping the transmission, changing the speed of the transmission, or making the impulse stronger or weaker. For example:
  - The **antidepressant** Prozac increases levels of the **transmitter substance** called **serotonin**.
  - Curare, used by South American Indians as an arrow poison, blocks the action of another type of transmitter molecule.

- **Beta blockers** are prescription drugs that block the transmitter molecule **adrenaline**, so they reduce the heart rate. They're used to treat people with problems with their heart rhythm, but some people use them to control anxiety during public performances.
- The drug **Ecstasy (MDMA)** works on serotonin, the same transmitter that Prozac affects.

- Following the transmission of a nerve impulse, the transmitter molecules should be removed from the synapse.
- MDMA blocks the sites on the neuron where MDMA is reabsorbed, increasing its concentration.
- MDMA therefore gives a feeling of well-being, because of increased levels of serotonin.
- After taking MDMA, the brain's serotonin is depleted, so the person is irritable and tired.

Improve your grade

Drugs

Foundation: Some chemicals affect how nerve impulses are transmitted across synapses. Give **two** examples of these chemicals, and state how these chemicals work. *AO1, AO2* [3 marks]

B6 Summary

A receptor is used to detect a stimulus; the response is produced by an effector.

Receptors and effectors can form part of complex organs.

Responding to change

The nervous system produces a quick, short-lived response to a stimulus.

The hormonal system produces a slower, longer-lasting response.

The nervous system is the central nervous system (brain and spinal cord) and the peripheral nervous system (the nerves).

The nervous system

The outer, folded layer of the brain is the cerebral cortex. It is concerned with intelligence, memory, language and consciousness.

Scientists map the brain using several techniques.

The CNS co-ordinates an animal's response to a stimulus.

Sensory neurons carry impulses from receptor cells to the CNS.

Motor neurons carry impulses from the CNS to effectors.

Sensory and motor neurons are linked by relay neurons.

The nervous system is made up of nerve cells or neurons; these transmit electrical impulses.

A neuron has a cell membrane and cytoplasm, which is extended into an axon. The myelin sheath (which surrounds some axons) insulates the nerve and speeds up the nerve impulse.

A reflex arc is a fixed nervous pathway that enables quick, 'automatic' responses independent of the brain.

Reflex actions

Simple reflexes include dropping a hot object and newborn reflexes, e.g. sucking.

A spinal reflex arc includes a receptor → sensory neuron → relay neuron → motor neuron → effector.

Neurons do not connect physically; impulses are transmitted across gaps called synapses.

Some toxins and drugs affect the transmission of nerve impulses across synapses.

Synapses

There are many different transmitters, each corresponding to a specific receptor.

An impulse arriving at the end of a nerve causes the release of a chemical transmitter.

The transmitter diffuses across the synapse, binds to receptor molecules, and sets up an impulse in the next neuron.

Simple organisms rely on reflexes for most of their behaviour; these aid survival.

More complex organisms show learned behaviours.

Learning and behaviour

Conditioning is a reflex response to a new (secondary) stimulus, learned by introducing it together with a main (primary) stimulus.

The human brain has billions of neurons that allow learning by experience.

When interacting with the environment, new neuron pathways form in the brain.

New skills can be learnt by repetition (this strengthens neuron pathways).

The number of pathways possible enables mammals to adapt to new situations.

Evidence suggests that children may only acquire some skills at a certain age.

Memory is the storage and retrieval of information.

Memory

You are more likely to remember something if you can see, or put, a pattern to it, repeat the information over a period of time, or associate a strong stimulus with it.

Memory can be short-term memory or long-term memory.

The multi-store model shows how short- and long-term memory are linked, but models of memory are limited for a number of reasons.

Atoms, elements and the Periodic Table

The history of the Periodic Table

- An **element** contains all the same type of **atoms**.
- The modern **Periodic Table** is based on the Russian chemist Dmitri Mendeleev's ideas.
- Mendeleev arranged elements into **groups** (vertical columns) and **periods** (horizontal rows) based on their **relative atomic masses** and patterns in their properties.
- Mendeleev left gaps for undiscovered elements and predicted properties of missing elements.
- Johann Döbereiner noticed 'triads' that linked patterns of the relative atomic masses for three elements.
- John Newlands noticed an 'octaves' pattern, where every eighth element had similar properties.
- Scientists rejected Döbereiner's triads and Newlands' octaves because most elements did not fit their 'patterns'.
- When new elements were discovered, they fitted Mendeleev's predictions.
- Data about properties of elements in the Periodic Table can be used to work out trends and to make predictions.

Lines of discovery

- When elements are heated they emit coloured flames. Some elements emit distinctive flame colours, e.g. lithium salts produce a red flame.
- The coloured light can be split into a **line spectrum** that is unique to each element.
- The discovery of some of elements, e.g. helium, happened because of the development of spectroscopy.
- Helium was discovered when chemists looked at the line spectrum from the sun.

The line spectrum of sodium is two single yellow lines, so close together they look like one.

Remember!
Every element has a different pattern of lines. The pattern of lines in a line spectrum can be used to identify the element.

Inside the atom

- Atoms have a tiny, central nucleus that contains **protons** and **neutrons**.
- **Electrons** travel around the outside of the atom in **shells**.

Particle	Charge	Mass
proton	+1	1
electron	−1	almost zero
neutron	0	1

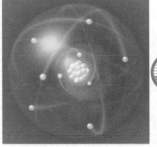

- All the atoms of a particular element have the same number of protons.
- Atoms have the same number of protons and electrons.
- The modern Periodic Table arranges atoms in order of their **proton number**.
- Number of protons + number of neutrons = relative atomic mass

A diagram of an atom showing the electrons orbiting the nucleus, which is made up of protons and neutrons (the nucleus is not to scale).

Higher

Ideas about science

You should be able to:
- separate data from explanations about the Periodic Table
- discuss why Mendeleev's table was an improvement on earlier ideas
- discuss how important it was that Mendeleev's predictions about elements were correct
- use data about elements to identify trends in properties and make predictions using the Periodic Table.

Improve your grade

The history of the Periodic Table

Foundation: Explain why Mendeleev's arrangement of elements was an improvement on Döbereiner's triads and Newlands' octaves.

AO1 [4 marks]

Electrons and the Periodic Table

Sorting electrons

- Electrons are arranged in shells around the nucleus.
- The first shell is closest to the nucleus and can hold 2 electrons.
- The second and third shells are further away from the nucleus.
- The second shell holds 8 electrons.
- The **electron arrangement** of oxygen can be written 2.6.
- The electron arrangement of chlorine can be written 2.8.7.

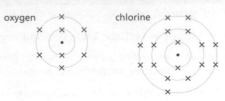

Electron arrangements of oxygen and chlorine atoms. Each cross represents an electron.

- The number of electrons in an atom is the same as the number of protons.
- The number of electrons in an atom is the same as the proton number.
- For the first 20 elements, the third shell holds 8 electrons.

Electron shell	Number of electrons
1	2
2	8
3	8

- Potassium, proton number 19, has an electron arrangement 2.8.8.1.

- Electrons in different shells have different **energy levels**.
- The closer the electron shell is to the nucleus, the lower the energy level.

Magnesium follows sodium in the Periodic Table, and has one extra proton and one extra electron in the third shell.

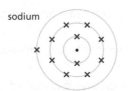
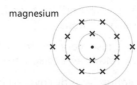

Finding elements in the Periodic Table

- In the Periodic Table, elements are arranged in order of proton number.
- A row across the Periodic Table is called a period.
- The number of electrons in the outer shell increases across a period.
- As you move from left to right along a period, elements change from **metals** to non-metals.

EXAM TIP

Make sure you know how to find the proton number of an atom from the Periodic Table. You need to be able to use the proton number to work out the electron structure of an atom by drawing 'dot and cross diagrams' (like the ones for oxygen and chlorine on this page) and by writing electron arrangements, e.g. 2.6 or 2.8.7.

- As you move from left to right along a period, each element has one more proton and one more electron.
- Properties, e.g. melting points, change across a period. These changes are called **trends**.
- Elements in Group 0 have full electron shells and are **inert** – this means they are very unreactive.

- The number of electrons in the outer shell of an atom is the same as its group number on the Periodic Table.
- Atoms of elements with up to 3 electrons in their outer shell are metals.
- Atoms of elements with 5 or more electrons in their outer shell are non-metals.
- Elements with full outer shells are the inert gases.

Improve your grade

Finding elements in the Periodic Table

Higher: An atom has the electronic arrangement 2.8.1.

Identify the element and explain why its electronic arrangement shows that it is likely to be a metal.

AO2 [3 marks]

Reactions of Group 1

Group 1 – the alkali metals

- A **group** is a vertical column in the Periodic Table.
- Group 1 is called the **alkali metals**.

- All elements in Group 1 are metals and have one electron in the outer shell of their atoms.
- Group 1 elements are soft metals that can be cut with a knife. The freshly cut surface is shiny but it **tarnishes** quickly in moist air by reacting with oxygen.
- The physical properties of Group 1, e.g. melting point, boiling point and density, show trends down the group.

- The reactivity of Group 1 elements is linked to the single electron in the outer shell. They all form **ions** with a 1+ charge by losing an electron.
- The outer electron is easiest to lose if the atom is bigger (because the electron is further from the nucleus). So reactivity increases down the Group 1 elements as the atoms get bigger.

Li lithium	Be beryllium
K potassium	Ca calcium
Rb rubidium	Sr strontium
Cs caesium	Ba barium
Fr francium	Ra radium

Group 1 in the Periodic Table – the alkali metals.

Reactions of Group 1 elements with water

- Group 1 elements all react with water, e.g. lithium and sodium fizz and move around on the surface of the water. The reaction gets more violent as you move down Group 1, e.g. potassium explodes and rubidium explodes more violently.
- In the reaction, hydrogen gas is formed (which 'pops' when lit).
- The reaction also makes a metal hydroxide, which is an alkali and turns **pH** indicator blue.
- Group 1 metals are flammable and their hydroxides are harmful and corrosive. When handling Group 1 metals they should be kept away from water and naked flames.

highly flammable · toxic · corrosive

harmful · explosive · oxidising

Take precautions when you see these hazard symbols.

- The equation for the reaction of Group 1 elements with water is:

metal + water ⟶ metal hydroxide + hydrogen

- If we use M to stand for any of the Group 1 metals, the general equation is:

$2M(s) + 2H_2O(l) \longrightarrow H_2(g) + 2MOH(aq)$

- For example, the **balanced symbol equation** for the reaction of sodium with water is:

$2Na(s) + 2H_2O(l) \longrightarrow H_2(g) + 2NaOH(aq)$

- For a balanced equation, the numbers of atoms of each element must be the same on both sides of the equation.

EXAM TIP

Exam questions often ask you to write or complete equations for the reactions of Group 1 metals with water and chlorine. Practise writing them. Make sure that you know the **formulae** of the compound and what numbers to use to balance the equations.

Reactions of Group 1 elements with chlorine

- Sodium reacts vigorously with chlorine to give a yellow flame; it makes a white solid (sodium chloride).
- The other Group 1 metals react in a similar way, and the reactions are faster down the group.

- The word and symbol equations for the reaction between sodium and chlorine are:

sodium + chlorine ⟶ sodium chloride

$2Na(s) + Cl_2(g) \longrightarrow 2NaCl(s)$

Remember!
Group 1 elements show a trend in reactivity. They get more reactive 'down the group'.

Improve your grade

Reactions of Group 1 elements with chlorine

Higher: Write word and symbol equations for the reaction of sodium with bromine. Compare the rate of reaction of sodium and potassium with bromine.

AO1 [3 marks]

Group 7 – the halogens

Which elements are halogens?

- The elements in Group 7 are called the halogens.
- The table shows the appearance of the halogens at room temperature and when they are warmed to form gases.

Halogen	Appearance at room temperature	Colour of gas
chlorine	pale green gas	pale green
bromine	red-brown liquid	reddish-brown
iodine	dark grey solid	purple

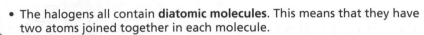

Group 7 of the Periodic Table, the halogens.

- The halogens all contain **diatomic molecules**. This means that they have two atoms joined together in each molecule.
- The formulae of the halogens are: chlorine: Cl_2; bromine: Br_2; iodine: I_2.
- The physical properties of the halogens show a trend down the group, e.g. melting points and boiling points increase.

Patterns in Group 7

- Group 7 elements are **corrosive** and **toxic**. They need to be used in a fume cupboard.
- Group 7 elements react with alkali metals and with other metals such as iron to form metal **halides**. Some examples are shown in the table.

Metal		Halogen		Metal halide
iron	+	chlorine	→	iron chloride
sodium	+	bromine	→	sodium bromide
potassium	+	iodine	→	potassium iodide

- Halogens are less reactive down the group. For example:
  - Sodium reacts vigorously in chlorine but reacts less violently with iodine.
  - Iron reacts vigorously in contact with fluorine but only reacts with iodine when heated.
  - Fluorine reacts even more vigorously than the other halogens.
- **Displacement reactions** happen when a more reactive halogen takes the place of a less reactive halogen in a compound.
  - Chlorine is more reactive than bromine and displaces bromine from potassium bromide solution.
  - Bromine is less reactive than chlorine and cannot displace chlorine from potassium chloride solution.

Remember!
The reactivity of Group 7 decreases (gets slower) down the group – this is the opposite pattern to Group 1.

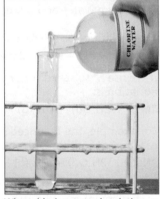

When chlorine water (a solution of chlorine in water) is added to potassium bromide solution, the chlorine displaces orange-coloured bromine.

- All halogen atoms have 7 electrons in their outer shell.
- The trend in reactivity is linked to the number of electron shells in the atom. For non-metals, the smaller the atom (the fewer the electron shells) the more reactive the element.

Improve your grade

Patterns in Group 7

Higher: Liz adds chlorine water to potassium bromide solution. The table shows what she sees and her explanation.

Halogen	Compound	Observations halide	Explanation
Chlorine	Potassium bromide	Solution turns brown	Bromine is made because chlorine displaces bromine. Chlorine is more reactive than bromine.

Predict what you will see when chlorine water is added to potassium iodide solution. Explain your reasoning.

AO2 [4 marks]

Ionic compounds

What are ionic compounds?

- Compounds of a Group 1 element and a Group 7 element (e.g. sodium chloride) are solids with high melting points.

- These compounds are **ionic compounds** because they contain charged particles, or ions, that are arranged in a regular pattern called a **crystal lattice**.

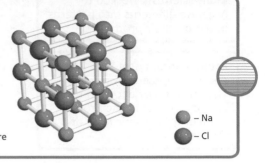

The ions in sodium chloride are arranged in a crystal lattice.

– Na
– Cl

Explaining properties

- Ionic compounds are **soluble** in water and conduct electricity when they are melted.

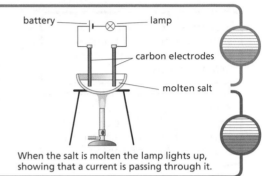

battery — lamp
carbon electrodes
molten salt

When the salt is molten the lamp lights up, showing that a current is passing through it.

- When an ionic crystal melts or **dissolves** in water the ions are free to move.

- An ionic compound conducts electricity when molten or in solution but not when solid. This is because the ions must move to the **electrodes** to complete the circuit.

Forming ions, formulae and charges

- An atom of a Group 1 element *loses* one electron to become a *positive* ion, e.g. Na becomes Na^+.

- An atom of a Group 7 element *gains* an electron to become a *negative* ion, e.g. Cl becomes Cl^-.

- When a Group 1 metal atom becomes an ion it loses one electron from its outer shell.

- All Group 1 metals become ions with a 1+ charge.

- When a Group 7 metal atom becomes an ion it gains one electron in its outer shell.

- All Group 7 metals become ions with a 1− charge.

- The ions have the same electron arrangement as an atom in Group 0. For example:

Electron arrangement	
atom	ion
Na: 2.8.1	Na^+: 2.8
Cl: 2.8.7	Cl^-: 2.8.8

same electron arrangement as a neon atom 2.8
same electron arrangement as an argon atom 2.8.8

- In the formula for an ionic compound the number of positive and negative charges balance, e.g. Na^+ and Cl^- have one positive and one negative charge and so form the compound NaCl.

- You can work backwards from the formula to work out the charge on the ions, e.g. in $CaCl_2$, there are 2− charges from the two Cl^- ions, so the calcium ion has a charge of Ca^{2+}.

sodium atom
Na
2,8,1

sodium ion
Na^+
2,8

chlorine atom
Cl
2,8,7

chlorine ion
Cl^-
2,8,8

EXAM TIP

Some of the formulae of compounds have more than one atom grouped together into an ion. The metal hydroxides in this topic contain hydroxide ions OH^-. Make sure you know their formula and charge – you might be asked to work out the formula of an unfamiliar hydroxide in the exam.

Higher

Improve your grade

Explaining properties

Foundation: Explain why sodium chloride conducts electricity when it is molten or dissolved in water but not when solid.

AO1 [4 marks]

C4 Summary

Many scientists worked to organise elements into patterns based on their properties.

Atoms contain protons, neutrons and electrons. The numbers of each can be worked out using the proton number and atomic mass of the element.

Atoms, elements and the Periodic Table

Mendeleev's ideas formed the basis for the modern Periodic Table.

Each element has a unique flame-test colour and line spectrum.

Electrons are arranged in shells around the nucleus and can be shown using electron arrangements.

A period is a row across the Periodic Table. There are trends in properties from left to right, e.g. metals to non-metals.

Electrons and the Periodic Table

Each shell holds a maximum number of electrons.

The position of an element on the Periodic Table gives information about its properties and its electron arrangement.

Group 1 metals are all soft, can be cut with a knife and tarnish in moist air.

The reactivity of Group 1 metals increases down the group and is linked to the electron arrangement of the atoms.

Reactions of Group 1

Group 1 metals react violently with water to give hydrogen and a metal hydroxide.

Group 1 metals react with chlorine to give a metal chloride.

Halogens react with metals and other less reactive halogen compounds (displacement reactions).

The appearances of the halogens are all different at room temperature and as gases. They all contain diatomic molecules, e.g. Cl_2.

Group 7 – the halogens

The halogens get less reactive down the group and this is linked to the electron arrangement of the atoms.

Positive ions form when atoms lose outer-shell electrons.
Negative ions form when they gain outer-shell electrons.
Ions have the same electron arrangement as Group 0 gases.

Ionic compounds

Ionic compounds conduct electricity when molten or in solution because the ions can move.

The formula of an ionic compound can be worked out by making sure that there are enough positive and negative ions so that the total charges balance.

Molecules in the air

Air

- **Dry air** contains non-metal elements, e.g. nitrogen, oxygen and argon.

- Air also contains small amounts of non-metal compounds, e.g. carbon dioxide and water vapour.

- The gases are non-metal elements and non-metal compounds. They are molecular because they contain atoms joined together into small molecules.

- The formulae and amounts of some of the gases in the air are shown in the table.

Substance	Formula	Percentage of dry air
nitrogen	N_2	78
oxygen	O_2	21
argon	Ar	1
carbon dioxide	CO_2	0.04

- Molecules can be shown either in 2-D or in 3-D.

- The atoms in the molecules are held together by **covalent bonds**.

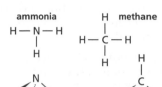

2-D and 3-D diagrams of some simple molecules.

- A covalent bond forms when atoms share a pair of electrons.

- The atoms are held together because the positively charged nuclei of both atoms are attracted to the negatively charged shared pair of electrons.

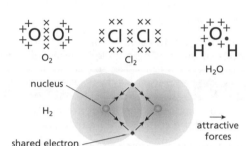

Simple molecular substances

- Simple molecular substances, such as molecules in the air, have very low melting points and boiling points.

Substance	Melting point (°C)	Boiling point (°C)
nitrogen	−210	−196
oxygen	−218	−183

EXAM TIP

A common question is to ask about data relating to melting points and boiling points of gases. Any substance with a melting point and boiling point below 25 °C is a gas at room temperature. Remember that many values for gases are negative. Bigger negative numbers mean lower boiling points and melting points.

- The melting points and boiling points are low because the attractive forces between small molecules are very weak. Very little energy is needed for molecules to overcome these forces and move apart.

- Molecules of elements and compounds have no electrical charge, so pure molecular substances cannot conduct electricity.

- For small covalent molecules, the forces *between* molecules are weak, but the forces *within* molecules (covalent bonds) are strong.

- When a molecular substance melts, the molecules are easily separated from one another but the molecules themselves are not broken up into separate atoms.

Improve your grade

Simple molecular substances

Higher: The table shows some data about oxygen.

Boiling point	State at room temperature	Density	Electrical conductivity
−218 °C	gas	very low	does not conduct

Use ideas about bonding and forces between molecules to explain the properties of oxygen.

AO2 [3 marks]

C5 Chemicals of the natural environment

Ionic compounds: crystals and tests

Ionic crystals

- The Earth's **hydrosphere** is all the water on Earth, including oceans, seas, lakes and rivers.

- The hydrosphere is mostly water, with some dissolved compounds called **salts**.

- When water evaporates, dissolved salts form solid **crystals**.

- Salts are **ionic compounds**. **Ions** have either a positive or a negative charge and are arranged in a giant 3-D pattern called a **lattice**.

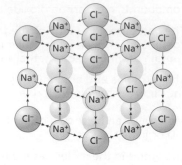

- The strong force of attraction between positively charged and negatively charged ions is called an **ionic bond**.

- Ionic compounds have high melting and boiling points because a large amount of energy is needed to overcome the forces between ions in the lattice.

- Ionic compounds do not conduct electricity when solid, because the ions are not free to move. When they are melted or dissolved in water, the ions can move and they conduct electricity.

The forces between positively and negatively charged ions pulls them together in a giant 3-D crystal lattice.

- In the formula for an ionic compound, the number of positive charges just balance and cancel out the negative charges, e.g. Na^+ and Cl^- make $NaCl$; Mg^{2+} and Cl^- make $MgCl_2$.

- Some ions, e.g. the sulfate ion SO_4^{2-}, contain groups of atoms. This is called a **molecular ion**.

Positively charged ions	Negatively charged ions
sodium (Na^+)	chloride (Cl^-)
potassium (K^+)	bromide (Br^-)
magnesium (Mg^{2+})	iodide (I^-)
calcium (Ca^{2+})	sulfate (SO_4^{2-})

Testing for ions

- Ions in a compound can be identified by their distinctive properties, e.g. compounds containing the copper ion are often blue.

- Solutions of some ionic compounds make a **precipitate** of an **insoluble** compound when they mix. The colour of the precipitate can be used to identify the ions in the compound.

- Adding an **alkali**, such as dilute sodium hydroxide, to different positive metal ions gives different colours of precipitate, as the table shows.

Ion	Observation
calcium, Ca^{2+}	White precipitate (insoluble in excess)
copper, Cu^{2+}	Light blue precipitate (insoluble in excess)
iron (II), Fe^{2+}	Green precipitate (insoluble in excess)
iron (III), Fe^{3+}	Red-brown precipitate (insoluble in excess)
zinc, Zn^{2+}	White precipitate (soluble in excess, giving a colourless solution)

- Negative carbonate ions are identified by adding dilute acid and looking for fizzing (**effervescence**).

- Negative chloride, bromide, iodide and sulfate ions are identified by adding dilute silver nitrate or dilute barium chloride and looking for precipitates.

EXAM TIP

You are not expected to know these tests 'off by heart'. In the exam you will have a sheet of tests and expected results to refer to, like the one on page 142. Practise using the sheet so it is familiar to you.

- Carbonates fizz when an acid is added because carbon dioxide gas is made in the reaction.

- Precipitates form when an insoluble solid is made in the reaction. For example, most metal hydroxides are insoluble, as is silver chloride from the reaction of silver ions and chloride ions.

- **Ionic equations** with **state symbols** show what happens when precipitates form. For example:
  - For positive ions reacting with hydroxide ions: $Cu^{2+}(aq) + 2OH^-(aq) \rightarrow Cu(OH)_2(s)$
  - For negative ions reacting with silver ions or barium ions: $Ag^+(aq) + Cl^-(aq) \rightarrow AgCl(s)$

Improve your grade

Testing for ions

Foundation: Sam tests a salt. He finds out it contains copper ions.

Describe what Sam does and what he sees.

AO2 [3 marks]

Giant molecules and metals

Metals, minerals and ores

- The **lithosphere** is the rigid outer layer of the Earth, made up of the crust and upper mantle. It contains rocks and **minerals**.

- Minerals are solids with atoms or ions arranged in a regular arrangement or lattice, e.g. carbon in the form of diamond or graphite.

- Silicon, oxygen and aluminium are very abundant elements in the Earth's lithosphere.

- Most of the silicon and oxygen on Earth are joined together in the Earth's crust as silicon dioxide, for example in the mineral quartz.

- Some minerals contain metals. Rocks that contain metal minerals are called **ores**.

- Some ores contain metal oxides.

- Copper, zinc and iron can be extracted from their ores by heating their metal oxides with carbon. Carbon **reduces** the metal oxide by taking away oxygen.

- The amount of minerals in ores varies. For some metals, e.g. copper, a huge amount of rock has to be mined to extract a small amount of metal.

- Extracting metals by heating their oxides with carbon is a **redox reaction**, because both oxidation and reduction happen.

- **Reduction** happens because the metal oxide *loses* oxygen.

- **Oxidation** happens because the carbon *gains* oxygen.

- For example, the extraction of zinc uses carbon to reduce zinc oxide to zinc:

  zinc oxide + carbon $\longrightarrow$ zinc + carbon dioxide

  $2ZnO(s)$ + $C(s)$ $\longrightarrow$ $2Zn(s)$ + $CO_2(g)$

- In this reaction the zinc has been *reduced* and the carbon has been **oxidised**.

Giant covalent structures

- Diamond and graphite contain many carbon atoms covalently bonded together in a regular pattern. This is called a **giant covalent structure**.

- Covalent bonds form when atoms share electrons.

- The bonds in diamond are very strong and need a large amount of energy to break them. This is why diamond has a very high melting and boiling points and does not dissolve in water.

- There are no free charged particles in diamond, so it does not conduct electricity when solid or when melted.

- Silicon dioxide also has a giant covalent structure, so it has similar properties to diamond.

- In diamond, each carbon atom is covalently bonded to four other atoms in a tetrahedral 3-D lattice.

- In graphite, each atom is strongly bonded to three others in sheets. The sheets are strong but there are is only a weak force between the layers, so they can slide over each other.

- There are free-moving electrons between the layers in graphite, so it conducts electricity.

The structure of diamond. Each carbon atom is joined to four others by covalent bonds.

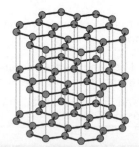

The structure of graphite. Each carbon atom is joined to three others by covalent bonds.

Remember!
Ionic compounds contain charged particles and so conduct electricity when they are molten. Covalent compounds have no charged particles and cannot conduct electricity. Graphite is the only exception to this rule.

Improve your grade

Giant covalent structures

Higher: Diamond is the hardest naturally occurring material on Earth.

Use ideas about the structure of diamond to explain why.　　　　*AO1* [3 marks]

Equations, masses and electrolysis

Equations

- Equations show the chemicals that react together (the reactants) and the chemicals that are made (the products).

- A balanced equation also shows the number of atoms of each element on each side of the equation. Equations are balanced because there is the same number of atoms of each element on both sides.

- State symbols show whether each chemical is solid (s), liquid (l), gas (g) or dissolved in water (aq).

- To balance equations, write the symbols for the elements and the formulae for the compounds. Write numbers in front of the formulae to give the correct number of atoms on each side.

$$2ZnO(s) + C(s) \rightarrow 2Zn(s) = CO_2(g)$$

Representing the balanced equation for the reaction zinc oxide with carbon.

Atomic mass and formula mass

- The **relative atomic mass** of an atom is the mass of an atom compared to the mass of an atom of carbon, which is given the value 12.

Remember!
The atomic mass of every element is in the Periodic Table – find any masses you need by looking there.

- The **relative formula mass** of a compound is the sum of the relative atomic masses of all the atoms or ions shown in its formula.

- For example, to find the relative formula mass of water, H_2O, use the relative atomic masses of hydrogen (1) and oxygen (16). The relative formula mass of water, H_2O is: $(2 \times 1) + 16 = 18$.

- The **gram formula mass** of an element or compound is its relative atomic mass or relative formula mass in grams, e.g. the gram formula mass of water is 18 g.

- Use this method to work out the percentage of a metal in a mineral:

$$\text{Percentage of metal in mineral} = \frac{\text{total mass of metal atoms}}{\text{gram formula mass}} \times 100\%$$

- To calculate the mass of a metal that can be extracted from a mineral, you need to work out the mass of metal in grams that can be extracted from 1 gram formula mass.

- For example, how much iron can be extracted from 80 g Fe_2O_3? One Fe_2O_3 formula unit can make 2Fe gram formula masses: 160 g can make $(2 \times 56) = 112$ g. So, 80 g can make 56 g.

Using electrolysis

- **Electrolysis** means passing an electric current through an ionic compound when it is either molten or dissolved in water.

- The compound is called the **electrolyte** because it conducts electricity.

- Electrolytes break down, or **decompose**, as the electricity passes through.

- Aluminium is extracted from aluminium oxide by electrolysis. Aluminium and oxygen are made in this process.

- Electrolyis is used to extract more reactive metals (e.g. aluminium) because their oxides cannot be reduced by carbon.

- Metals form at the negative **electrode** because positive metal ions are attracted to the negatively charged electrode (**cathode**).

- Negative ions such as chloride (Cl^-) and oxide (O^{2-}) move to the positively charged electrode (**anode**). Non-metals, such as chlorine and oxygen, form at the anode.

- At the negative electrode, metals ions gain electrons and become neutral metal atoms, e.g. aluminium ions gain 3 electrons to form aluminium atoms: $Al^{3+} + 3e^- \rightarrow Al$

- At the positive electrode, non-metal ions lose electrons and become neutral non-metal atoms, e.g. 2 oxygen ions lose 2 electrons each to form oxygen gas: $2O^{2-} \rightarrow O_2 + 4e-$

Improve your grade

Using electrolysis

Higher: An electric current is passed through molten potassium chloride.

Write ionic equations to help you to explain what happens at each electrode and name the products that form.

AO1 [5 marks]

Metals and the environment

- Metals are very useful because they are strong, **malleable** (they can be hammered into shape), have high melting points and are good conductors of electricity.

- Atoms in metals are held together by **metallic bonds**. The atoms are arranged in a regular pattern in a giant lattice.

- Metallic bonds are strong so a lot of energy is needed to melt or reshape them.

- Metal atoms lose their outer-shell electrons to form positive ions. In solid metals, the outer-shell electrons form a 'sea of electrons', which can move freely.

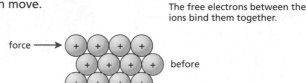

'sea' of electrons

positive metal ions

forces of attraction pull ions together

The free electrons between the ions bind them together.

- The attraction between the positive ions and the sea of electrons is very strong, so that metals have high melting points and high strength.

- Metals conduct electricity because the electrons can move.

- In pure metals, all the atoms are the same size and can roll over each other. This means that metals can be reshaped (they are malleable) even though the bonds are strong.

force → before

after

Remember!
Metals and ionic compounds conduct electricity in different ways. Ionic compounds only conduct when they are molten or in solution because the ions need to move.

Applying a force to change the shape of a metal does not affect the arrangement of the atoms, so it remains strong.

Metals in the environment

- Some metals are poisonous, e.g. lead, mercury and cadmium.

- Waste poisonous metals from mines destroys habitats and damages soil and water sources.

- Extracting metals makes pollutant gases that cause acid rain.

- Large amounts of waste rock need to be processed to produce very small amounts of some metals such as copper.

- Processing large amounts of rock uses a lot of energy.

- Some minerals contain compounds of metals with sulfur. During extraction of the metal, sulfur dioxide is made. Sulfur dioxide gas forms acid rain and damages plants and fish.

- We need large amounts of copper for electrical wiring and circuits, pipes and building materials. Some waste copper is recycled.

- Lead is a **toxic** metal that was used to make batteries for vehicles.

- Modern batteries are made using lithium. Lithium is not toxic but it is difficult to extract enough lithium to meet the demand for batteries.

EXAM TIP

You do not need to learn facts about individual metals and their effects on the environment, but it is important that you can discuss data and information that you are given in questions. Typical questions will ask you about the costs and benefits of extracting metals – you need to be able to give the benefits of all of the uses of metals to people, as well as discussing the harm that extraction causes to the environment.

Ideas about science

You should be able to:

- discuss the costs and benefits to people and the environment of extracting metals.

- explain what sustainability means and apply ideas about sustainability to extracting metals

- discuss why people may have different views about the impact of mining on their lives and the environment.

Improve your grade

Metals in the environment

Foundation: Old car batteries are made from lead. Lead is toxic. Modern batteries use alternative, non-toxic metals.

Use ideas about cost and benefit to explain why manufacturers could not stop using lead to make batteries until alternative batteries were developed.

AO3 [3 marks]

C5 Summary

Dry air contains 78% nitrogen, 21% oxygen and about 1% argon and other gases such as carbon dioxide.

Gases in the air contain small molecules. The atoms in the molecules are held together by covalent bonds.

The melting points and boiling points of gases in the air can be explained by using ideas about weak forces between their molecules.

Molecules in the air

Ionic salts are compounds containing positive and negative ions in a 3-D lattice.

The negative ions in a salt can be identified by testing with a dilute acid, dilute silver nitrate or dilute barium chloride.

Ionic compounds: crystals and tests

Ionic compounds have very high melting points and conduct electricity when they are molten or dissolved in water but not when they are solids.

The positive ions in a salt can be identified by adding dilute sodium hydroxide and looking at the colour of the precipitate.

The lithosphere contains rocks and useful minerals. Metal minerals are called ores.

The properties of giant covalent structures are related to the way that the atoms are held in strong, 3-D lattices of covalent bonds.

Giant molecules and metals

Metals can be extracted from ores by reactions that involve reduction and oxidation.

Diamond, graphite and silicon dioxide all have giant covalent structures.

Balanced equations show the numbers of atoms that react. The same numbers of atoms are on both sides.

Ionic equations show how metal ions gain electrons at the negative electrode and non-metal ions lose electrons at the positive electrode.

Equations, masses and electrolysis

Relative atomic mass, relative formula mass and gram formula mass can be used to work out how much metal can be extracted from a mineral.

Electrolysis is used to extract reactive metals from their compounds. The compound breaks down when an electric current passes through.

The properties of metals make them useful for a wide range of purposes.

Metals and the environment

Metals contain positive ions in a sea of free-moving electrons. This structure gives metals high melting points, high strength, malleability and good electrical conductivity.

Metal extraction causes environmental harm because large amounts of waste rock are produced and some metals are very toxic.

Making chemicals, acids and alkalis

Making chemicals

- Hazard symbols (see page 79) are used to show that chemicals are hazardous.
- Chemists and engineers must assess the risks before using chemicals to make a new product.
- **Chemical synthesis** means using simple substances to make new, useful chemical compounds.
- The chemical industry uses chemical synthesis to make chemicals for food additives, fertilisers, dyes, paints, pigments and pharmaceuticals (medicines).

Acids and alkalis

- **Indicators** turn different colours in **acids** and **alkalis**. Litmus is red in acids and blue in alkalis. Universal indicator is orange or red in acids and green to blue in alkalis.
- **Pure** acid compounds can be solids, liquids or gases. These compounds dissolve in water to form dilute acids that can be tested using indicators.

State when pure	solid	liquid	gas
Examples of acid	citric acid; tartaric acid	sulfuric acid; nitric acid; ethanoic acid	hydrochloric acid

- Sodium hydroxide, potassium hydroxide and calcium hydroxide are common alkalis.
- The **pH** scale is a measure of how strong an acid or an alkali is.
- The pH can be measured using universal indicator or a pH meter. The colour of the universal indicator can be compared to a colour chart to find the pH number of a sample.
- Neutral solutions have a pH of 7. pH numbers for acids are below 7 and for alkalis are above 7.

> **Remember!**
> The lower the pH of an acid, the stronger the acid. The higher the pH of an alkali, the stronger the alkali.

Universal indicator shows a range of colours across the pH scale from 1 (red) to 14 (dark blue).

Reactions of acids

- Acids react with many metals and metal compounds to make a **salt**.
- Acids react with many metals to form a salt and hydrogen gas, e.g.:

  calcium + hydrochloric acid $\longrightarrow$ calcium chloride + hydrogen

- Acids react with metal oxides and hydroxides to form a salt and water, e.g.:

  magnesium oxide + sulfuric acid $\longrightarrow$ magnesium sulfate + water

  sodium hydroxide + nitric acid $\longrightarrow$ sodium nitrate + water

- Acids react with metal carbonates to form a salt, water and carbon dioxide gas, e.g.:

  calcium carbonate + hydrochloric acid $\longrightarrow$ calcium chloride + water + carbon dioxide

- Salts are **ionic compounds** and contain a positively charged metal **ion** and a negative ion from the acid.

Acid	hydrochloric	sulfuric	nitric
Formula	HCl	H_2SO_4	HNO_3
Negative ion	chloride (Cl^-)	sulfate (SO_4^{2-})	nitrate (NO_3^-)

- The reactions of acids can be shown using symbol equations with **state symbols**.
- The state symbols can be either solid (s), liquid (l), gas (g) or solution in water (aq), e.g.:

  – $Ca(s) + 2HCl(aq) \longrightarrow CaCl_2(aq) + H_2(g)$ – $Mg(OH)2 + H_2SO_4(aq) \longrightarrow MgSO_4(aq) + H_2O(l)$

  – $NaOH(aq) + HNO_3(aq) \longrightarrow NaNO_3(aq) + H_2O(l)$ – $CaCO_3(s) + 2HCl(aq) \longrightarrow CaCl_2(aq) + H_2O(l) + CO_2(g)$

- To work out the formula of a salt, the number of positive charges must equal the number of negative charges. For example, in potassium sulfate, two potassium ions (K^+) are needed to balance the charge on the sulfate ion (SO_4^{2-}), so the formula is K_2SO_4.

Metal	sodium	potassium	magnesium	calcium	copper	zinc	iron(II)	aluminium	iron(III)
Ion	Na^+	K^+	Mg^{2+}	Ca^{2+}	Cu^{2+}	Zn^{2+}	Fe^{2+}	Al^{3+}	Fe^{3+}

- A balanced equation has the same number of each type of atom on each side of the equation.

Improve your grade

Reactions of acids

Higher: Write a word equation and a balanced symbol equation to show the reaction between sodium hydroxide and sulfuric acid.

AO1 [3 marks]

C6 Chemical synthesis

Reacting amounts and titrations

Reacting amounts

- The **formula** of a compound is the simplest ratio of the numbers of atoms that are in the compound.

- The **relative atomic masses** of atoms can be found on the **Periodic Table**.

- The **relative formula mass** of a compound is the sum of the relative atomic masses of all the atoms in the formula. For example, the relative formula mass of magnesium chloride, $MgCl_2$, is worked out like this:
  RAM Mg = 24; RAM Cl = 35.5
  Relative formula mass $MgCl_2$ = 24 + (2 × 35.5) = 95

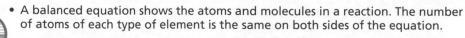

The relative atomic mass is the number above the symbol of the element. This number compares the mass of the atom with the mass of a carbon atom, which has the value 12.

- A balanced equation shows the atoms and molecules in a reaction. The number of atoms of each type of element is the same on both sides of the equation.

- Relative atomic mass (for atoms) and relative formula mass (for compounds) can be used to work out the amounts of **reactants** and **products** in the reaction.

EXAM TIP

Only work out the relative formula masses for the chemicals mentioned in the equation. In the example below you can ignore the HCl and the H_2O because the question does not ask about them.

- The balanced equation for a reaction can be used to calculate the minimum quantity of reactants needed to make a particular amount of a product.

- For example, what mass of magnesium chloride can be made by reacting 1 tonne of magnesium oxide with hydrochloric acid?

	MgO	+ HCl $\rightarrow$ MgCl$_2$	+ H$_2$O
RFMs	MgO = 24 + 16 = 40	MgCl$_2$ = 24 + (2 × 35.5) = 95	
Reacting masses	40 tonnes	95 tonnes	
	1 tonne	95/40 = 2.38 tonnes	

Titrations

- When an acid reacts with an alkali it becomes **neutral**. This is called a **neutralisation** reaction.

- A **titration** is used to measure the volume of acid and alkali that exactly react together.

- An indicator is added so that you can see when neutralisation happens. The indicator suddenly changes colour at the **end-point** of the titration.

- A titration is repeated to check that the results are close together. Variations between readings are small differences that happen due to small experimental errors.

- The **range** of readings is the spread of readings from the highest to the lowest.

- The **true value** should fall within the range of the readings.

- An estimate of the true value can be worked out by calculating a **mean** (an average) of the results.

- An **outlier** is a reading that is very different to most of the others. Outliers should be left out when you calculate the mean, because they are usually the result of errors in measurement.

- The volumes obtained from a titration will always be in the same proportion if the same concentrations of the same solutions are used.

Carrying out a titration. The end-point is reached when one drop from the burette changes the colour of the indicator.

Ideas about science

You should be able to:

- work out a best estimate for a titration by working out the mean of the accurate results

- explain why repeating measurements leads to a better estimate from a set of repeated titration measurements

- make a sensible suggestion about the range within which the true titration value probably lies

- identify any outliers in data from titrations.

Improve your grade

Reacting amounts

Foundation: What is the relative formula mass of calcium hydroxide, $Ca(OH)_2$?

Use the Periodic Table to help you.

AO2 [2 marks]

Explaining neutralisation & energy changes

Explaining neutralisation

- When an acid reacts with an alkali, a salt and water are always made.

- All acids contain hydrogen ions (H^+) when they are dissolved in water. The pH of an acid is related to the concentration of H^+ ions in the acid solution.

- All alkalis contain hydroxide ions, OH^-, when they are dissolved in water.

- In neutralisation reactions the hydrogen ions and the hydroxide ions join up to form water molecules:

$$H^+(aq) + OH^-(aq) \longrightarrow H_2O(l)$$

- This is the **ionic equation** for all neutralisation reactions. The negative ion from the acid and the positive ion from the alkali are left in the solution to form the salt.

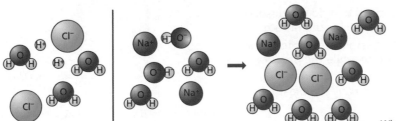

hydrochloric acid solution　　　sodium hydroxide solution　　　sodium chloride + water

Remember!
Hydrochloric acid always forms chloride salts, sulfuric acid forms sulfate salts, and nitric acid forms nitrate salts.

When the ions in an acid and an alkali are mixed, water molecules are formed.

- The positive ion from the alkali and the negative ion from the acid make the salt.

- The formula of the salt can be worked out by looking at the charges on the two ions. For example:

sodium hydroxide + hydrochloric acid ⟶ sodium chloride + water

Salt: sodium chloride. Positive ion: Na^+; negative ion: Cl^-;

so formula: NaCl.

Higher

Energy changes

- **Exothermic** reactions give out heat energy. The temperature of the surroundings rises.

- **Endothermic** reactions take in heat energy. The temperature of the surroundings falls.

- An **energy level diagram** summarises the energy changes in a reaction.

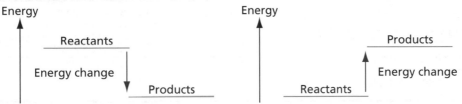

In exothermic reactions, the reactants give out energy, so the reactants are always at a higher energy value than the products.

For endothermic reactions, the reactants take in energy and so are always at a lower energy level than the products.

EXAM TIP

Make sure that you can draw energy level diagrams for exothermic and endothermic reactions. Notice that the arrows to represent the energy change always go from the reactants to the products.

- Energy changes for large scale-reactions in industry need to be carefully controlled, because very extreme temperature changes could cause overheating, explosions or fires.

◉ Improve your grade

Energy changes

Foundation: Sam does some experiments. She does four different reactions in test tubes and takes a note of the temperature before mixing and 60 s after mixing. Sam's results are shown in the table.

Experiment	Temperature before mixing (°C)	Temperature 60 s after mixing (°C)
1	17	21
2	18	11
3	16	18

Which reaction is the most exothermic? Explain how you can tell.　　　*AO3* [3 marks]

Separating and purifying

The importance of purity

- A **pure** substance has nothing else mixed with it.

- In industry, pure substances need to be separated from impurities, such as left-over reactants or other products, before they are used. Some impurities may be harmful.

- **Filtration** can be used to separate a solid from a liquid or from a solution.

Crystallisation

- Crystallisation is used to purify impure solid crystals. The process has several steps:

  1 *Dissolving* – **dissolve** the product in a small amount of hot water (use only the minimum amount of water necessary to dissolve the product).

  2 *Filtering* – filter off any solid impurities that do not dissolve. The solution that comes through the filter is the **filtrate**.

  3 *Evaporating* – the filtrate starts to crystallise as some of the water **evaporates** off. Cool the filtrate while the product continues to **crystallise**.

  4 *Filtering* – filter off the crystals, leaving any soluble impurities in the solution.

  5 *Drying* – dry the crystals in a **dessicator** or oven.

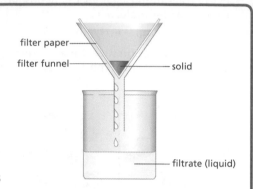

Filtering off solid impurities during crystallisation.

Percentage yield

- The percentage yield at the end of an experiment is worked out from the actual yield and the theoretical yield.

- The actual yield is the mass of product measured at the end of the experiment.

- The theoretical yield is the predicted yield. It is calculated from the amount of reactants used and the equation for the reaction.

- Percentage yield $= \dfrac{\text{actual yield}}{\text{theoretical yield}} \times 100\%$

- For example, an experiment reacts 2.4 g magnesium with hydrochloric acid. The actual yield of magnesium chloride is 5.7 g. What is the percentage yield?

Step 1: work out the theoretical yield:

Equation: $Mg + 2HCl \longrightarrow MgCl_2 + H_2$
Relative masses: 24   95
Reacting masses: 2.4 g   9.5 g
Theoretical yield of magnesium chloride = 9.5 g

Step 2: work out the percentage yield:

Percentage yield $= \dfrac{5.7}{9.5} \times 100\% = 60\%$

> **Remember!**
> The theoretical yield is worked out by looking at the equation for the reaction. The actual yield is what you measure at the end of your experiment. You need both values to work out the percentage yield.

> ### EXAM TIP
> Make sure you know how to work out theoretical yields from reacting amounts. At Higher tier, you will probably be asked to work out a theoretical yield first and then a percentage yield.

Improve your grade

Percentage yield

Higher: Ben prepares some copper sulfate crystals. He talks to Liz about his method.

Ben: 'First I added solid copper carbonate to sulfuric acid until the fizzing stopped.

Then I filtered off some unreacted copper carbonate.

Next I heated the filtrate to evaporate some of the water and left the solution to cool for a few minutes until some crystals formed.

Then I filtered off the crystals and weighed them. I made 1.4 g of crystals. I worked out that my theoretical yield is 1.6 g.'

Liz: 'Your percentage yield will be really inaccurate because you have missed some steps out of your method.'

Calculate Ben's percentage yield. Explain why his percentage yield is likely to be inaccurate. *AO2* [4 marks]

Rates of reaction

Measuring rates of reaction

- The **rate of reaction** is the amount of a product produced or the amount of reactant used up in a certain time. It is usually measured as the amount per second.

- Chemical engineers look for ways to control reactions. By speeding the reaction up they can make them more economical. They also need to ensure that reactions occur at a safe rate.

- In industry, chemical engineers aim to produce the most amount in the minimal time. They change conditions to make reactions faster to make the process as economical as possible, but they also consider the cost of the energy and safety.

- If the reaction makes a gas, the rate can be followed by:
  - measuring the volume of gas made at set times, e.g. every 30 seconds.
  - measuring the decrease in mass of the flask as the gas leaves the reaction.

- If the reaction makes a solid, the rate can be followed by measuring the time taken until you cannot see a cross underneath the flask or beaker.

- A colorimeter can be used to follow the rate of a colour change.

- Rate of reaction graphs show the change in the amount of reactant or product against time.

- The gradient of the curve at any point gives the rate of reaction.

Remember!
The steeper the gradient, the faster the reaction. The reaction stops when the line becomes horizontal.

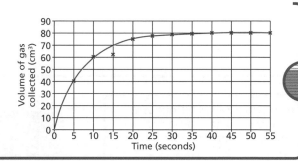

Changing rates of reactions

- For reactions to happen, particles must collide. The more collisions the faster the reaction.
- A **catalyst** is a substance that speeds up a chemical reaction but is not used up.

- Reactions are faster when:
  - the temperature of the reactants increases
  - the size of solid particles are smaller (this increases the surface area)
  - the concentration of reactants in solution increases (concentration is measured in grams per dm^3 of solution).
- To investigate the effect of changing one of these factors on the rate of the reaction, it is important that all other factors are kept constant.

The greater the concentration of reactants, the greater the collision frequency.

low concentration

- The rate of reaction increases when the frequency of collisions increases.
- Rate of reaction increases with a larger surface area of solids and with a higher concentration of solutions because the frequency of collisions increases.
- Chemical engineers control rates by controlling the factors that affect the rate, e.g. temperature, concentration, particle size or using a catalyst.

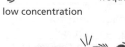
high concentration

Ideas about science

You should be able to:

- discuss how rates of reaction are controlled so that the benefits of faster reactions outweigh the costs

- discuss the costs to the environment of changing reaction conditions to increase the rate of industrial processes, e.g. the effects of using large amounts of energy to run a process at a high temperature

- discuss how using a catalyst makes industrial chemical reactions more sustainable.

Improve your grade

Changing rates of reactions

Foundation: Jack investigates the rate of the reaction between a large lump of calcium carbonate and dilute hydrochloric acid. He measures the volume of gas given off every 30 seconds.

The reaction takes place very slowly.

Suggest what changes Jack could make to his experiment to make the reaction faster. *AO1 [3 marks]*

C6 Summary

The chemical industry uses large-scale chemical synthesis to make useful chemical compounds.

The pH of acids and alkalis can be tested using indicators or a pH meter.

Acids react with metals, metal oxides, metal hydroxides and metal carbonates to give salts and other products. These reactions can be summarised in equations.

Making chemicals, acids and alkalis

Relative formula masses can be worked out using relative atomic masses.

Reacting amounts and titrations

Balanced equations can be used to work out masses of reactants and products in reactions.

Titrations can be used to measure the exact amount of acids and alkalis that react together.

Acid and alkali reactions can be shown as an ionic equation between hydrogen ions and hydroxide ions.

$H^+ + OH^- \longrightarrow H_2O$

Explaining neutralisation and energy changes

Exothermic reactions give out heat energy; endothermic reactions take in heat energy.

Energy changes in reactions can be shown on energy level diagrams.

Products of chemical reactions must be separated from impurities.

The percentage yield for preparing a salt can be calculated from the theoretical yield and the actual yield.

Separating and purifiying

Solids can be separated from liquids and solutions by filtration. Crystals can be purified by crystallisation.

Rate of reaction depends on temperature, concentration of solutions and surface area of solids.

Rates of reaction

Rate of reaction increases when the frequency of particle collisions increases.

Catalysts increase the rates of reactions but are not used up.

Speed

Calculating speed

- The **speed** of an object tells us how far it will travel in a certain time.
- Speed (m/s) = $\dfrac{\text{distance travelled (m)}}{\text{time taken (s)}}$
- Most things do not travel at constant speed, so we usually calculate **average speed**.
- Speed is measured in metres per second (m/s) or kilometres per hour (km/h).
- For example, you walk 3 km to school, and it takes you an hour.
  Speed = 3 km ÷ 1 hour = 3 km/h or: Speed = 3000 m ÷ 3600 s = 0.83 m/s

Remember!
Make sure you convert time into seconds before you calculate speed in metres per second.

- If you measure average speed over a very short time interval you get very close to a value for **instantaneous speed**.

- The **displacement** of an object is the distance from its start point in a straight line. When you run once all the way around a running track, the distance you run is 400 m, but your displacement is zero.
- The displacement is expressed as a distance with a direction.
- Displacement is a vector quantity – it has a size (magnitude) and a direction.
- The **velocity** of an object is its speed in a certain direction. It is also a vector quantity.

Average velocity (m/s) = $\dfrac{\text{displacement (m)}}{\text{time taken (s)}}$

Distance–time graphs

- A **distance–time graph** can be used to visualise a journey.
- The time for the journey is plotted on the x-axis (horizontal).
- The distance travelled is plotted on the y-axis (vertical).
- A straight line means that the vehicle is travelling at a constant speed. On the graph, line A shows a constant speed.
- A horizontal line means that the vehicle is stationary; speed is zero. On the graph, line B shows that the bus has stopped for 40 seconds after travelling 300 m.

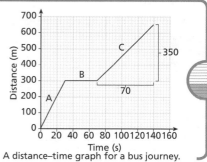

A distance–time graph for a bus journey.

- The **gradient** of the line is equal to the speed.
- The steeper the gradient of the line, the faster the speed. For example, on the graph, line C is less than line A so the bus is travelling slower.
  Gradient of line A = 300 ÷ 30 = 10 m/s
  Gradient of line C = 350 ÷ 70 = 5 m/s

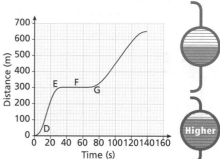

- A curved line means that the speed is changing.
- If the curve is getting steeper it means that the vehicle's speed is increasing.

Displacement–time graphs

- Return journeys can be visualised on a **displacement–time graph**.
- When the vehicle has returned to its starting point its displacement will be zero.

Displacement–time graph: the object moves away at constant speed for 30 s, stops for 40 s and then travels back in the opposite direction for 100 s. It ends up 200 m behind its starting point.

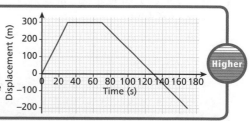

- The displacement would be negative if the vehicle travelled *behind* its starting point.

Improve your grade

Calculating speed

Higher: Henry is planning a train journey from Ipswich to Birmingham. The train journey is in two parts:

Departure time		Arrival time		Distance travelled (km)
12:00	Ipswich	13:00	Ely	80
13:15	Ely	15:45	Birmingham	120

a Which train is faster? Show your calculations.

b Taking into account the wait for the connection at Ely, what is the average speed of the journey from Ipswich to Birmingham?

AO1, AO2 [5 marks]

Acceleration

Speeding up and slowing down

- The rate at which the speed of an object increases is known as **acceleration**.
- Acceleration is measured in metres per second squared – m/s².

$$\text{Acceleration (m/s}^2) = \frac{\text{change in speed (m/s)}}{\text{time taken (s)}}$$

- So, for a car that speeds up from rest to 25 m/s in 5 seconds, acceleration is: $\frac{(25 - 0)}{5} = 5$ m/s²

Remember!
Always calculate the change in speed as final speed minus initial speed.

- When an object slows down it has a negative acceleration, sometimes called *deceleration* or *retardation*.
- There has to be net force acting on an object to cause acceleration (or deceleration).
- When the net or overall force is zero, the acceleration is zero.

Changing direction

- A vehicle travelling at constant speed around a corner is changing its velocity.
- Acceleration is defined as the rate of change of velocity.

$$\text{Acceleration (m/s}^2) = \frac{\text{change in velocity (m/s)}}{\text{time taken (s)}}$$

Speed–time graphs

- A **speed–time graph** is used to show the changes in speed during a journey.
- Speed is plotted on the *y*-axis (vertical) and time on the *x*-axis (horizontal).

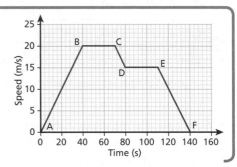

A speed–time graph for a short car journey.

- A horizontal line means that the speed of the object is constant – it is at steady speed. Between points B and C on the graph, the car is travelling at a constant speed of 20 m/s.
- If the horizontal line is along the *x*-axis, the speed is zero – the object is stationary.
- A straight line going up shows acceleration. On the graph, line AB shows acceleration of: $(20 - 0) \div 40 = 0.5$ m/s²
- A straight line going down shows deceleration. On the graph, line EG shows deceleration of: $(0 - 15) \div 30 = -0.5$ m/s²
- The steeper the line, the greater the size of the acceleration.
- The instantaneous speed of a vehicle in a certain direction is its instantaneous velocity.

- A **velocity–time graph** also shows the direction in which an object is travelling.
- A positive velocity means that the object travels in a certain direction and a negative velocity means that the object is travelling in the opposite direction.
- The gradient of the velocity–time graph is equal to the acceleration. For the first 60 seconds, the acceleration of the shuttle train is: $(12 - 0) \div 60 = 0.2$ m/s²

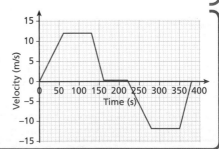

Higher

Improve your grade

Speeding up

Higher: A cyclist rides around a velodrome track. When he sets off he takes 1 minute to reach a top speed of 14 m/s. Then he cycles 10 laps round the oval track at a steady speed.

Calculate the cyclist's initial acceleration, and explain why he continues to accelerate after that.

AO1 [5 marks]

Forces

Forces between objects

- A **force** is a push or a pull which acts between two objects.

- Forces always act in pairs, e.g. you push against a wall; the wall pushes back on your hand.

- A **repulsive** force pushes objects apart; an **attractive** force pulls objects towards each other.

Remember!
The two forces in an interaction pair are equal in size, opposite in direction and act on different objects.

- The size of the force is always the same on both objects.

- The force on each object acts in the opposite direction to each other, e.g. when a ship floats on water, the upthrust (buoyancy) pushes up on the ship and the ship pushes down on the water. This provides a repulsive force pair.

- When you stand you push downwards on the floor. The force is equal to your weight.

- At the same time, the floor exerts an upwards force on you – called the **reaction force** (or the reaction of the surface). It stops gravity from pulling you through the floor.

- The reaction force is equal and opposite to your weight, so you are not accelerating up or down. The reaction force balances your weight.

- If you jump upwards, you need to push harder on the floor – the reaction force increases. Now the forces are unbalanced and you will accelerate upwards.

Rockets (and jets) push gas out of the back of the engine with a large force (yellow arrow). The gas pushes back on the engine with an equal but opposite force (red arrow), propelling the rocket forward.

- Forces are vector quantities, i.e. they have a magnitude (size) and a direction.

- A force is represented by an arrow on a diagram.

- On a scale diagram, the length of the arrow represents the magnitude and the direction of the arrow shows which way the force acts.

The car pushes backwards on the road through the force of friction between the tyres and the road (yellow arrow). The road pushes the car forward with an equal but opposite force (red arrow).

Friction

- Friction is a force which acts between two surfaces. As the two surfaces slide over one another, friction acts to oppose the motion.

- The size of the friction force depends on:
  - the roughness of the surfaces (rougher surfaces give more friction)
  - how hard the surfaces are pushed together (the heavier the object the more friction).

- When you try to push an object along a surface, friction will be equal to the applied force and acts in the opposite direction, so the object will not move.

- As you increase the applied force, the friction will increase too.

- Eventually the friction reaches a maximum value and the object will start to move – this is called **limiting friction**.

- The kinetic energy of the moving object is transferred to heat energy in both surfaces.

- Lubrication (oil) is used to reduce the friction between moving parts of machinery to stop them getting too hot and wearing out.

- You need friction in order to walk. When you walk your feet push against the friction on the floor, so the floor pushes your foot forwards.

- The **resultant** force (or overall force) acting between your foot and the floor is a combination of friction and the reaction of the surface.

- These forces combine to push the foot in a diagonal upwards direction.

The reaction force (red) and the friction force (blue) combine to push the foot in the direction of the resultant force (black).

Improve your grade

Forces between objects

Higher: James kicked his football towards the goal. There was a force on the ball when it was kicked. This force was part of an interaction pair.

Describe the partner force of the kicking force in the interaction pair.

AO1, AO2 [3 marks]

The effects of forces

Adding the effects of forces

- There are usually several forces acting on an object at the same time.
- The resultant force is the total or overall force acting on an object.
- When you add forces together you must account for both the size and the direction of the force.

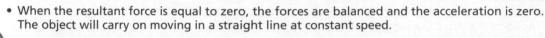

Adding two 5 N forces together can equal 10 N, or 0 N or anything in-between.

- When the resultant force is equal to zero, the forces are balanced and the acceleration is zero. The object will carry on moving in a straight line at constant speed.
- In a frictionless space, once an object starts moving it should keep moving at the same speed.
- When the forces are unbalanced, there is a net force on the object and it will speed up, slow down or change direction.

Terminal velocity

- Objects falling accelerate towards the ground at 9.8 m/s², due to gravity. The force of gravity always acts towards the centre of the Earth.
- The atmosphere creates an upwards force that slows down falling objects. This is known as **air resistance** or **drag**.
- Drag acts in the opposite direction to the speed (or velocity) of the object.
- Drag force increases as the speed of the object increases.
- The larger the surface area of the object, the larger the drag force.

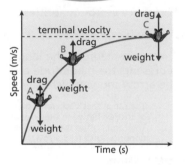

- The constant maximum speed reached by a falling object is known as its **terminal velocity**. As the diagram opposite shows:

  A At first the force of gravity is larger than the drag force, so the object accelerates.

  B As speed increases so does drag; the acceleration decreases.

  C When drag equals the force due to gravity there is no resultant force and the acceleration is zero. The object continues at terminal velocity.

- The ideas about terminal velocity work in the same way for vehicles, with friction and drag acting in the opposite direction to the driving force.

Collisions and momentum

- Large forces are exerted during collisions. The size of the force depends on:
  - the mass of the object (the heavier the object, the larger the force)
  - the speed (or velocity) of the object (the faster the object, the larger the force)
  - the duration of the impact (the longer the time to stop, the lower the force).
- Most car safety devices, e.g. air bags, crumple zones, seat belts, crash helmets, are designed to increase the impact time, thus reducing the force in a collision.

- **Momentum** (kg m/s) = mass (kg) × velocity (m/s)
- A resultant force will change an object's momentum. The larger the force exerted, the larger the change in momentum.

  Change of momentum (kg m/s) = resultant force (N) × time for which it acts (s)

- Force is equal to the rate of change of momentum.

  $$\text{Force (N)} = \frac{\text{change in momentum (kg m/s)}}{\text{time taken (s)}}$$

- For example, a car of mass 1200 kg crashes into a wall at a speed of 20 m/s. The collision stops the car in a time of 1.5 s.

  Change in momentum = (1200 × 20) − 0 = 24 000 kg m/s

  Force = 24 000 ÷ 1.5 = 16000 N

Improve your grade

Terminal velocity

Higher: Explain how air bags reduce injury in a car crash. Include ideas about momentum in your answer. *AO1, AO2* [3 marks]

Work and energy

Work

- The **energy** used by the movement of a force is known as the **work** done.
- Energy and work are both measured in joules (J). Energy is defined as the ability to do work.
- Work done (J) = Force (N) × distance moved in the direction of the force (m)

- When work is done *on* an object, energy is transferred to that object.
- When work is done *by* an object, energy is transferred from the object to something else.
  Amount of energy transferred (J) = Work done (J)

- All forms of energy have the potential to do work.
- Energy from food is transferred in our bodies so we can do exercise.
- Not all energy is transferred as work – some is always dissipated as heat.

Potential energy

- When you lift an object, you do work against gravity.
- 1 joule of work will lift a weight of 1 Newton a distance of 1 metre.
- The work is transferred to **gravitational potential energy** (GPE) of the object.

- As an object is raised, its gravitational potential energy increases. As an object falls, its gravitational potential energy decreases.
- Change in gravitational potential energy (J) = weight (N) x vertical height difference (m)
  For example, if your weight is 700 N and you climb stairs a height of 3 m:
  Gain in gravitational potential energy = weight × height = 700 N × 3 m = 2100 J

Kinetic energy

- When you push an object to get it moving (increase its velocity) you do work. The work is transferred to the moving object as **kinetic energy** (KE).
- The greater the mass of the object and the faster its speed, the greater its kinetic energy.

- Kinetic energy (J) = ½ × mass (kg) × [velocity]^2 ([m/s]^2)
  For example, for a car of mass 800 kg travelling at a speed of 12 m/s:
  KE = ½ × mass × velocity^2 = ½ × 800 × 12^2 = 400 × 144 = 57 600 J

  > **EXAM TIP**
  > Always remember to square just the velocity – do that step first in your calculation.

- The work done by an applied force is the same as the change in the kinetic energy of the object.
  For example, the driving force of a car is 8 kN and it moves a distance of 7.2 m:
  Work done = force × distance moved = 8000 x 7.2 = 57 600 J = change in kinetic energy

Energy transfers

- As a roller coaster travels round its track, going up and down, its energy changes from kinetic energy to gravitational potential energy.
- Its total energy at any time is the sum of its kinetic energy and its gravitational potential energy.
- When there are no resistive forces the total energy remains constant. This is known as the principle of **conservation of energy**.

- Energy can be neither created nor destroyed; it can only transfer between objects or change its form.
- Usually some energy is used up doing work against friction and air resistance – this means that some energy is dissipated as heat.

- When an object falls, its potential energy is transferred to kinetic energy.
- Ignoring energy transferred due to friction and air resistance:
  Loss in gravitational potential energy = gain in kinetic energy

⦿ Improve your grade

Energy transfers
Foundation: A spacecraft is returning to Earth. It has a gravitational potential energy of 8 MJ on re-entry.
a What is its maximum possible increase of kinetic energy as it falls?
b Explain why the actual increase of kinetic energy will be less than this value.   *AO1, AO2* [3 marks]

P4 Summary

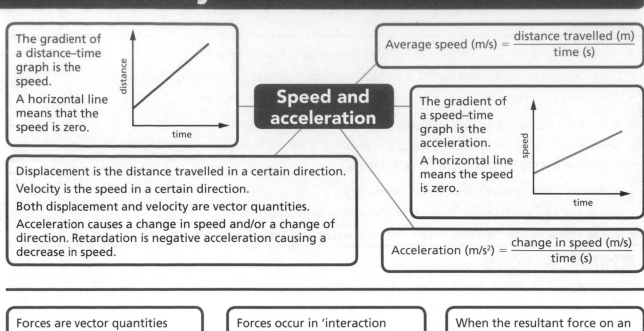

The gradient of a distance–time graph is the speed.

A horizontal line means that the speed is zero.

Average speed (m/s) = $\dfrac{\text{distance travelled (m)}}{\text{time (s)}}$

Speed and acceleration

The gradient of a speed–time graph is the acceleration.

A horizontal line means the speed is zero.

Displacement is the distance travelled in a certain direction.

Velocity is the speed in a certain direction.

Both displacement and velocity are vector quantities.

Acceleration causes a change in speed and/or a change of direction. Retardation is negative acceleration causing a decrease in speed.

Acceleration (m/s²) = $\dfrac{\text{change in speed (m/s)}}{\text{time (s)}}$

Forces are vector quantities – they have both size and direction.

When more than one force acts on an object the resultant force is the sum of all the individual forces.

Forces occur in 'interaction pairs' which always:
- are the same size
- act in opposite directions.
- act on different objects.

When the resultant force on an object is zero:
- the forces are balanced.
- there is no acceleration. The object carries on moving in a straight line at constant speed.

Forces and motion

Friction forces act between two surfaces in the opposite direction to motion.

Reaction forces act on objects upwards from surfaces.

When an object is travelling in a straight line, if the driving force is:
- greater than the drag forces, the object will speed up
- equal to the drag forces, the object will continue to move at constant speed – sometimes known at 'terminal velocity'
- smaller than the drag forces, the object will slow down.

Momentum (kg m/s) = mass (kg) × velocity (m/s)

Momentum

To decrease the size of forces in collisions, the time taken to stop can be increased by using: air bags; seat belts; crumple zones; crash helmets.

A resultant force will cause a change in momentum.

Force (N) = $\dfrac{\text{change in momentum (kg m/s)}}{\text{time taken (s)}}$

When a force moves an object it does work.

Work done (J) = force (N) × distance moved (m)

When work is done energy is transferred.

Work done (J) = energy transferred (J)

The energy of a moving object is its kinetic energy.

The faster the motion and the heavier the object, the larger the kinetic energy.

Kinetic energy (J) = ½ × mass (kg) × [velocity]² ([m/s]²)

Energy is always conserved – it can neither be destroyed nor created, only transferred.

In any energy transfer, some energy is dissipated or wasted as heat.

Work and energy

As you raise an object, its gravitational potential energy (GPE) increases.

GPE (J) = weight (N) × height (m)

Electric current – a flow of what?

Static electricity

- An **atom** is made up of charged particles. It has a positive **nucleus** with negative **electrons** orbiting it. The nucleus is made up of **neutral neutrons** and positive **protons**.

- Neutral objects have no overall charge.

- There are **electrostatic forces** between charged objects, e.g. hair stands up when it is attracted to a charged comb.

- Like charges repel; unlike charges attract.

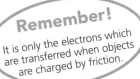

electron: negative charge, –

proton: positive charge, +

neutron: neutral

- There is an electrostatic force of attraction between the positively charged nucleus and the negatively charged electrons in an atom.

- The outermost electrons are less strongly attracted to the nucleus and can be removed by rubbing.

- When two insulating objects are rubbed together they become charged, because electrons are transferred from one object to the other.
  - The object which has lost electrons will become positively charged.
  - The object which has gained electrons will become negatively charged.

Remember!
It is only the electrons which are transferred when objects are charged by friction.

- When you brush your hair, individual hairs become similarly charged and repel each other, making your hair stick up.

- When you take off a nylon or polyester top, there can be a spark or a crackle over your head. This is caused by the electrons moving through the air from the negatively charged clothing to your positively charged hair.

- During a thunderstorm, charge builds up in the clouds. When the amount of charge becomes large enough to break down the insulation of the air, the charge flows between the cloud and the Earth as a flash of lightning.

Conductors and insulators

- Metals are good electrical **conductors** because they have **free electrons**. This means there are lots of charges free to move.

- Plastics are electrical **insulators**. There are few free electrons in plastics, so there are few charges free to move.

free electrons from outer **shells** of metal **ions**

metal ions

Free electrons in the metal cannot pass through the plastic layer.

plastic layer with no free electrons

Moving charges

- When the bulb is lit in a circuit, there is an **electric current**.

- The moving electrons, or electric current, transfer energy to light the bulb.

cell: supplies energy to electrons

electrons

electrical energy transferred to light energy

- In a complete circuit, there are free electrons in all the metal components and connecting wires. The cell (or battery) supplies energy to the electrons. The electrons carry a negative charge, so they will flow from the negative terminal of the cell towards the positive terminal.

- The flow of charge is the electric current.

- The bulb is converting the energy carried by the electrons into light (and heat) energy.

- Electric current is the **rate of flow of charge**, or the charge flowing per second.

- Electric current is measured in **Ampères**, or amps (A) for short.

- The more energy the charged particles receive from the power supply, the greater the current.

- In an electric circuit, charge is conserved and energy is transferred.

Improve your grade

Static electricity

Higher: Bella was rubbing a nylon comb with a duster. When she put the comb near her head, her hair moved towards the comb. Explain why this happened. *AO1, AO2* [5 marks]

Current, voltage and resistance

Measuring current and voltage

- An **ammeter** is used to measure current. An ammeter is connected in series. The circuit symbol for an ammeter is: —Ⓐ—

- A **voltmeter** is used to measure **voltage**. A voltmeter is connected in parallel across a **component** in a circuit. The circuit symbol for a voltmeter is: —Ⓥ—

- The unit of voltage is the **volt** (V).

- The larger the voltage of the battery of in a circuit, the bigger the current.

- The voltage across a power supply is a measure of how much energy is supplied to the circuit.

- The voltage across a component is a measure of how much energy is transferred in the component.

> **EXAM TIP**
>
> Always talk about the current *through* a component and the voltage *across* a component.

- In a circuit, the charges (free electrons) are the energy carriers. They collect energy at the power supply and transfer energy at a component.

- **Power** is the rate at which energy is transferred. It is measured in watts (W).
  Power (W) = voltage (V) × current (A)

- A voltmeter measures the difference in energy between the terminals of a battery or bulb.

- The difference in energy per unit charge is known as the **potential difference (p.d.)**. This is the scientific term for voltage.

- A potential difference of 1 volt means that 1 joule of energy is transferred into or out of electrical form for each unit of charge.

Electrical resistance

- The more **resistance** in a circuit, the lower the current.

- All electrical components, e.g. lamps, motors, have some resistance to the flow of charge through them.

- The greater the voltage across a resistor, the larger the current.

- The circuit symbol for a resistor is: —▭—

- Resistance is a measure of how much a conductor opposes the current. Its unit is the ohm (Ω).

- Copper wires have such a low resistance that we can ignore it.

- A **variable resistor** is a device that allows you to control the current by changing the amount of resistance wire in a circuit.

- The symbol for a variable resistor is: —▭⟋—

- Resistance (in ohms) = $\dfrac{\text{voltage (in volts)}}{\text{current (in amps)}}$ or: $R = \dfrac{V}{I}$

Ohm's law

- A graph of voltage against current will give a straight line through the origin. This means that the current through a fixed resistor is directly proportional to the voltage across it (at constant temperature).

- The higher the resistance, the lower the gradient.

Remember!
Resistance is equal to 1/gradient of the line *only* if the voltage is along the x-axis and current is on the y-axis.

fixed resistor with small resistance
fixed resistor with large resistance
Current (*I*)
Voltage (*V*)

Ideas about science

You should be able to:

- suggest why using experimental values of current or voltage may not give the true value of resistance in a circuit.

Improve your grade

Electrical resistance

Higher: Harry was recording values of current and voltage across a resistor so he could calculate the resistance.

a He recorded a current of 0.06 A when the voltage across the resistor was 4 V. Calculate the resistance.

b Harry then repeated the experiment with a voltage of 6 V. What current reading did he expect to get? Explain why he might not get this exact value.
AO1, AO2 [5 marks]

Useful circuits

Series and parallel circuits

- Components connected **in series** are in a line.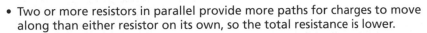

- The current is the same through all the components connected in series.

- The more cells connected in series, the greater the potential difference.

- The potential difference across the components adds up to the p.d. across the battery.
  Supply p.d. = p.d. across R_1 + p.d. across R_2 + p.d. across R_3

- The p.d. across each component will be in proportion to its resistance.

- The overall resistance will be the sum of all the individual resistances: $R_{total} = R_1 + R_2 + R_3$

- Components **in parallel** are each connected separately to the power supply (see right).

- The charge has a choice of pathways, so the current is shared between each branch. The current to and from the power supply is the sum of the current through all branches. $I_{total} = I_1 + I_2 + I_3$

- Two or more resistors in parallel provide more paths for charges to move along than either resistor on its own, so the total resistance is lower.

- The current through each resistor is *inversely proportional* to its resistance, i.e. it is largest through the component with the smallest resistance.

Remember!
You can use the relationship $R = V/I$ for each branch of the circuit.

- Work is done by the power supply to provide energy to the charged particles. A bulb uses the energy to do work to provide heat and light; a resistor uses the energy to do work to provide heat.

- In a series circuit, the work done on each unit of charge by the battery must equal the work done on it by the circuit components.
  - More work is done by the charge moving through a large resistance than through a small one.
  - Two or more resistors have more resistance than one on its own, because the battery has to move charges through both of them.
  - A change in the resistance of one component, e.g. variable resistor, will cause a change in the potential differences across all the components.

- Cells connected in parallel will have the same potential difference as one cell on its own, but the amount of energy in the circuit will increase.

- The potential difference across components connected in parallel is always the same, and is equal to the pd of the battery.

Thermistors and LDRs

- A **thermistor** is a semiconductor whose resistance changes with temperature.

- An **light dependent resistor (LDR)** is a semiconductor whose resistance changes as the amount of light falling on it changes. In bright light the resistance will be low.

- The left graph shows the variation of resistance with temperature. At high temperature the resistance is lower.

- The right graph shows the variation of resistance with light intensity for an LDR.

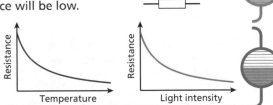

Metals and semiconductors

- The graph opposite shows how the current through a bulb varies with increasing potential difference. As the wire in the bulb gets hotter, its resistance increases.

- In semiconductors, as the temperature or light intensity increases there are more free electrons to carry the current, so the current is higher.

- The positive ions in the metal structure have more energy and vibrate more. The free electrons will collide more often. This means that they cannot move as fast, so the current decreases.

Improve your grade

Series circuits and parallel circuits

Foundation: Susan and Mark were discussing adding resistors to a circuit. Susan said that if you added more resistors the total resistance would increase. Mark told her that sometimes adding more resistors would decrease the total resistance. Explain why both Susan and Mark are correct.   *AO1, AO2* [5 marks]

Making an electric current

- A **magnetic field** is a space around a magnet in which a magnetic force acts.
- The magnetic field is strongest where the field lines are closest together.

- A voltage is **induced** when a magnet is moved near a piece of wire (or when a wire is moved near a magnetic field). If the piece of wire is part of a circuit, a current will flow.
- A voltage is always induced when there is relative movement between a magnet and a coil of wire.
- The direction of the current is reversed when the motion of the wire is reversed, or the magnet is turned round.
- The current will increase if the speed of motion increases, a stronger magnet is used or there are more turns of wire in the coil.

Generators

- A continuous supply of electricity is produced when there is continuous relative motion between a magnet and a coil of wire.
- The coils of wire continuously 'cut' the magnetic field lines so a voltage is induced. This is called **electromagnetic induction**.

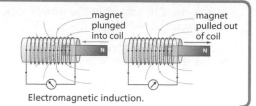
magnet plunged into coil   magnet pulled out of coil

Electromagnetic induction.

- A larger voltage is induced if:
  - the strength of the magnet is increased
  - an iron core is used inside the coil
  - the number of turns in the coil is increased
  - the rate at which the coil is turned is increased.
- The faster the rate of cutting field lines, the larger the induced voltage.
- Mains electricity is produced by generators in power stations that induce an alternating voltage.

Higher

- As a coil rotates in a uniform magnetic field it cuts the lines of magnetic field at different rates:
  - When the coil is at right angles to the field lines, it cuts no field lines, so the induced voltage is zero.
  - When the coil is parallel to the field lines, its rate of cutting field lines is at a maximum, so the inducted voltage is at a peak.
  - As the coil rotates, it cuts field lines in a different direction, so the direction of the voltage alternates.

Distributing mains electricity

- **Direct current (d.c.)** always flows in the same direction. Batteries produce d.c. electricity.
- **Alternating current (a.c.)** changes direction at regular intervals.
- In the UK, mains electricity is a.c. and is generated at 230 V and at a frequency of 50 Hz. It is transmitted along cables held up by pylons at very high voltages.

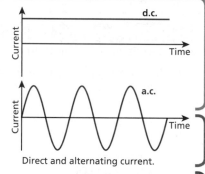
Direct and alternating current.

- Transformers (see page 105) are used to step up the voltage at the power station and to step down the voltage near our homes. Transformers only work with a.c.

Higher

- The UK's electricity supply is a.c. because it is easier to generate a.c. electricity in large amounts. Also, many different fuels can be used in power stations.
- It is more efficient to transmit electricity at very high voltages.

Ideas about science

You should be able to:
- identify factors that might affect an outcome
- make predictions about the outcome of changing a variable.

Improve your grade

Generators

Foundation: Jenny uses a dynamo to power her bicycle lights.

a Explain why the lights are dim when she cycles slowly.

b Suggest one advantage and one disadvantage to using dynamo lights instead of battery-powered lights.

AO1, AO2 [5 marks]

Electric motors and transformers

The magnetic effect of a current in motors

- **Motors** are used in many electrical appliances that make things move, e.g. hairdryer, DVD player.

- There is a circular magnetic field around a wire carrying electric current.

- If the wire is made into a coil, the magnetic field pattern becomes similar to that of a bar magnet. This is called an **electromagnet**.

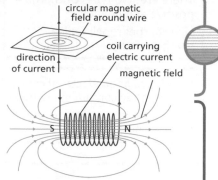

Electric currents generate magnetic fields.

- When a current flows in a wire that is in a magnetic field, the wire experiences a force. If the wire is free to move, it moves. This is called the **motor effect**.

- The force is largest when the current is at right angles to the magnetic field lines. The direction of the force is always at right angles to both the current in the wire and the magnetic field lines.

- No force is experienced when the current is parallel to the magnetic field lines.

- The direction of the force is reversed if either the current or the magnetic field is reversed.

- When a simple motor (see opposite) is placed in a uniform magnetic field, one side of the rectangular current-carrying coil is forced upwards, while the other is forced downwards to produce rotation.

- The motor will turn faster if: the current is increased; the number of turns on the coil is increased; the magnetic field is stronger; there is a soft iron core in the coil.

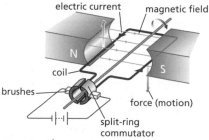

An electric motor.

- The motor effect works because of the interaction between the magnetic field around the current carrying wire and the magnetic field of the permanent magnets.

- In the diagram, the two magnetic fields reinforce each other above the wire and cancel each other out below the wire, so the wire is forced down.

- The coil of the motor is connected to the power supply using a commutator. The commutator swaps contacts with the coil every half turn to reverse the current through the coil. This keeps the motor turning.

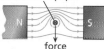

Combining magnetic fields.

Transformers

- A transformer changes the voltage of an a.c. power supply.

- It consists of two separate coils around an iron core.

- The input voltage is fed into the **primary coil**.

- The output voltage is across the **secondary coil**.

- The larger of the two voltages will be across the coil with the most turns.

- A **step up transformer** converts a low voltage input to a higher voltage output. The primary coil will have fewer turns than the secondary coil.

- A **step down transformer** converts a high voltage input to a lower voltage output. The primary coil will have more turns than the secondary coil.

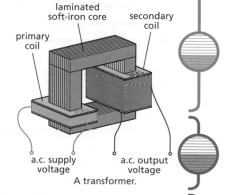

A transformer.

- The alternating current in the primary coil creates an alternating magnetic field around it. The magnetic, soft iron core channels the magnetic field through the secondary coil.

- The alternating magnetic field will continuously cut through the wires in the secondary coil and an a alternating voltage will be induced across the secondary coil.

- If the number of turns in the secondary coil is doubled, the output voltage will be doubled.

- The turns ratio is equal to the voltage ratio:

$$\frac{voltage\ across\ primary\ coil}{voltage\ across\ secondary\ coil} = \frac{number\ of\ turns\ in\ primary\ coil}{number\ of\ turns\ in\ secondary\ coil}$$

Improve your grade

Transformers

Higher: The mains supply at home is at 230-V a.c. A computer needs a supply at 23 V. Describe how the voltage of the mains is converted to the lower voltage. *AO1* [5 marks]

P5 Summary

Electric charge

There are two types of charge – positive and negative.
Like charges repel; unlike charges attract.

Electrostatic effects are caused by a transfer of electrons.
A positively charged object has lost some electrons.
A negatively charged object has gained some electrons.

Metals are good electrical conductors because they contain many free electrons.
Plastics are insulators because there are few free electrons.

Electric current is the rate of flow of electric charge.
Electric current is measured in amps (A).

Circuit electricity

Energy is supplied to a circuit by a battery or cell.
The voltage or potential difference (p.d.) across the cell is a measure of the amount of energy per unit of charge.
The energy is transferred to the component in the circuit.

Resistors oppose the motion of the charged particles, and get hot.
Resistance is measured in ohms (Ω).
The higher the resistance, the lower the current. The higher the voltage, the higher the current.
Ohm's law states:

$$\text{Resistance } (R) = \frac{\text{voltage } (V)}{\text{current } (I)}$$

When resistors are connected in series, the total resistance is increased.
The current around a series circuit is the same everywhere.
The p.d. across the components adds up to the p.d. across the power supply.
$$V_{total} = V_1 + V_2 + V_3$$
The p.d. is split in proportion to the resistance.

An LDR is a semiconductor whose resistance decreases in brighter light.
A thermistor is a semiconductor whose resistance usually decreases as temperature increases.
The resistance of metals increases with increasing temperature.

When resistors are connected in parallel, the total resistance is less than any of the resistors individually.
The p.d. across each component is equal to the p.d. across the power supply.
The current has a choice of pathways.
The total current through the power supply is the sum of the currents through each component. $I_{total} = I_1 + I_2 + I_3$

Generating electricity

Alternating current (a.c.) changes direction at regular intervals.
Direct current (d.c.) always flows in the same direction.

Mains electricity supply in the UK is 230 V at 50 Hz a.c.

Electricity generators work by electromagnetic induction.
When a wire moves relative to a magnetic field, a voltage is induced.

Transformers change the voltage of a.c. electricity. Step up transformers increase voltage; step down transformers decrease it.
A transformer consists of two coils around an iron core. A varying current in the primary coil produces a varying magnetic field; the varying magnetic field induces a voltage in the secondary coil.

$$\frac{voltage\ across\ primary\ coil}{voltage\ across\ secondary\ coil} = \frac{number\ of\ turns\ in\ primary\ coil}{number\ of\ turns\ in\ secondary\ coil}$$

In a generator, a magnet is rotated within a coil of wire to induce a voltage. The size of the induced voltage can be increased by: increasing the speed of rotation; the strength of the magnetic field; the number of turns on the coil; placing an iron core inside the coil.

Electric motors

Motors are used in all electrical appliances that convert electrical energy to movement, e.g. vacuum cleaners, CD players.

The motor effect:
A wire carrying a current experiences a force when in a magnetic field. The force is at maximum value when the wire is at right angles to the magnetic field lines. Continuous rotation of a wire coil can be produced if the current is reversed in the coil every half turn, with the use of a commutator.

Nuclear radiation

The nuclear atom

- An **atom** is the smallest part of an element.

- **Neutrons** and **protons** form the **nucleus** of an atom. **Electrons orbit** the nucleus at high speed.

- Electrons are the smallest particle and are negatively charged.

- Protons have mass 2000 times that of an electron and are positively charged.

- Neutrons have the same mass as protons and have no charge.

- An atom is neutral, so has an equal number of electrons and protons.

- Nearly all the mass of the atom is concentrated in the nucleus.

An atom of helium

- Evidence about the structure of the atom was obtained from the 1909 Rutherford-Geiger-Marsden alpha scattering experiment, during which alpha particles were fired at gold foil.

- The observations recorded were:
  - Most **alpha particles** passed straight through the gold foil undeviated.
  - A few particles were deflected through small angles.
  - Even fewer bounced straight back from the foil.

- The conclusions were:
  - The atom is mostly empty space because most particles went straight through.
  - The mass and charge of the atom is concentrated in a small area in the centre of the atom (the nucleus).
  - The nucleus was positive because the positive alpha particles were repelled.

The Rutherford-Geiger-Marsden scattering experiment.

- A **strong nuclear force** holds the protons and neutrons in the nucleus together. It has to balance the repulsive electrostatic force between the protons.

- The number of protons in a nucleus determines the element and its chemical properties.

- **Isotopes** are atoms of the same element with the same number of protons but differing numbers of neutrons in the nucleus. For example, the three isotopes of hydrogen are hydrogen (1 proton, 1 neutron), deuterium (1 proton, 2 neutrons) and tritium (1 proton, 3 neutrons).

Radioactive elements

- **Radioactive** elements are unstable and constantly emit **ionising radiation**. This makes them become more stable.

- Ionising radiation knocks out electrons from atoms, forming a positive **ion**.

- Ionising radiation can be either high-energy particles or high-energy electromagnetic waves.

- **Background radiation** is low-level ionising radiation that is all around us.

- Some background radiation comes from outer space as cosmic rays, but most comes from rocks and soil.

Remember!
Not all background radiation comes from natural sources – some comes from industry and medical uses.

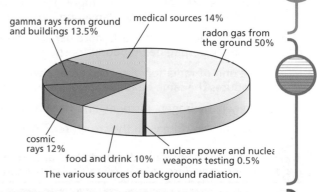

The various sources of background radiation.

- Radioactivity is a random process. We never know when an unstable nucleus will decay and emit ionising radiation.

- The amount of radiation emitted is dependant only on the amount of the radioactive element present. The behaviour of radioactive materials is not affected by physical or chemical processes, e.g. temperature or atmospheric conditions.

Improve your grade

Radioactive elements

Foundation: Radioactive materials emit ionising radiation. Explain what is meant by 'ionising radiation'.

AO1 [3 marks]

Types of radiation and hazards

Types of ionising radiation

- There are three types of ionising radiation emitted by radioactive materials.

	Charge	Mass			
Alpha (α)	Positive (2⁺)	Heavy	Deflected by magnetic and electric **fields**	Highly ionising	Low penetration
Beta (β)	Negative (1⁻)	Very light	Deflected by magnetic and electric fields (in opposite direction to alpha)	Not as ionising as alpha	Medium penetration
Gamma (γ)	Neutral	No mass	Not deflected by magnetic or electric fields.	Low ionising power	High penetration

- As alpha, beta and gamma radiation go through air they ionise air molecules and lose energy.

- Alpha particles are massive and can easily knock off electrons, so lose energy quicker and travel less far (few cm in air).

- Beta particles have a range in air of about a metre, but can be stopped by about 3 mm of aluminium.

- Gamma rays are not stopped by air, but just spread out and become less intense. Thick lead is used to absorb gamma rays.

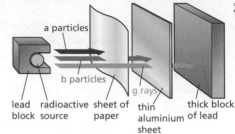

Alpha, beta and gamma have different penetrating properties.

- An alpha particle is a helium nucleus. It consists of 2 protons and 2 neutrons, so it has an atomic mass of 4 and a charge of +2. It is written as $^4_2$He.

Higher

- A beta particle is a fast-moving electron, so it has an atomic mass of zero and a charge of −1. It comes from the nucleus when a neutron changes into a proton and an electron. It is written as $^0_1$e.

- Gamma rays are very-high-frequency electromagnetic waves.

Hazards of ionising radiation

- Ionising radiation can damage living cells. The damage depends on the type and intensity of the radiation.

- Ionising radiation collides with living cells and knocks elections out of the atoms, leaving positive ions.

- High intensity radiation may kill the living cells and cause tissue damage, and could lead to radiation sickness. Some high intensity radiation causes cells to become **sterile**.

- Lower intensity radiation can affect the cell's genetic makeup, causing **mutations** that could lead to cancer.

- Alpha particles are large and highly ionising, but do not pass through the skin. Inside the body alpha particles would be highly damaging, but are relatively safe outside the body.

- Beta and gamma are much more penetrating and will pass through skin, so are more dangerous outside the body despite being poorer ionsisers.

- The unit of radiation absorbed **equivalent dose** is the **Sievert (Sv)**. One Sievert of alpha, beta or gamma produces the same biological effect, and is a measure of the possible harm done to your body.

- Oxygen, hydrogen, nitrogen and carbon are highly susceptible to ionisation and are abundant in the body.

- Ions can interfere with the structure of DNA, causing it to behave incorrectly and damage living cells.

Higher

- Once atoms are ionised by radiation and form ions, these charged ions can break and make chemical bonds, therefore changing molecular structure.

Ideas about science

You should be able to:

- suggest ways of reducing the risks from ionising radiation

- interpret and discuss information on the size of risks, presented in different ways.

Improve your grade

Hazards of ionising radiation

Higher: Explain why it is more dangerous to inhale (breathe in) an alpha emitter than a beta emitter.

AO1, AO2 [4 marks]

Radioactive decay and half-life

Radioactive decay

- When a radioactive nucleus emits an alpha or beta particle, the number of protons and neutrons changes and it becomes a new element.

- The radioactivity of a material decreases over time because the amount of radioactive nuclei decreases. This is called **radioactive decay**.

- The **activity** of a radioactive sample only depends on the number of unstable nuclei present.

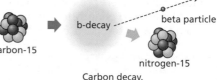

alpha particle

a-decay

uranium-238

thorium-234

Alpha decay.

- Many large nuclei are unstable and emit alpha particles. For example, uranium-238 decays to the element thorium with the emission of an alpha particle.

- Carbon-15 is an unstable form of carbon that emits a beta particle and becomes nitrogen.

- Emitting gamma rays does not change one element to another.

- The new element is known as the **daughter product**, and may or may not be radioactive.

- Many radioactive elements belong to a **decay chain**, in which the first radioactive element decays into a second element, which then decays into a third element, and so on.

beta particle

b-decay

carbon-15

nitrogen-15

Carbon decay.

- An unstable nucleus is in an *energetic state* – it has too much energy. It needs to lose energy to become more stable.

- When nuclei emit alpha or beta particles they change into more stable nuclei, but may still be in an energetic state. So they often emit gamma rays as well, to reduce their energy.

- Scientists represent nuclei in the form $^M_A X$, where X is the chemical symbol for the element, M is the mass number (number of protons + number of electrons) and A is the atomic number (number of protons).

- An equation for alpha decay is:

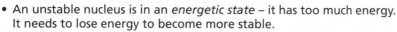

$$^{238}_{92}U \xrightarrow{\alpha\text{-decay}} {}^4_2He + {}^{234}_{90}Th$$

238 – 4
92 – 2

- An equation for beta decay is:

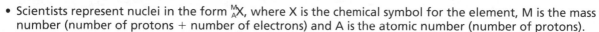

$$^{15}_{6}C \xrightarrow{\beta\text{-decay}} {}^0_{-1}e + {}^{15}_{7}N$$

15 – 0
6 + 1

- The total mass and the total atomic number (or charge) must be the same on both sides of the equation.

Half-life

- The **half-life** of a radioactive element is the time taken for half the nuclei in a sample to decay. The half-life is specific to each radioactive element.

- Half-lives can vary from fractions of a second to millions of years.

- The activity of a radioactive source (the amount of radiation emitted) is a measure of its rate of decay.

- When there are plenty of radioactive nuclei present at the beginning, the rate of decay is faster than when most of the nuclei have already decayed.

- The activity can never reach zero – it just continuously decreases to a negligible value.

- Scientists can find the half-life of a sample by recording the radioactive count rate over time and plotting a graph. They read off the time it takes the count rate to drop from 80 to 40, 40 to 20, 20 to 10, etc., and calculate the average.

- So, if you started with a count rate of 120 counts per minute (cpm), after three half-lives the count rate would be: $\frac{1}{8} \times 120 = 15$ cpm

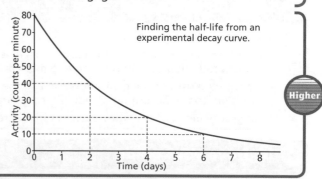

Finding the half-life from an experimental decay curve.

Improve your grade

Half-life

Higher: Radon-220 decays by alpha emission with a half-life of 52 seconds. The initial activity is 640 counts per second. How long will it take for the activity to become 80 counts per second?

AO1, AO2 [3 marks]

Uses of ionising radiation and safety

Uses of ionising radiation

1 *Treating cancer* – ionising radiation can kill cells, so can be used to kill cancerous cells. This is known as **radiotherapy**. Gamma radiation is usually used. Some healthy tissue around the tumour can be damaged, so the radiation must be focused on the tumour.

2 *Sterilising medical instruments* – these can be irradiated with gamma radiation to kill bacteria.

3 *Sterilising food* – as soon as fresh food is picked and ready to transport, micro-organisms will start the decay process. If the food is irradiated with gamma radiation the micro-organisms will be killed. This makes the shelf life of the food much longer.

- For sterilisation, the radiation must penetrate the packaging and be capable of killing bacteria – so gamma emitters are used.

4 *Detecting tumours*: brain and other tumours can be detected using a **radioactive tracer**. A gamma emitter with a half-life of a few hours is injected; the radiation is detected from outside to build up a computer image of the tumour.

EXAM TIP
Always explain which type of radiation you need and whether you need a long or short half-life for all uses of radioactivity.

- Radioactive tracers are usually beta or gamma emitters, as they must be able to penetrate skin and tissue. The half-life needs to be a few hours, so that it has time to reach the affected parts of the body in sufficient quantities, but not last so long that it damages the body.

Keeping people safe

- Exposure to radiation is called **irradiation**.

- People are exposed to radiation all the time and the risk to health is usually insignificant. However, it does depend on the level of radiation and the length of exposure.

- The dose equivalent in Sieverts can be used to evaluate the level of risk from radiation and what harm may be done.

Remember!
A **hazard** is anything which may cause harm. The **risk** is the chance, high or low, that somebody could be harmed by the hazard.

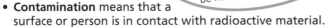

From the sky About 400 000 cosmic rays pass through us each hour

From the air 30 000 atoms of radioactive gases breathed in disintegrate in our lungs each hour

From food 15 million potassium-40 atoms disintegrate inside our bodies each hour

From soil and building materials More than 200 million gamma rays pass through us each hour

We are exposed to ionising radiation all the time.

- **Contamination** means that a surface or person is in contact with radioactive material.

- Radioactive waste can be contained to prevent contamination. If radioactive waste cannot be contained it must be diluted to safe concentrations.

- High-level contamination, such as fallout from a nuclear explosion, will need more intervention, e.g. administering iodine to affected people.

- People who work with radioactive sources, e.g. radiographers and workers in nuclear power stations, regularly need to have their level of exposure monitored.

- A film badge can monitor radiation. The level of exposure is measured by assessing how black the film has become.

- Lead shielding in the form of aprons or protective walls is used to protect radiographers.

Ideas about science

You should be able to:

- identify examples of risks which arise from a new scientific or technological advance, such as using gamma radiation to sterilise food and medical instruments
- identify the risks of the contamination and the benefits of using radioactive materials in a given situation, to the different individuals and groups involved
- suggest benefits of activities, such as radiotherapy, that are known to have risk.

Improve your grade

Uses of ionising radiation

Higher: Radioactive tracers can be used in many different applications. Which of the following radioactive isotopes would be most suitable for studying plant nutrition?

Explain your choice.

Isotope	Type of radiation	Half-life
Phosphorus-32	Beta decay	14 days
Nitrogen-16	Beta decay	7 seconds
Bismuth-210	Alpha decay	5 days

AO1 [3 marks]

Student book pages 272–275

Nuclear power

Energy from the nucleus

- **Nuclear fission** releases energy by a heavy nucleus, e.g. uranium, splitting into two lighter nuclei.

- **Nuclear fusion** releases energy by two light nuclei, e.g. hydrogen, combining to create a larger nucleus.

- Materials that can provide energy by changes in the nucleus are known as **nuclear fuels**.

fusion

energy

- Nuclear fission and fusion release much more energy than chemical reactions involving a similar mass of material. This is because the energy that holds nuclei together (**binding energy**) is much larger than the energy that holds electrons in place.

fission

energy

- In nuclear fission, a neutron is fired at a uranium or plutonium nucleus to make it unstable. The nucleus breaks down into two smaller nuclei of similar size, and releases some more neutrons.

- The neutrons released in the fission reaction can go on to initiate more fission reactions. This is known as a **chain reaction**.

- Only one neutron from each fission needs to go on to initiate the next fission.

- It is necessary to have a critical mass of nuclear fuel for the chain reaction to be viable.

- More and more neutrons will be released in each subsequent reaction, and the chain reaction will get out of control unless the number of neutrons is controlled.

$^{141}_{56}Ba$

$^{135}_{92}U$

$^{1}_{0}n$

$^{1}_{0}n$

$^{1}_{0}n$

$^{92}_{36}Kr$

energy

Ba

U

Kr

U

Ba

Kr

U

Ba

Kr

n

n

n

n

n

n

A chain reaction results from the neutron emissions.

- Energy released in nuclear fission and fusion is calculated using $E = mc^2$, where E is the energy in joules, m is the mass in kg and c is a constant equal to the speed of light in a vacuum or 3×10^8 m/s.

Nuclear power generation

- About a sixth of the UK's electricity is generated in nuclear power stations. They all use nuclear fission.

- Nuclear fission produces radioactive waste, which must be disposed of carefully.

- Nuclear wastes are categorised according to their level of risk:
  - **Low level waste**, e.g. contaminated paper, clothing, is not dangerous to handle but should be disposed of with care. It is burnt and sealed in containers before being buried in landfill.
  - **Intermediate level waste**, e.g. chemical sludges, reactor parts, is more radioactive and needs shielding. Waste with a longer half-life is buried deep underground.
  - **High level waste**, e.g. spent fuel rods, is highly radioactive. Some of this waste is mixed with molten glass and contained in stainless steel drums before careful storage.

- In a nuclear reactor, the **fuel rods** contain pellets of uranium. Neutrons cause the fuel to undergo fission. The energy is released as kinetic energy of particles (heat).

- A **coolant** (gas or liquid) circulated around the reactor absorbs the heat and transfers it to a steam generator. Electricity is then generated in the same way as a conventional power station.

- **Control rods** (usually made of boron) absorb some of the neutrons. The control rods can be raised or lowered to control the fission rate.

Harnessing fusion energy

- Fusion produces a lot more energy per kg than fossil fuels. Its by-products are not radioactive and it does not release carbon dioxide into the atmosphere.

- Isotopes of hydrogen are readily available and only small amounts are needed.

- Despite its great potential, more energy is consumed producing fusion reactions than is released by it.

Improve your grade

Energy from the nucleus

Foundation: Explain the difference between nuclear fission and nuclear fusion.   *AO1, AO2* [4 marks]

Atoms have neutrons and protons in the nucleus, which are held together by a strong nuclear force. The nucleus is surrounded by electrons.

The Rutherford-Geiger-Marsden scattering experiment gave evidence for the structure of the atom.

An isotope is an atom of an element with the same number of protons but a different number of neutrons in the nucleus.

Ionisation occurs when radiation collides with an atom and knocks electrons out of orbit, leaving a positively charged ion.

Nuclear radiation

When an unstable nucleus emits alpha or beta particles it decays to become more stable – it becomes a new element.

An equation for alpha decay is: $^{238}_{92}U \longrightarrow {}^{4}_{2}He + {}^{234}_{90}Th$

An equation for beta decay is: $^{14}_{6}C \longrightarrow {}^{0}_{-1}e + {}^{14}_{7}N$

The activity of a radioactive source decays over time.

The half-life is the time taken for the activity of a radioactive sample to halve, or the time taken for the number of radioactive nuclei to halve.

Unstable nuclei are radioactive and emit ionising radiation.

Background radiation comes from natural and man-made sources.

Radioactivity is completely random and is not affected by chemical or physical changes.

There are three types of ionising radiation:

- Alpha – helium nucleus, massive, 2+ charge, highly ionising, low penetration. $^{4}_{2}He$
- Beta – fast-moving electron, low mass, 1+ charge, medium ionising power, medium penetration. $^{0}_{-1}e$
- Gamma – high frequency electromagnetic waves, low ionising power, high penetration.

Ionisation can cause damage to living cells – they could be killed or become cancerous.

Ions (formed by radiation colliding with atoms) can react with other chemicals.

Uses and safety

Exposure to ionising radiation is called irradiation.

Contamination occurs when you are in contact with radioactive materials.

Background radiation causes irradiation and contamination all the time.

Radiation dose (measured in Sieverts) is a measure of the possible harm done to our bodies. The higher the radiation dose, the greater the risk.

People who work with radioactive materials must monitor their exposure carefully.

Ionising radiation is used to treat cancer, sterilise medical equipment and food, and as a tracer in the body.

Choosing the radioactive source involves thinking about both the type of radiation emitted and the half-life.

The longer the half-life, the longer the source will be considered dangerous.

Nuclear fuels can release a lot more energy than chemical fuels.

Some of the binding energy is released as there is a change of mass.

Energy released $E = mc^2$.

Nuclear power

Nuclear power stations produce radioactive waste.

Low level waste, intermediate level waste and high level waste are disposed of in different ways.

Nuclear fusion occurs when two light nuclei, e.g. hydrogen, come close enough together to fuse and form a heavier product, e.g. helium. This releases energy.

Nuclear fission occurs when a large unstable nucleus, e.g. uranium or plutonium, splits into two similar-sized smaller nuclei. This gives off large amounts of energy and some neutrons.

The neutrons go on to initiate further fission of more nuclei in a chain reaction.

In nuclear power stations, the number of neutrons which go on to cause more fission in the fuel rods is managed by control rods.

The fuel rods are surrounded by a coolant which transfers heat energy to water to produce steam. The steam turns a turbine to generate electricity.

Page 4 Studies on twins

Higher: How do studies of identical twins help us to understand the effect of the environment on the phenotype for a characteristic? *AO2* [3 marks]

Identical twins are born with identical sets of genes, so any differences in their phenotypes for a characteristic must be the result of an effect of their environment.

Answer grade: C. This answer is correct, but could be extended. It's important to spell out each stage of your answer, so the first part must refer to the genotypes of identical twins being the same.

For a B grade, you need to raise a more subtle point. Identical twins are usually brought up in a very similar environment, so studies of identical twins that have been separated at birth or at an early age are especially useful, as their environments may be very different.

Page 5 Variation in offspring

Foundation: Explain how variation occurs in the offspring produced by humans. *AO1* [5 marks]

Variation occurs because we all have different genes, which we inherit from our parents.

Answer grade: D. This answer is correct but lacks detail. For full marks at grade C, explain that it is because of *sexual reproduction* that we show variation. Be specific and say that we inherit half our genes from our mother and half from our father. However, siblings differ because they inherit a different *combination* of genes from their parents. Show how these combinations might occur, using diagrams of chromosomes and genes. You should also mention that the environment we live in also leads to variation in humans.

Page 6 Genetic diagrams

Higher: A disease called cystic fibrosis is caused when two recessive alleles of a gene are present. The diagram shows the occurrence of cystic fibrosis in the family.

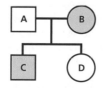

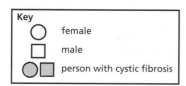

```
Key
○  female
□  male
◐◼  person with cystic fibrosis
```

What are the genotypes of:

a father A?

b daughter D?

Explain your answer. *AO3* [4 marks]

Father A is CC or Cc, because he does not have cystic fibrosis. As the daughter D is normal, she must be Cc.

Answer grade: B. Sentence 1 is partly correct, but it is possible to say with certainty what the genotype is. In sentence 2, the daughter's genotype given is correct, but the explanation is incomplete.

For full marks at grade A/A*, you need to *explain* that father A does not have cystic fibrosis, so must have at least one dominant gene. As the son has cystic fibrosis, the father must be Cc, and not CC (if he was CC, there would be no chance of producing a child with cystic fibrosis). The mother (B) has cystic fibrosis, so she must be cc. This means that daughter D must have inherited a c allele from her mother, but as she is normal, daughter D must be Cc.

Page 7 Gene disorders

Foundation: Write down the name of one dominant genetic disorder, and one recessive genetic disorder. For each disorder, list three symptoms the person will show. *AO1* [4 marks]

Huntington's disease: The symptoms are tremors, uncontrollable shaking and memory loss.

Cystic fibrosis: The symptoms are thick gluey mucus affecting the lungs and difficulty breathing.

Answer grade: E. In the question, there are 2 marks for naming the disorders and 1 mark for each list of three symptoms. The student has not said which is the dominant genetic disorder (Huntington's disease) and which is the recessive (cystic fibrosis). These are required to get the 2 marks.

For full marks, you must write down **three** symptoms for each disorder, as stated in the question, to get one mark for each. For Huntington's disease, these also include an inability to concentrate and mood changes. Note that 'tremors' and 'uncontrollable shaking' are the same thing, so three symptoms weren't given in the answer. For cystic fibrosis, symptoms also include chest infections and difficulty in digesting food.

Page 8 Stem cells

Higher: Discuss the future use of stem cells in medicine. *AO1, AO2* [4 marks]

Stem cells are unspecialised, so they can be used to produce different types of body cells. So stem cells can be used to renew damaged or destroyed cells, for example in spinal injuries, heart disease, Alzheimer's disease and Parkinson's disease.

Answer grade: C. This answer is correct; the student has referred to the fact that stem cells are unspecialised, so have the potential to develop into different cell types. However, the student has just focused on the possible replacement of damaged or destroyed cells.

For a B grade, it's important to mention other uses too. Stem cells can be used to improve our understanding of how cells become specialised. This occurs in the early stages of a person's development, by the switching on and off of particular genes. Stem cells can also be used in the testing of new drugs.

Page 10 Our defence system

Higher: A bacterium enters your blood stream. Describe the series of events leading to the bacterium being destroyed by your immune system.　　　*AO1* [5 marks]

Antibodies are produced. The antibodies lead to the destruction of the bacterium.

> **Answer grade: B.** The answer correctly states that it is the production of antibodies that leads to the bacterium being killed, but it does not describe the series of events that lead to its destruction.
>
> For full marks, explain that the antibodies are made by white blood cells and how this is done. You could explain that the invading bacterium (antigen) becomes attached to a white blood cell with a matching antibody; the white blood cell then divides and the white blood cells all make many antibody molecules, which attach to the invading bacterium and destroy it.

Page 11 Vaccination programmes

Higher: The graph shows the number of cases of measles, and deaths from measles, in England and Wales from 1940 to 2006.

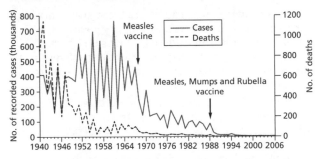

Discuss what the data suggest about the effectiveness of the measles vaccines.　　　*AO3* [4 marks]

The graph shows how the introduction of the measles vaccine reduced measles cases. When the combined measles, mumps and rubella vaccine was introduced, it reduced these further.

> **Answer grade: C.** The answer says, correctly, that the vaccines reduced the cases of measles, but it lacks detail, and doesn't describe how the *deaths* from measles changed over the time period.
>
> For full marks, you would need to say that the numbers of cases of measles showed fluctuations (some very large) across this time period (and even after the introduction of the vaccines). The *deaths* from measles have shown a continuous decrease, even *before* the vaccines were introduced. You could also comment on the small peak around 1994. Remember, the data may show more than one trend, so you may need to describe and explain these one by one.

Page 12 Trialling new treatments

Foundation: When testing a chemical on humans to see if it would be suitable as a new antibiotic, a placebo is sometimes used.

Explain why placebos are important, and how ethical issues with using placebos are overcome. *AO1* [4 marks]

Placebos are important because patients often feel an improvement even if the treatment does not have the active component. Ethical problems arise if the group provided with the new treatment begins to improve.

> **Answer grade: D.** The answer says why placebos are important, but does not explain fully what a placebo is. An ethical problem with the use of placebos is given but there is no indication as to how this problem is overcome.
>
> For full marks, explain that a placebo is a tablet or solution made to look just like the new drug, but without the active ingredient. Patients sometimes show an improvement when given any kind of 'treatment', so the use of a placebo will suggest if this is the case, and help to distinguish this from improvement resulting from the 'real' treatment. The ethical problem given can be overcome if members of the group using the placebo are switched to the treatment so they can benefit from it also.

Page 13 The heart and circulatory system

Foundation: The heart needs its own blood supply to live. Describe how the heart receives this and what happens in a person with coronary heart disease.　　　*AO1* [5 marks]

The coronary arteries supply the heart with blood. In coronary heart disease, the coronary arteries become blocked with fatty deposits. The person has a heart attack.

> **Answer grade: D.** This answer begins by stating that the coronary arteries are involved, but there is no mention of how these keep the *whole* of the heart supplied with oxygen (they run over the surface of the heart, so all the heart receives blood). Also, while the answer states that the coronary arteries become blocked and heart attack results, there is no description of how this happens. You should explain that, when the coronary arteries get blocked, the heart is deprived of oxygen (or part of it will be), so the heart muscle will die.

Page 14 Heart rate and blood pressure

Foundation: Explain the term 'blood pressure' and describe how blood pressure affects health.
　　　AO1 [2 marks]

Blood needs to be under high pressure to be pumped around the body, otherwise it would reach all parts of the body. High blood pressure increases the chance of heart attack and stroke.

> **Answer grade: E.** The first part clearly explains why blood needs to be under high pressure to be pumped around the body. The next sentence states the effects of high blood pressure, but does not mention the effects of low blood pressure. Low blood pressure will result in dizziness and fainting.

Page 15 The kidneys

Higher: Explain the effect of Ecstasy (MDMA) on the water balance of a person's body.　*AO1* [3 marks]

Ecstasy increases the production of anti-diuretic hormone (ADH) by our pituitary gland. So Ecstasy causes our body to retain water.

> **Answer grade: B.** This answer *describes* the effects of Ecstasy rather than *explaining* them. To get full marks you would need to explain that by increasing the amount of ADH – a hormone that prevents urine production – Ecstasy reduces the amount of urine produced. You could use a diagram to illustrate this.

Page 17 Extinction

Foundation: The harlequin ladybird originated in Asia and arrived in Britain in 2004. It eats the same food as our native ladybird (aphids – greenfly and blackfly). It also predates on many insects (other ladybirds, and the eggs of butterflies and moths) and eats fruit.

Suggest why the spread of the harlequin ladybird might affect food webs. *AO2* [4 marks]

As it eats the same food as our native ladybirds, it might out-compete them for food. Our native ladybirds will die. As it is a predator, it will prey on many other species of insect. It may also eat fruit.

Answer grade: D. The answer correctly refers to competition for food and predation of native species, and ladybirds eating fruit, but does not explain how these affect food chains. To gain full marks, you would need to explain that invasive species can also bring disease (affecting native ladybirds and perhaps other insects). By eating fruit the harlequin ladybird may deprive other animals of food, and this may affect the reproduction of the plants.

Page 18 Indicators of environmental change

Foundation: Lichens are sensitive to sulfur dioxide in the air. The distribution of three different types of lichen was measured in a city centre, and at different distances from it. The results are shown below.

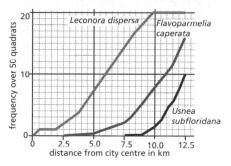

a Suggest what the graph tells you about the resistance of the lichens to pollution. *AO3* [1 mark]

b Why did the scientists measure the frequency of the lichens in 50 quadrats? *AO2* [2 marks]

a *The data suggest that* Leconora dispersa *is the most resistant to pollution.*

b *To improve the reliability of the results.*

Answer grade: E. The answer to part **a** tells us that *Leconora dispersa* is the most resistant to pollution. To get full marks, put the lichens into order. *Leconora dispersa* is the most resistant, followed by *Flavoparmelia caperata*, with *Usnea subfloridana* being the least resistant.

Part **b** states, correctly, that taking counts from 50 quadrats will provide reliable results, but this answer is vague and needs to be explained further. You need to say that taking counts from a larger number of quadrats (than say, 10) is better, as using 10 quadrats, or fewer, may not truly reflect how the lichen is distributed. This will increase the confidence you have in the results.

Page 19 Variation and natural selection

Higher: Scientists have observed that the number of cases in which disease-causing bacteria are resistant to antibiotics has increased over the last 30 years. Explain why some people think that this is evidence of natural selection. *AO2* [4 marks]

The bacteria have genes that makes them resistant to antibiotics. This is evidence for natural selection because only the bacteria that have these genes are able to survive and reproduce.

Answer grade: B. While these points are correct, the answer lacks detail and does not link the points together or describe the sequence of events. For full marks, you should say that in any population of bacteria there is natural variation. When exposed to the antibiotic, most of the bacteria will be killed, but some may survive, because they have genes that make them resistant. These bacteria will reproduce, so these 'resistant genes' will spread throughout the population. The bacteria that survive have therefore been *selected* by the antibiotic, and over a short time, the species of bacterium will become resistant to the antibiotic. You could illustrate this process nicely with a diagram.

Page 20 Evidence for evolution

Higher: The following are sequences of very short sections of DNA from four species of primate – a human, Neanderthal man, a chimpanzee and a gorilla.

Human	CTGGGCGCGTGCGGTTGTCCTGGTCCTGCT
Neanderthal	CCGGGCGCGAGCGGTTGTCCTGGTCCTGCA
Chimpanzee	CCGGGCGCGTGCGGTTCACCAGGTCCTGCA
Gorilla	CAGGGCGCGGGAGGTTTACCACATGCTTCA

Look at the sequences and suggest what the evolutionary relationships of the animals are based on the data. Explain the level of confidence you have in your conclusion, and suggest how this could be increased. *AO2, AO3* [5 marks]

The Neanderthal is closest to humans, followed by the chimpanzee, then the gorilla.

The comparison is of only 30 bases, of the millions found in each organism, so many more need to be analysed.

Answer grade: B. This answer is correct, but the reasoning in the first sentence could be quantified, and the second sentence makes just one point.

For full marks, explain that the sequences of the human and Neanderthal have three differences, the human and the chimpanzee have five differences, and the human and the gorilla have eleven differences. Explain that the similarities not only indicate the closeness, but also how long ago the branches may have occurred. You could extend the comparison in the second sentence by looking at other evidence, such as that from the fossil record, and from similarities in other patterns of the organisms, e.g. their skeletons.

Page 21 Conserving biodiversity

Foundation: When investigating species extinctions, explain why we can use real data of human populations but need a computer model for numbers of extinctions. *AO2* [4 marks]

It is possible to record the numbers of the human population, but not species of organisms.

Answer grade: D. The answer is correct, but to get full marks you should say that the actual number of species of organisms on the planet is unknown – many are still undiscovered. It's not possible to record the extinction of a species if we don't know that it exists, so it's only possible to estimate these. And of course we cannot do this for the past.

Page 23 The Earth's atmosphere

Foundation: The early atmosphere was mainly composed of water and carbon dioxide. Suggest how these were gradually removed. *AO1* [4 marks]

Water condensed to form oceans. Carbon dioxide was removed by plant photosynthesis and by forming fossil fuels.

Answer grade: D/C. Both sentences are correct, but the answer lacks detail. For full marks, you need to explain that water condensed because the Earth cooled, and that fossil fuels form when plants and animals die and are buried under certain conditions. Carbon dioxide is also removed by dissolving in oceans and forming sedimentary rocks.

Page 24 How has human activity changed air quality?

Foundation: How do humans affect air quality? *AO1* [4 marks]

Fuels such as coal give out waste gases. Cars give out gases that make asthma worse. Carbon dioxide is linked to acid rain. People cut down trees and burn them, adding carbon dioxide to the air.

Answer grade: E/D. This answer gains 1 mark. The first sentence gains a mark for linking pollutant gases to burning fuels, and the final sentence gains a mark for linking burning trees to carbon dioxide release. Sentence 2 is correct – cars emit both sulfur dioxide and nitrogen dioxide, which are linked to asthma – but this is not clearly stated. Sentence 3 has incorrect science. For full marks, you would need to correct sentences 2 and 3 and state that nitrogen oxides from cars are also linked to acid rain.

Page 25 What happens when fuels burn?

Foundation: Methane (CH_4) reacts with oxygen to form carbon dioxide and water.

Finish the diagram to show this reaction. Use ● to represent a carbon atom, ● to represent a hydrogen atom and ○ to represent an oxygen atom.

methane + oxygen ⟶ carbon dioxide + water
AO2 [3 marks]

Answer:

methane + oxygen ⟶ carbon dioxide + water

Answer grade: C/D. The formula representations for oxygen, carbon dioxide and water are all correctly drawn, which gains 2 marks. However, the answer is incomplete. For full marks, the equation needs to be balanced. Do this by counting each atom and making sure that all of them are the same on both sides. To balance the equation, two molecules of water and two molecules of oxygen need to be drawn.

Page 26 How pollutants are formed

Foundation: Describe the damage acid rain causes, and explain why the UK is being blamed for acid rain damage in Europe. *AO1* [4 marks]

Acid rain is made when sulfur dioxide dissolves in water vapour in clouds. The rain clouds are blown by winds. Acid rain kills trees and wildlife in lakes.

Answer grade: C/B. The first sentence uses correct scientific terms but does not explain the link between acid rain and pollution from power stations in the UK. The next two sentences state a fact but do not explain the causes of these facts. For full marks you need to explain why acid rain is a problem. For example, you could say that acid rain kills trees on land and wildlife in lakes by gradually changing the pH to such an extent that animals and plants cannot tolerate it.

Page 27 Reducing air pollution from transport

Foundation: Suggest ways of reducing air pollution from cars. *AO1* [4 marks]

You could walk to school instead of getting someone to drive you. You could also take a bus instead of using a car. Finally, switch off your engine if you've stopped for a few minutes.

Answer grade: D/E. The first sentence gives a sensible suggestion and gains 1 mark. Sentence 2 is the same idea about not driving, so is too similar to get a separate mark. Sentence 3 is a different idea worth 1 mark. For full marks you would need to make some additional suggestions, such as make public transport cheaper and use cleaner fuels or clean exhaust gases.

Page 29 Comparing materials and measuring properties

Foundation: Sub aqua divers should always leave a marker buoy at the dive site to warn other boats to stay away, reducing the risk of divers being hit when they surface.

Describe the main properties that the marker buoy would need. *AO1* [4 marks]

It will need to float and should be brightly coloured.

Answer grade: E. This answer gives two sensible, simple properties, but these two ideas alone will not gain all the marks available for this question. For full marks, you need to suggest extra properties, such as being waterproof, hard-wearing or staying in shape.

Page 30 Natural and synthetic materials

Higher: What are synthetic materials, and what advantages do they have over natural materials? *AO1* [4 marks]

Some natural materials are found in the Earth's crust formed from plants and from animals. Limestone is a natural material than can be used to make a synthetic material. Synthetic materials are man-made. Cement is an example. It can be made to order but goes off if you leave it a long time or if it gets damp.

Answer grade: E/D. Sentence 3 gains 1 mark, but while the other statements are true, they do not answer the question. For full marks, you would need to explain that synthetic materials are cheaper, can be made to order, are manufactured by chemical reactions, and enable properties to be designed to suit a particular purpose.

Page 31 Investigating boiling points

Higher: Explain why during the distillation of crude oil, small hydrocarbon molecules rise to the top of the tower. *AO1* [4 marks]

The temperature drops with height up the tower. Small gases have lower boiling points. They can cool a lot more before becoming a liquid.

Answer grade: C. All three sentences are worthy of 1 mark each. For full marks, you would need to add that small-sized molecules have weak intermolecular forces.

Page 32 Improving polymers

Higher: A company is making rotor blades for a radio-controlled toy helicopter.

A polymer needs to be made stronger but more flexible. What could be done, and how will it change the properties? *AO2* [3 marks]

To make the plastic harder, cross-links could be made. To make it flexible, use a plasticiser. Plasticisers are small molecules inserted to disrupt the chains.

Answer grade D/C. Sentences 2 and 3 are both correct and worth 1 mark. In Sentence 1, cross-linking will make the polymer harder, but this would prevent the plasticiser being added and would not make it flexible. For full marks, you need to add that increasing chain length to improve strength and using a plasticiser for flexibility would work.

Page 33 Nanotechnology

Foundation: Explain what nanoparticles are, and suggest why some act as catalysts. *AO1* [4 marks]

Nanoparticles are small groups of atoms joined together. Some can be used as catalysts as they have a large area. They provide more sites for reactants to meet and react.

Answer grade: E/D. Sentences 1 and 3 are worth 1 mark each. Sentence 2 is incorrect – it is not a large area. For full marks you need to add large surface area, and explain that nanoparticles work as catalysts by providing sites for reactants to meet. You would also need to include the idea of size – up to a thousand atoms or 100 nm.

Page 34 Are nanoparticles safe?

Higher: Fresh Crop is a company selling mixed salads. They are considering adding silver nanoparticles to their food packaging to help prevent bacterial decay.

Explain why some people believe this may have risks. *AO2* [3 marks]

The silver nanoparticles might get onto the food and be eaten. If it gets inside you, you might get ill. The nanoparticles could kill 'good' bacteria in your body.

Answer grade: D/C. This answer is worth 2 marks for the ideas of getting inside and causing a possible problem. For full marks, the idea of uncertainty is needed, linked to the ideas that long-term evidence is not available.

Page 36 Limestone, coal and salt

Foundation: Coal, limestone and salt are major raw materials for industry. Choose one and state how it is formed. *AO1* [4 marks]

Salt is made when water evaporates from the sea. Coal is made when dead plants decompose. Limestone is made from sediments.

Answer grade: F/G. The first sentence is worth 1 mark. The question says choose *one* raw material, so normally only the first answer will be marked. For full marks, you would need to state how salt is formed, i.e. warm sea drying up, rock dust combining, and being buried by sediments.

Page 37 Extracting and using salt

Higher: Describe how salt is obtained by solution mining, and suggest why the chemical industry prefers this method of extraction. *AO1* [4 marks]

Solution mining uses water to dissolve underground salt deposits. It comes out of the ground in solution, and this is easier for industry to use.

Answer grade: C. Each sentence is worth 1 mark each, but the answer lacks detail. For full marks, you need to explain that water needs to be pumped in under pressure, and salt solution returns up another pipe. Industry prefers this method as the salt obtained is purer, so there is less cost for removing impurities.

Page 38 About alkalis

Foundation: Use the two tables below to write a word equation for making potassium nitrate.

Soluble hydroxides	sodium hydroxide – NaOH
	potassium hydroxide – KOH
	calcium hydroxide – Ca(OH)$_2$
Soluble carbonates	sodium carbonate – Na$_2$CO$_3$
	potassium carbonate – K$_2$CO$_3$

Acid	Salt
hydrochloric – HCl	chloride – Cl
sulfuric – H$_2$SO$_4$	sulfate – SO$_4$
nitric – HNO$_3$	nitrate – NO$_3$

AO2 [4 marks]

Potassium hydroxide + sulfuric acid ⟶ potassium nitrate

Answer grade: E/F. This answer gains 1 mark for choosing potassium hydroxide as the alkali needed. For full marks, an acid needs choosing which will give nitrate, so the only one is nitric acid. Adding the equation: acid + base ⟶ salt + water would be useful, to show full understanding, and water needs to be added as a product.

Page 39 The benefits and risks of adding chlorine to drinking water

Foundation: Worldwide, over 100 000 people die from a disease called cholera each year, through drinking dirty water. In Britain, cholera is rare.

Explain these statements, and suggest advice for people who drink water directly from rivers.

AO2 [4 marks]

In the UK we add chlorine to water to kill cholera. The best advice for people in Third World countries is do not go to the toilet close to water.

Answer grade: D/E. Both sentences are worth 1 mark each. To improve to a C grade, you would need to add the following points: cholera breeds in dirty water; some countries do not have an enclosed water supply or cannot afford to chlorinate; the best advice is to boil drinking water.

Page 40 Should we worry about PVC?

Higher: Cling-film is used to wrap food. It may be made from thin sheets of PVC, which contain plasticiser molecules called phthalates to increase flexibility. These molecules have passed safety tests. Despite this, some people are still worried about their safety.

Suggest reasons for some people's reluctance to accept the risk. *AO2* [4 marks]

Phthalates might be able to get out of the cling-film on to the food. When the food is eaten, they might cause harm to your body. The safety test may not have been carried out correctly. If they think there is a risk, it's best to avoid it.

Answer grade: B/A. Sentences 1, 2 and 4 are worth 1 mark each. To improve to an A* grade, you could explain that safety tests might not test if phthalates can escape. You could also explain what phthalates might do to your body over time, and that it would be difficult to calculate the actual risk.

P1 Improve your grade

Page 42 The solar system

Foundation: Describe the motion of moons and planets in the solar system. *AO1 [4 marks]*

The planets go in circles around the Sun. Each planet is a ball of rock. There are moons around some of the planets. There are also smaller lumps called comets and asteroids. These also go around the Sun.

> **Answer grade: F.** This answer is only worth 2 marks. It contains lots of information which was not asked for. For full marks you should say that moons orbit planets, and that the Sun is at the centre of the planet orbits.

Page 43 Fusion of elements in stars

Higher: Explain how most of the material in and around you was created by stars. *AO1 [4 marks]*

Stars give out energy because they can fuse hydrogen atoms into helium atoms. When they run out of hydrogen, the helium is fused to make heavier atoms such as carbon and oxygen. This needs a higher temperature and pressure in the star. In turn, these atoms then fuse to make even heavier atoms. This only happens in stars that are big enough for gravity to make the pressure and temperature high enough.

> **Answer grade: B.** This answer is worth 3 marks. To gain full marks you would need to explain how the atoms get out of the star at the end of its lifetime – that it is the explosion of a star as a supernova that sends material into space where it can form a new solar system.
>
> Many students assume that fusion can take place throughout a star, so all of the hydrogen has to be used up before helium fusion can start. In fact, fusion only takes place in the very middle. Even at the end of its lifetime, a star still contains a lot of hydrogen.

Page 44 Continental drift

Foundation: Explain why Wegener's theory of continental drift was not accepted when it was first published. *AO1 [4 marks]*

Wegener was not a geologist, so nobody paid attention to his ideas. He couldn't explain why the continents should move and their speed was too small to be measured with the instruments that they had then.

> **Answer grade: B.** This answer sticks closely to the question and doesn't waste space by describing Wegener's theory. The three points made here are relevant and gain 1 mark each.
>
> To push this answer to an A grade, you need to say a bit more about why geologists rejected Wegener's theory. You would need to explain that because geologists already had theories which explained a lot of Wegener's observations, they didn't see why they should accept his theory.

Page 45 Tectonic plates

Higher: Explain why there are volcanoes at plate boundaries. *AO1 [4 marks]*

Volcanoes allow rock to get out from the Earth's core onto the land. They can do this at plate boundaries because that is where the ground splits open to let the lava through.

> **Answer grade: F.** This answer gains only 1 of the 4 marks. To get full marks, you need to discuss what happens when plates meet head-on and subduction occurs, as well as what happens where plates are moving apart. You also need to use the correct scientific terms, such as crust and mantle.

Page 46 Finding out about waves

Higher: Sound in steel has a speed of 2 km/s. What is the wavelength of a sound wave in steel which has a frequency of 80 000 Hz? *AO2 [2 marks]*

$Speed = wavelength \times frequency$,

so $wavelength = \dfrac{speed}{frequency}$

$Wavelength = \dfrac{2}{80\ 000} = 0.000025\ m$

> **Answer grade: B.** The answer successfully rearranges the equation, but the student has forgotten to convert the speed into m/s before substituting it into the equation. For full marks, the answer should be 2 000 / 80 000 = 0.025 m.
>
> Many students calculate the wrong answer because they put the numbers straight into their calculator. If you write down the calculation first, then you would be able to do the sum again to check it.

Page 48 Ionisation

Higher: Bacteria are single-cell organisms that can pollute drinking water. Explain why exposing the water to ultraviolet light removes the bacteria, but exposing it to visible light does not affect the bacteria.

AO2 [4 marks]

The ultraviolet kills the bacteria because it is an ionising radiation. Light is not an ionising radiation, so does not kill the bacteria. Many bacteria absorb light to make their food, so light will probably make the pollution worse.

Answer grade: D/C. The first two sentences correctly explain why ultraviolet light removes bacteria from drinking water but visible light does not, and so gain 1 mark each. The last sentence gains no marks because it is not relevant to the question. To earn full marks, you would need to describe what happens to molecules when they are ionised (electrons are removed) and explain why only ultraviolet light can do this (its photons transfer enough energy).

Page 49 Microwaves

Higher: Explain the risks of cooking food with microwaves. *AO2* [3 marks]

Microwaves are strongly absorbed by water in our cells. This damages cells by ionising them, making them into cancer cells.

Answer grade: E. This is a well-ordered answer which states causes and effects in a logical order. However, there is an important factual error – microwaves are not an ionising radiation so are unlikely to lead to cancer. To get full marks you need to say that the microwaves transfer energy as heat in the water, and that too much heat will kill the cell.

Students often assume that all radiation damages people by giving them cancer. Microwaves and infrared don't have enough energy to ionise materials, so they transfer their energy as heat – and so can damage people by burning them.

Page 50 Global warming

Foundation: Explain how increasing carbon dioxide in the atmosphere results in global warming.

AO2 [4 marks]

The extra carbon dioxide stops infrared radiation from escaping into space. Global warming is going to melt all the ice and make the sea rise up and flood lots of land, making it difficult for us to grow all the food we need.

Answer grade: F. This answer gains 1 mark only – it has provided just one relevant scientific fact (sentence 1), but then wastes time describing some consequences of global warming. To get full marks you need to discuss the incoming energy from the Sun as light, as well as the outgoing radiation from the Earth as infrared.

Only answer the question you have been asked to! If you provide extra, unnecessary information, this will waste valuable time in an exam. This could cost you marks in later questions when time becomes short.

Page 51 Analogue and digital

Foundation: Describe the difference between analogue and digital signals used for radio broadcasts.

AO1 [4 marks]

Analogue sets the size of the wave, but digital turns it on and off. So an analogue signal looks like a wave which is getting gradually bigger and then smaller all the time. A digital wave is just there or not there.

Answer grade: D. This answer earns 3 marks without wasting any time in writing down unasked-for details (such as why the performance of digital is so much better than analogue). To earn full marks you would need to point out that digital performs better than analogue because it isn't affected so much by other signals such as noise and interference.

Page 53 Power

Foundation: A kettle comes with this warning:
This kettle must be used with a 230 V, 50 Hz supply. The current in the leads will be 8.7 A.

How much energy in kilowatt-hours is transferred from the supply when the kettle is used for 15 minutes?
AO2, AO3 [3 marks]

The power of the kettle = 230 V × 8.7 A = 2001 W

This is 2.001 kW.

The energy transferred = 2.001 × 15 = 30 kWh

> **Answer grade: E.** The student has correctly calculated the power of the kettle, and earns 2 marks for this. They wisely selected the data required to do this, ignoring the 50 Hz.
>
> To get full marks, you need to convert the 15 minutes into 0.25 hours for the second calculation. To help you avoid errors such as this, develop the habit of including the units as well as the numbers when you write down calculations.

Page 54 Efficiency

Foundation: Explain why governments have passed legislation which forces people to use energy-efficient lamps in their homes. *AO2 [4 marks]*

Energy-efficient lamps transfer less energy wastefully to the environment than the old sort, for the same amount of energy transferred to light. This means that the nation needs to use less electricity.

> **Answer grade: D.** The explanation of efficiency is excellent in this answer, and gains 3 of the 4 marks. To gain the final mark you would need to develop the argument to its logical conclusion, by explaining that using less electricity reduces our impact on the environment.

Page 55 How power stations work

Foundation: Describe how a power station transfers energy in coal to electricity. *AO1 [5 marks]*

The coal is burnt to transfer heat energy to water, boiling it into high pressure steam. This is used to make a magnet rotate inside a wire, making electricity.

> **Answer grade: E.** This answer earns 3 out of the 5 marks, one for each basic correct fact. However, one stage of the process (the turbine) is missing, and the description of the generator is incomplete (it should be a coil of wire). You would need to include both of these in your answer to gain full marks.

Page 56 Renewable energy sources

Foundation: The majority of the electricity in the UK is generated from fossil fuels. Explain the advantages and disadvantages of using wind and hydroelectric technology instead. *AO1 [4 marks]*

Fossil fuels are non-renewable, so won't last forever. Wind and hydroelectric technologies are renewable energy sources, so can be used forever. Wind power has the problem that it only works while the wind is blowing and lots of people think that wind turbines are ugly. Hydroelectric power stations can only be built in mountains, a long way from where people want to use electricity. So perhaps we need both renewable and non-renewable sources for our electricity.

> **Answer grade: D.** This answer correctly states the advantage of using wind and hydroelectric technologies to generate electricity. However, only the first disadvantage for wind technology earns a mark; the second disadvantage gains no marks because it is subjective and contains no science. Sentence 3 about hydroelectric power stations also earns no marks (distance is not a disadvantage because the National Grid transfers energy from faraway places to where it is needed). To gain full marks, you would need to state a disadvantage for the use of hydroelectric technology, such as its effect on the environment as it floods land or the expense of building it.

Page 57 Dealing with future energy demand

Foundation: It has been suggested that every person on Earth should only be allowed to produce a certain amount of carbon dioxide each year. Explain what impact this could have on your lifestyle. *AO2 [4 marks]*

I would have to use less energy from fossil fuels because these make smoke when they are burnt. I could use energy from renewable sources of energy such as solar and wind instead, but it might be difficult to make enough electricity this way.

> **Answer grade: D.** This answer is quite good and gains 3 of the 4 marks. The candidate has stated three pieces of science, and has organised them to make a clear logical argument. Also, they have not wasted time talking about specific details, such as having to use a more efficient washing machine or only showering once a week, and have kept to general statements. To earn full marks you would need to mention carbon dioxide instead of smoke.

Page 59 Enzymes

Higher: Explain what happens to enzyme activity as the temperature is increased. *AO1 [5 marks]*

As the temperature increases, enzyme activity increases up to a certain temperature. In most enzymes, activity begins to decrease above about 40 °C, and ceases at about 60 °C, because the shape of the enzyme is changed.

Answer grade: B/A. This answer gains marks for describing the pattern of enzyme activity correctly and accurately stating that this pattern is shown by most enzymes (some enzymes found in bacteria can work well between 80 and 100 °C). The answer does not give reasons why enzyme activity increases, however, which is owing to an increase in the reaction rate at higher temperatures. Furthermore, the answer says that enzyme activity stops at around 60 °C, but does not explain why.

To gain full marks you would need to explain that enzyme activity stops at around 60 °C because, at this temperature, the heat changes the shape of the active site (by breaking bonds in the enzyme molecule), so the substrate will no longer fit into it. At this point, the enzyme is said to be denatured.

Page 60 Glucose: making it and using it

Foundation: Explain how the products of photosynthesis are used by the plant. *AO1, AO2 [5 marks]*

The products of photosynthesis are used in respiration. They are also used to produce the chemicals required for growth.

Answer grade: D/C. The first sentence gains 1 mark but does not mention that glucose is the main product of photosynthesis that is used for respiration. The second sentence also gains 1 mark, but lacks detail. For full marks, you need to discuss the chemicals used for growth that glucose is used to produce. These include starch for storage in cells and structures such as seed, cellulose for cells walls, and proteins for growth.

Page 61 Moving chemicals in and out of plants by osmosis

Higher: Explain how plant roots take up water, and how this water moves across a plant root. *AO2 [5 marks]*

The plant roots take up water by osmosis, because water is in a higher concentration in the soil than in a root cell. This dilutes the water in the first plant root cell, so water moves across the root.

Answer grade: C/B. The first sentence is correct, but could have begun with a definition of osmosis, for example: 'Osmosis is the net water movement from an area where it is low concentration to where it is in high concentration.'

The second sentence is also correct, but does not explain fully why water should move. For full marks, you would need to describe how movement of water into the first root (hair) cell dilutes the cell contents, so the water concentration is now higher than in the cell next to it, deeper into the root. Water will move into this second cell by osmosis, whose water concentration is now increased, so water will move into the third cell, and so on. This process could be illustrated in an annotated diagram.

Page 62 Investigating the effects of light on plant growth

Foundation: Describe and explain how an ecologist would compare how a plant is distributed in two meadows. *AO1, AO2 [5 marks]*

Throw a quadrat ten times in the first meadow, then in the second, and count the numbers of the plant in each quadrat. For each meadow, calculate the mean number of plants per quadrat.

Answer grade: D/C. The answer is correct, but the description lacks some important detail. For full marks, you would need to suggest an appropriate size of quadrat. For example, for a meadow an appropriate area of quadrat would be 1 x 1 m or 0.25 x 0.25 m, depending on the size of the meadow. It should also be emphasised that the quadrat should be thrown or placed (using coordinates) at random. Though not essential, for effective comparison it is best to calculate the mean number of plants per metre square in each meadow.

Page 63 Fermentation

Foundation: Explain what the graph below tells us about changes in concentration of sugar and ethanol as yeast grows.

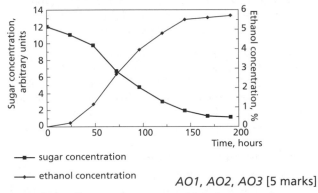

- ■— sugar concentration
- ◆— ethanol concentration

AO1, AO2, AO3 [5 marks]

Yeast is feeding on the sugar, which decreases in concentration. It is producing ethanol by the process of fermentation, so its concentration is increasing.

Answer grade: C/B. Sentence 1 is correct, though it is better to say that the yeast is using the sugar for respiration. Sentence 2 is also correct but lacks detail. It does not mention that yeast produces ethanol under anaerobic conditions. The rise in ethanol concentration for the first 24–25 hours is slow. This is because, at this stage, anaerobic respiration is taking place only very slowly (oxygen is available to the yeast but later runs out).

For full marks you would also need to explain that the graph shows that ethanol concentration (and sugar concentration) eventually levels off, because the ethanol is inhibiting the growth of the yeast and/or the sugar is running out.

Page 65 Cell specialisation in animals

Foundation: Compare how cells become specialised in animals and plants. *AO2 [5 marks]*

In animal cells, specialised cells are produced from stem cells. In plants, they are produced following cell division in regions called meristems.

> **Answer grade: D/C.** Both of these statements are correct, but the student does not go on to add detail. For full marks, you should mention embryonic and adult stem cells in animals. You need to explain that cells become specialised in the embryo after the eight-cell stage, when cells produced by embryonic stem cells differentiate. You should also say that some stem cells remain in the adult (adult stem cells), and that these can differentiate into a limited number of cell types. Finally, you need to state that in plants, when meristem cells divide into two, one of the new cells produced by the meristem can differentiate.

Page 66 Plant clones

Foundation: Explain how and why plant breeders who have produced a new variety of plant take many cuttings from it. *AO1, AO2 [5 marks]*

Plant cuttings are taken from the new variety by cutting off a shoot from the plant and placing it in compost that is kept moist (or sometimes water). After around two weeks, the cutting will produce roots, and will grow into a new plant. Plants grown from cuttings are identical to the parent plant.

> **Answer grade: D/C.** Sentence 1 is correct, but does not mention that the cutting is usually dipped in hormone rooting powder to help it to root. The final paragraph is also accurate, in that the plants are identical to the parent, but it does not relate this to the question, i.e. its importance when producing many identical copies of the new variety of plant.
>
> For full marks, you need to say that the root cutting is dipped in hormone rooting powder, in order to promote the growth of roots from the portion of the cutting under the surface of the compost. You also need to say that as the plants grown from cuttings are identical to the parents, the characteristics of the new variety will be present in all the plants produced.

Page 67 Mitosis and meiosis

Higher: Explain why gametes (sex cells) are produced by meiosis and not by mitosis. *AO1, AO2 [4 marks]*

Meiosis is used to produce gametes in order to keep the chromosome number constant (46 in humans) from generation to generation.

> **Answer grade: C/B.** The sentence is correct, but offers little in the way of explanation, or comparison with mitosis, so only gains 1 of the 4 available marks.
>
> For full marks, you first need to say that human cells contain 46 chromosomes, which can be assembled as 23 pairs (each one of the pair carrying the same type of genes). If the cells that produce gametes divided by mitosis, at fertilisation, the zygote/offspring would have 92 chromosomes, so the chromosome number would double every generation. These cells therefore divide by a process called meiosis, so each gamete in humans contains 23 chromosomes; one from each pair. This means that the chromosome number is restored on fertilisation, with the zygote having 23 pairs of chromosomes.

Page 68 Protein synthesis

Higher: Describe the process by which proteins are produced. *AO1 [5 marks]*

Proteins are assembled on a ribosome. The order of the amino acids in the protein is specified by the genetic code of the DNA.

> **Answer grade: B.** Both of these statements are correct, but the description lacks some important detail.
>
> For full marks, you should be begin by saying that in the nucleus of the cell, messenger RNA is synthesised using the DNA of the gene (the gene that codes for this protein) as a template. The mRNA passes into the cytoplasm and attaches to a ribosome. The amino acids are ferried in to the ribosome and the amino acids are bonded together (in the order specified by the genetic code).

Page 69 Stem cell research and therapy

Higher: Scientists have reprogrammed skin cells to function as stem cells. Explain, in principle, how this technique is carried out, and why this might be preferable to using embryonic stem cells. *AO2 [5 marks]*

Skin cells have been changed into stem cells using a chemical treatment. This method is preferable to using embryonic stem cells, as when these are removed from the embryo, the embryo is destroyed.

> **Answer grade: C/B.** Sentence 1 is correct, although it does not explain what the chemical treatment does. Sentence 2 is also correct, and points out the main ethical problem with using embryonic stem cells, but does not point out the deficiencies of using the transformed cells.
>
> For full marks at grade A, you need to say that the chemical treatment reactivates genes that have become inactive, so that the transformed cells can develop into different cell types. You also need to explain that these treatments are in the early stages of their development, and although a number of cell types have been produced, cells have not yet been produced that will develop into all cell types.

Page 71 Neurons

Higher: Explain how nerve cells (neurons) are adapted to transmitting nerve impulses. *AO2* [5 marks]

Nerves are the longest cells in the body as they have to reach all parts of the body. They have a long extension to the cell called the axon. The axon is insulated by a fatty covering called the myelin sheath.

Answer grade: D/C. Both of these statements are correct, but the student has missed some important points.

For full marks, begin by saying that the nerve impulse is an electrical impulse (which explains why it's important to be insulated). To extend the answer to an A grade, say that the presence of the myelin sheath not only insulates the neuron, but also enables much greater transmission speeds, as the nerve impulse jumps from one gap in the sheath to the next.

For a C grade it's important to mention that the neuron has extensions called dendrites, which enable it to communicate with other neurons. A more subtle point, at A/A* grade, is that the end of the axon contains chemical transmitter molecules that enable it to communicate with other nerve cells and other effectors.

Page 72 Synapses

Higher: Describe how a nerve impulse is transmitted from a sensory nerve to a nerve close to it in a spinal cord. *AO1, AO2* [5 marks]

As the nerve impulse reaches the end of the nerve, a chemical transmitter is released. This passes across the synapse, and sets up a nerve impulse in the nerve on the spinal cord.

Answer grade: B. The answer is correct but misses some detail. Also, although the student has said correctly that the nerve passes across a synapse, they have not defined what a synapse is.

For full marks, you need to say that nerves are not connected together physically; instead a chemical transmitter is released from the first nerve and passes across a gap called a synapse. Point out that the type of chemical transmitter used is dependent on the location and type of nerve. Finally, you need to describe how, after the impulse has passed, the remaining chemical transmitter in the synapse is reabsorbed into the first nerve (or alternatively broken down by an enzyme).

Page 73 Instinctive and learned behaviour

Higher: A bird eats a poisonous, brightly coloured caterpillar. It is sick, but survives. In future, it avoids eating this type of caterpillar. Explain how this is an example of a conditioned reflex and how it might help the bird's survival. *AO2* [5 marks]

After being sick, the bird learned to avoid the poisonous, brightly coloured caterpillars. It had associated the bright colours of the caterpillar with the unpleasant experience. This is called a conditioned reflex.

Answer grade: C. While this answer is correct, the student has not defined the two stimuli involved. For full marks, you need to define the poisonous/distasteful nature of the caterpillar as the primary stimulus, and the bright colours of the caterpillar as the secondary stimulus.

The student has also not explained how the response involved in the conditioned reflex – avoiding brightly coloured caterpillars – has no direct connection with their distastefulness or poisonous nature. You need to explain that, after tasting the caterpillars once or possibly a few times, the bird would come to associate the bright colours with distastefulness.

Finally, you need to mention how this can help the bird's (and the caterpillar's) survival. In being sick, the bird removed the poisonous caterpillar from its gut, but on another occasion may have eaten sufficient or kept it in its gut for long enough to kill it. So in not eating the caterpillar again, poisoning would be avoided.

Page 74 Brain structure

Foundation: Describe how scientists have mapped the areas of the brain to see how it works. *AO1* [3 marks]

Neuroscientists have studied people with brain injuries and investigated how people react when their brains are stimulated using electrodes.

Answer grade: D. Both of these statements are correct, but the answer lacks detail. For full marks, you should refer to invasive and non-invasive techniques, and describe these. It's also important to say how the effects of brain injury are studied.

The answer also gives no information on non-invasive techniques, e.g. scanning techniques such as MRI scanning. You need to explain that these are used to compare the structure and activity of the brains of healthy people and people with brain disease, and when a person is stimulated by music, language, etc.

Page 75 Drugs

Foundation: Some chemicals affect how nerve impulses are transmitted across synapses. Give **two** examples of these chemicals, and state how these chemicals work. *AO1, AO2* [3 marks]

Prozac increases levels of a chemical transmitter substance that carries the impulse between nerves. Toxins can block certain chemical transmitters.

Answer grade: E. The first sentence is correct, and is complete, as the question only says 'state' and doesn't ask for a description. The second sentence is also correct, but does not give an example, just a type of chemical that affects transmission. For full marks, you need to provide an example of a toxin that blocks a chemical transmitter, e.g. curare, which is a poison used on the tips of arrows by South American Indians, or botulin toxin ('botox').

Page 77 The history of the Periodic Table

Foundation: Explain why Mendeleev's arrangement of elements was an improvement on Döbereiner's triads and Newlands' octaves. *AO1 [4 marks]*

Mendeleev's arrangement was better because it used the properties of elements and put them into groups. All of the element properties fitted, but elements in triads and octaves did not all fit. Triads and octaves only worked for some elements.

Answer grade: D/C. A good feature of this answer is that it talks about Döbereiner and Newlands, as the question asks. However, the student only discusses one aspect of the table – the idea that all of the element properties fit the table. The most important reasons that Mendeleev's table was an improvement are because he left gaps and he predicted the properties of new elements. When they were discovered, the 'missing' elements fitted Mendeleev's predictions.

Page 78 Finding elements in the Periodic Table

Higher: An atom has the electronic arrangement 2.8.1.

Identify the element and explain why its electronic arrangement shows that it is likely to be a metal. *AO2 [3 marks]*

The element is sodium. It is a metal because sodium is a metal.

Answer grade: C. The answer scores only 1 mark, for identifying the metal. You can do this by working out that the total number of electrons in the atom is 11, which is the same as the proton number of sodium. However, the answer does not explain what the electron arrangement shows. For the other 2 marks you would need to say that atoms with one electron in the outer shell are likely to be metals, and that they will be in Group 1, which only contains metals.

Page 79 Reactions of Group 1 elements with chlorine

Higher: Write the word and symbol equations for the reaction of sodium with bromine. Compare the rate of reaction of sodium and potassium with bromine. *AO1 [3 marks]*

sodium + bromine $\longrightarrow$ sodium bromide

$Na + Br \longrightarrow NaBr$

Potassium reacts faster because it is further down the group.

Answer grade: C/B. The word equation is correct but the formula for bromine is wrong – it should be Br_2. If you are aiming at grades A or B you need to be able to write equations for the reactions with bromine and iodine as well as chlorine. They follow the same pattern: just swap 'Br' for 'Cl' or 'I' in the equations. The correct equation is $2Na + Br_2 \longrightarrow 2NaBr$. The last point is correct, the reactivity increases down the group, so potassium reacts faster.

Page 80 Patterns in Group 7

Higher: Liz adds chlorine water to potassium bromide solution. The table shows what she sees and her explanation.

Halogen	Compound	Observations	Explanation
Chlorine	Potassium bromide	Solution turns brown	Bromine is made because chlorine displaces bromine. Chlorine is more reactive than bromine.

Predict what you will see when chlorine water is added to potassium iodide solution. Explain your reasoning. *AO2 [4 marks]*

You would see the solution go brown because iodine is made and it looks brown.

Answer grade: C/B. This answer gets 2 marks. The observations are correct, and it is correct that iodine is made, but you need to 'model' your answer on the explanation in the table. Look at the number of marks – there are 4 in total. To gain the other 2 marks available you need to mention that chlorine displaces iodine and explain that this is because chlorine is more reactive than iodine.

Page 81 Explaining properties

Foundation: Explain why sodium chloride conducts electricity when it is molten or dissolved in water but not when solid. *AO1 [4 marks]*

Sodium chloride conducts because it is an ionic compound and the ions need to move to be able to conduct electricity.

Answer grade: D. There are 4 marks available and several parts to the question, so you need to give an 'in-depth' answer here.

First, you need to explain why sodium chloride conducts electricity. This answer gains 1 mark by saying that sodium chloride is an ionic compound. However, this is a 'why' question so a higher-level answer is needed. The answer goes on to correctly say that the ions must be able to move, and gets 1 mark for this.

Notice that the question also asks about 'when molten' and 'when dissolved in water' and 'not when solid'. The answer has not mentioned any of these, so is only worth 2 marks. A better answer would go further to say that ions can only move when the compound is molten or when dissolved in water, but that ions cannot move in the solid.

C5 Improve your grade

Page 83 Simple molecular substances

Higher: The table shows some data about oxygen.

Boiling point	State at room temperature	Density	Electrical conductivity
–218 °C	gas	very low	does not conduct

Use ideas about bonding and forces between molecules to explain the properties of oxygen.

AO2 [3 marks]

Oxygen has a low boiling point and it is a gas. It has a low density and does not conduct. This is because it has covalent bonds.

Answer grade: C/B. This answer only gets 1 mark. The answer copies the information in the table (this is a common mistake) but does not explain any of the properties in terms of bonding and forces between molecules, which the question asks you to do. To gain full marks you need to link the low boiling point, density and state to the fact that the forces between molecules are very low, and the low electrical conductivity to the fact that the molecules have no charge.

Page 84 Testing for ions

Foundation: Sam tests a salt. He finds out it contains copper ions.

Describe what Sam does and what he sees.

AO2 [3 marks]

He adds some sodium hydroxide and he looks for a precipitate.

Answer grade: D/C. This answer makes two of the main three points. To get the last mark, it is important to make sure you give the colour of the precipitate.

Page 85 Giant covalent structures

Higher: Diamond is the hardest naturally occurring material on Earth.

Use ideas about the structure of diamond to explain why.

AO1 [3 marks]

Diamond is hard because the bonds are very strong. Each carbon atom in diamond is bonded to four others in a giant, 3-D structure, which holds each atom tightly in place.

Answer grade: B/A. This is a good answer which gains 2 of the 3 available marks. Notice how the answer uses these important terms – 'bonds', 'strong', 'giant structure' and '3-D'. However, a key term is missing from this answer – you need to describe the bonds as 'covalent'. It is very important to try to include the correct level of language when answering questions on the Higher tier.

Page 86 Using electrolysis

Higher: An electric current is passed through molten potassium chloride.

Write ionic equations to help you to explain what happens at each electrode and name the products that form.

AO1 [5 marks]

$2KCl \longrightarrow 2K + Cl_2$

The potassium chloride breaks down and it makes potassium and chlorine.

Answer grade: C. This response does not fully answer the question. The question does not ask for an overall equation, it asks for an equation at each electrode. The full answer needs to show each ionic equation: $K^+ + e \longrightarrow K$ and $2Cl^- \longrightarrow Cl_2 + 2e$. Also, saying that the potassium chloride 'breaks down' is not enough. You need to explain that the potassium ions are positive and so go to the negative electrode to gain electrons, and the negative chloride ions move to the positive electrode and lose electrons.

Page 87 Metals in the environment

Foundation: Old car batteries are made from lead. Lead is toxic. Modern batteries use alternative, non-toxic metals.

Use ideas about cost and benefit to explain why manufacturers could not stop using lead to make batteries until alternative batteries were developed.

AO3 [3 marks]

People could not go without cars so they needed to invent a new battery. Lead batteries are toxic but only the people at the garage touch them.

Answer grade: D/C. This answer gets 2 of the available 3 marks. It gives one benefit (needed for cars) and also talks about one cost to a group of people who were affected by using lead batteries. A better answer would identify more costs, such as the problems with dumping old car batteries or the environmental damage caused by toxic metals, or identify more benefits of continuing to use the old batteries, such as it is very expensive to develop new technology.

Page 89 Reactions of acids

Higher: Write a word equation and a balanced symbol equation to show the reaction between sodium hydroxide and sulfuric acid. *AO1* [3 marks]

sodium hydroxide + sulfuric acid ⟶ sodium sulfate + water

$$NaOH + H_2SO_4 \longrightarrow NaSO_4 + H_2O$$

Answer grade: C/B. This answer gains only 1 of the 3 available marks. The word equation is correct (hydroxides react with acids to give a salt and water), but the symbol equation is incorrect for two reasons. Firstly, the formula of the salt is incorrect: sodium ions have a +1 charge (Na^+) and sulfate ions have a –2 charge (SO_4^{2-}), so two Na^+ ions are needed to balance the charge of each SO_4^{2-} ion, i.e. the formula of sodium sulfate is Na_2SO_4. Secondly, the equation must be balanced. Counting the numbers of atoms on each side of the equation, the equation balances if a '2' is put in front of the NaOH, giving $2NaOH + H_2SO_4 \longrightarrow Na_2SO_4 + H_2O$.

Page 90 Reacting amounts

Foundation: What is the relative formula mass of calcium hydroxide, $Ca(OH)_2$?

Use the Periodic Table to help you. *AO2* [2 marks]

RAM Ca = 39

RAM O = 16

RAM H = 1

Relative formula mass = 39 + 16 + 1 = 56

Answer grade: D. This gets 1 of the 2 available marks. It is a really good idea to write down the masses as you find them on the Periodic Table – this gets 1 mark. If you had made a small error you would still get 'error carried forward' if you have set your work out clearly. However, the relative formula mass is wrong. In the formula $Ca(OH)_2$ the 2 at the end shows that there are two oxygen atoms and two hydrogen atoms in the formula. So the correct answer is $39 + (2 \times 16) + (2 \times 1) = 73$.

Page 91 Energy changes

Foundation: Sam does some experiments. She does four different reactions in test tubes and takes a note of the temperature before mixing and 60 s after mixing. Sam's results are shown in the table.

Experiment	Temperature before mixing (°C)	Temperature 60 s after mixing (°C)
1	17	21
2	18	11
3	16	18

Which reaction is the most exothermic? Explain how you can tell. *AO3* [3 marks]

Reaction 2 because it is the biggest temperature change.

Answer grade: F. This answer is incorrect. Exothermic reactions give out energy, and in Experiment 2 the reaction is endothermic (takes in heat energy). For full marks you would need to say that the most exothermic reaction is Experiment 1, because this gives the largest temperature increase.

Page 92 Percentage yield

Higher: Ben prepares some copper sulfate crystals. He talks to Liz about his method.

Ben: 'First I added solid copper carbonate to sulfuric acid until the fizzing stopped.

Then I filtered off some unreacted copper carbonate.

Next I heated the filtrate to evaporate some of the water and left the solution to cool for a few minutes until some crystals formed.

Then I filtered off the crystals and weighed them. I made 1.4 g of crystals. I worked out that my theoretical yield is 1.6 g.'

Liz: 'Your percentage yield will be really inaccurate because you have missed some steps out of your method.'

Calculate Ben's percentage yield. Explain why his percentage yield is likely to be inaccurate.

AO2 [5 marks]

$$Percentage\ yield = \frac{actual\ yield}{theoretical\ yield} \times 100\%$$

$$= \frac{1.4}{1.6} \times 100\% = 87.5\%$$

It will be inaccurate because Ben only did it once. He should have done it lots of times and taken an average. He didn't purify his crystals either.

Answer grade: B. This answer gets 3 of the available 5 marks. It gains 2 marks for correctly working out percentage yield. The point that Ben should have purified his crystals gains 1 mark. However, while repeating an experiment is a good idea, this takes too long when preparing salts, so this point does not score a mark.

Liz points out that the method has some missing steps – a better answer would identify these. To gain an additional 2 marks you could mention any two of the following steps: waiting for all the crystals to form; the use of crystallisation to purify the crystals; drying the crystals in an oven or dessicator; weighing the dried crystals.

Page 93 Changing rates of reactions

Foundation: Jack investigates the rate of the reaction between a large lump of calcium carbonate and dilute hydrochloric acid. He measures the volume of gas given off every 30 s. The reaction takes place very slowly.

Suggest what changes Jack could make to his experiment to make the reaction faster. *AO1* [3 marks]

You could try heating it up because the reaction goes faster when it is at a higher temperature so that would make the gas come off faster.

Answer grade: E/D. This answer gets 1 mark for the idea of raising the temperature. However, if you look at the question, it clearly asks for *changes* – not just one change. There are 3 marks available, so to gain full marks you need to say three things. Two other changes Jack could make are: he could use a more concentrated acid, and he could use smaller pieces of calcium carbonate instead of one big lump.

P4 Improve your grade Explaining motion

Page 95 Calculating speed

Higher: Henry is planning a train journey from Ipswich to Birmingham. The train journey is in two parts:

Departure time		Arrival time		Distance travelled (km)
12:00	Ipswich	13:00	Ely	80
13:15	Ely	15:45	Birmingham	120

a Which train is faster? Show your calculations.

b Taking into account the wait for the connection at Ely, what is the average speed of the journey from Ipswich to Birmingham?

a The first train does 80 km in an hour. The second train does 120 km in an hour and a half, so they are both the same speed.

b The total time is 3.45 hours, so the average speed = 200 ÷ 3.45 = 58 km/h.

Answer grade: D. The answer to part **a** is incorrect, as the students has said that the journey from Ely to Birmingham only takes 1½ hours not 2½ hours. In part **b** the student has said that 3 hours 45 minutes is 3.45 hours, not 3.75. However, the working is clear so some marks will be awarded.

Page 96 Speeding up

Higher: A cyclist rides around a velodrome track. When he sets off he takes 1 minute to reach a top speed of 14 m/s. Then he cycles 10 laps round the oval track at a steady speed. Calculate the cyclist's initial acceleration, and explain why he continues to accelerate after that. *AO1 [5 marks]*

Acceleration = $\frac{\text{change in speed}}{\text{time}} = \frac{14}{1} = 14$ m/s²

He is still accelerating afterwards because his speed changes as he goes round the curves.

Answer grade: D/C. The student has used the correct equation to find acceleration, for which they gain 1 mark. However, they have forgotten to change 1 minute to 60 seconds. The correct calculation is 14 ÷ 60 = 0.23 m/s².

The explanation of why the cyclist continues to accelerate is accurate but incomplete. To gain more marks, you need to use the word velocity and explain that as the cyclist changes direction the direction changes, which means that the velocity changes.

Page 97 Forces between objects

Higher: James kicked his football towards the goal. There was a force on the ball when it was kicked. This force was part of an interaction pair.

Describe the partner force of the kicking force in the interaction pair. *AO1 [3 marks]*

When James kicked the ball, he applied a force on the ball so it pushed back on his foot.

Answer grade: C/D. This answer is basically correct. When answering a question like this, make sure that it is clear that the force is *from* the football and on the foot. This answer is just about clear enough.

However, the student has forgotten to mention that the two forces in an interaction pair are equal and opposite. You would need to do this to get full marks.

Page 98 Terminal velocity

Higher: Explain how air bags reduce injury in a car crash. Include ideas about momentum in your answer. *AO1 [3 marks]*

The air bag cushions the driver so he takes more time to stop. His momentum has reduced so the force is less.

Answer grade: C/D. This answer goes some way to explain how air bags reduce injury in a car crash, but is only partly correct. There is some confusion about change of momentum. The change of momentum does not depend on the time for the collision (only the mass and velocity change). To gain full marks, explain that the force is reduced because the change in momentum is slower.

Page 99 Energy transfers

Foundation: A spacecraft is returning to Earth. It has a gravitational potential energy of 8MJ on re-entry.

a What is its maximum possible increase of kinetic energy as it falls?

b Explain why the actual increase of kinetic energy will be less than this value. *AO1 [3 marks]*

a 8 MJ

b Because not all the potential energy becomes kinetic energy.

Answer grade: F/E. The student has correctly calculated the answer to part **a**, and gains 1 mark for this. However, no marks are awarded for the answer to part **b**, because it does not explain *why*. For full marks, you need to say that the spacecraft slows down in the atmosphere due to air resistance.

Page 101 Static electricity

Higher: Bella was rubbing a nylon comb with a duster. When she put the comb near his head, her hair moved towards the comb. Explain why this happened.

AO1 [5 marks]

When Bella rubbed the comb it got charged. When it went near her hair there was an electrostatic force of attraction so the hair was attracted towards the comb. The comb and the hair must have had opposite charge.

Answer grade: B/C. The first sentence does not explain how the comb becomes charged, so gains no marks. The second sentence makes good use of the key term *electrostatic force* and correctly describes the hair as being attracted towards the comb, gaining 2 marks. The student also correctly identifies the comb and hair has having opposite charge, and for this the answer gains an additional 1 mark.

For full marks, you would need to include a more detailed explanation about electrons being rubbed off the comb by friction, making the comb positively charged.

Page 102 Electrical resistance

Higher: Harry was recording values of current and voltage across a resistor so he could calculate the resistance.

a He recorded a current of 0.06 A when the voltage across the resistor was 4 V. Calculate the resistance.

b Harry then repeated the experiment with a voltage of 6 V. What current reading did he expect to get? Explain why he might not get this exact value.

AO1, AO2 [5 marks]

a *Resistance* $= \frac{4}{0.06} = 66.7$

b *Current* $= \frac{6}{66.7} = 0.09$ A

He might have connected it up wrong.

Answer grade: D/C. The student's answer to part **a** of this question gains 1 out of the 2 available marks for correctly calculating the resistance. To gain the second mark, you would need to include the unit for resistance ( ).

In part **b**, the student has correctly calculated the current and has used the right unit, so gains 2 marks. However, the explanation about why the experiment might not give the exact value gains no marks. To gain full marks, you need to give reasons why there might be lower value for current, such as there might be a dirty connection or a broken lead.

Page 103 Series circuits and parallel circuits

Foundation: Susan and Mark were discussing adding resistors to a circuit. Susan said that if you added more resistors the total resistance would increase. Mark told her that sometimes adding more resistors would decrease the total resistance. Explain why both Susan and Mark are correct.

AO1 [5 marks]

When you add more resistors in series they are all in a line and you can add up all the resistors to give a bigger value for resistance, so Susan is right. Mark is talking about adding resistors in parallel – then the more resistors you add, the lower the total.

Answer grade: E/D. This answer is basically correct, giving a simple explanation of why both Susan and Mark are right, for which it gains 3 of the 5 marks available.

To gain full marks, the answer needs more detail. You would need to explain that adding resistors in series makes it harder for the charged particles to flow through the circuit. You also need to say that in parallel circuits the charged particles have a choice of pathway and it is easier to flow if there are more pathways.

Page 104 Generators

Foundation: Jenny uses a dynamo to power her bicycle lights.

a Explain why the lights are dim when she cycles slowly.

b Suggest one advantage and one disadvantage to using dynamo lights instead of battery powered lights.

AO1 [5 marks]

a *When you cycle slower the magnet doesn't spin as fast so there is a lower voltage.*

b *The advantage of using dynamo is that she doesn't need to buy batteries, and the disadvantage is that the light is dimmer.*

Answer grade: D. The student's answer to part **a** of this question is clear and accurate, gaining 2 of the 3 marks available. For full marks, you would need to give a more detailed explanation by saying that the lower induced voltage means lower current will flow through the bulb.

In part **b** of this question the student has correctly identified one advantage to using a dynamo, and gains 1 mark for this. However, a dynamo light is not necessarily dim. To gain full marks, you need to give a clear disadvantage, for instance, the light would go out when you had to stop to give way.

Page 105 Transformers

Higher: The mains supply at home is at 230 V a.c. A computer needs a supply at 23 V. Describe how the voltage of the mains is converted to the lower voltage.

AO1 [5 marks]

You need a step down transformer to reduce the voltage. The transformer has more coils on the primary coil and fewer on the secondary. The input voltage produces a magnetic field, which induces a lower voltage on the other coil.

Answer grade: C/B. This answer covers the basics and is accurate, so each sentence gains 1 mark.

For full marks, you would need to go into greater detail. First, you need to calculate the ratio of turns in the transformer (the primary coil needs 10 times as many turns as the secondary coil). Second, you need to say that the input creates a varying magnetic field to induce the output voltage.

Page 107 Radioactive elements

Foundation: Radioactive materials emit ionising radiation. Explain what is meant by 'ionising radiation'.
AO1 [3 marks]

Ionising radiation turns atoms into ions.

Answer grade: E/F. While this answer is basically correct, it needs more detail. You need to say that when alpha or beta particles collide with an atom, some of the electrons are knocked off, leaving a positively charged ion.

Page 108 Hazards of ionising radiation

Higher: Explain why it is more dangerous to inhale (breathe in) an alpha emitter than a beta emitter.
AO1 [4 marks]

Alpha particles are more ionising, so they can cause a lot of cell damage, but they can't pass through skin.

Answer grade: C/B. This answer is essentially correct, but if fails to explain why alpha radiation is more ionising than beta radiation. Also, while the last part of the answer is correct physics, it does not answer the question.

To gain full marks, you need to say that an alpha emitter is more massive and has double the charge of a beta particle, so it is easier for it to ionise other atoms.

Page 109 Half-life

Higher: Radon-220 decays by alpha emission with a half-life of 52 seconds. The initial activity is 640 counts per second. How long will it take for the activity to become 80 counts per second? *AO1 [3 marks]*

640 is 8 × 80 so it is 8 half-lives.

8 × 52 seconds = 416 seconds

Answer grade: C/D. The first statement is correct (640 is 8 x 80), but the student has used the incorrect value to work out the time for the half-life. Half-life is the time it takes to halve the activity, so 640 – 320 – 160 – 80 is 3 half-lives. You would then multiply the half-life (52 seconds) by 3 to get the answer (156 s).

Page 110 Uses of ionising radiation

Higher: Radioactive tracers can be used in many different applications. One application is for research into plant nutrition.

Which of the following radioactive isotopes would be most suitable for studying plant nutrition? Explain your choice.

Isotope	Type of radiation	Half-life
Phosphorus-32	Beta decay	14 days
Nitrogen-16	Beta decay	7 seconds
Bismuth-210	Alpha decay	5 days

AO1 [3 marks]

The nitrogen, because it has a very short half-life so will not cause too much damage. It is also beta decay which is better than alpha, which would not be detected outside the plant.

Answer grade: B/C. The arguments the student puts forward for beta decay are correct. However, the student has failed to think through the consequences of their decision – a half-life of 7 seconds is too short to study plant nutrition, as all the radioactivity will run out before the plant grows. For this reason, a beta emitter with a longer half-life (phosphrus-32) would be more suitable.

Page 111 Energy from the nucleus

Foundation: Explain the difference between nuclear fission and nuclear fusion. *AO1 [4 marks]*

Nuclear fission is when an atom splits up and fusion is when two atoms combine to form a new atom.

Answer grade: E/F. This answer is basic and accurate, so gains 2 marks. Note that there are 4 marks available for this question, so to gain full marks you need to give a more detailed response. You would need to explain that fission occurs with large unstable nuclei, such as uranium, and that fusion occurs with small nuclei such as hydrogen.

Ideas About Science

Understanding the scientific process

As part of your Science assessment, you will need to show that you have an understanding of the scientific process – Ideas about Science.

Science aims to develop explanations for what we observe in the world around us. These explanations must be based on scientific evidence, rather than just opinion. Scientists therefore carry out experiments to test their ideas and to develop theories. The way in which scientific data is collected and analysed is crucial to the scientific process. Scientists are sceptical about claims that cannot be reproduced by others.

You should be aware that there are some questions that science cannot currently answer and some that science cannot address.

Collecting and evaluating data

You should be able to devise a plan that will answer a scientific question or solve a scientific problem. In doing so, you will need to collect and use data from both primary and secondary sources. Primary data is data you collect from your experiments and surveys, or by interviewing people.

While collecting primary data, you will need to show that you can identify risks and work safely. It is important that you work accurately and that when you repeat an experiment, you get similar results.

Secondary data is found by research, often using ICT (the Internet and computer simulations), but do not forget that books, journals, magazines and newspapers can also be excellent sources. You will need to judge the reliability of the source of information and also the quality of any data that may be presented.

Presenting and processing information

You should be able to present your information in an appropriate, scientific manner, using clear English and the correct scientific terminology and conventions. You will often process data by carrying out calculations, drawing a graph or using statistics. This will help to show relationships in the data you have collected.

You should be able to develop an argument and come to a conclusion based on analysis of the data you collect, along with your scientific knowledge and understanding. Bear in mind that it may be important to use both quantitative and qualitative arguments.

You must also evaluate the data you collect and how its quality may limit the conclusions you can draw. Remember that a correlation between a factor that's tested or investigated and an outcome does not necessarily mean that the factor caused the outcome.

Changing ideas and explanations

Many of today's scientific and technological developments have benefits, risks and unintended consequences.

The decisions that scientists make will often raise a combination of ethical, environmental, social and economic questions. Scientific ideas and explanations may change as time passes, and the standards and values of society may also change. It is the job of scientists to discuss and evaluate these changing ideas, and to make or suggest changes that benefit people.

Glossary

A

absorb to take in energy from electromagnetic radiation; this is transferred to the particles of the material 48, 49, 50, 52

abundance a measure of how common a species is in an area 62

acceleration the rate at which the velocity of an object changes 96, 98, 100

accuracy how near a reading is to the true value 21, 40

acid a chemical compound which when dissolved in water gives a pH reading of under 7 and turns litmus red 24, 25, 26, 38, 41, 84, 88–94

acid rain rainwater which is made more acidic by pollutant gases 24–26, 28, 87

active site part of an enzyme where a substrate can fit neatly into it 59, 64

active transport the movement of chemicals into or out of a cell from areas of low concentration to high concentration, where the cell controls the direction in which chemicals move rather than the difference in concentration 61, 63, 64

activity (radioactivity) the amount of radiation emitted from a material 109, 112

adaptation the way in which a species changes over time to become better able to survive in its environment 17, 20, 22

adrenaline a hormone that helps prepare your body for action in the 'fight or flight' response 75

adult stem cells unspecialised body cells that can develop into other, specialised cells that the body needs 8, 9, 65, 69, 70

aerobic respiration respiration that requires oxygen 60, 63–64

air resistance the upwards force exerted by air molecules on an object 98–99

alkali a chemical compound which when dissolved in water gives a pH reading of over 7 and turns litmus blue 38, 41, 79–80, 84, 89–91, 94

alkali metal very reactive metal in Group 1 of the Periodic Table, for example sodium 79–80

alkanes a family of hydrocarbons (C_nH_{2n+2}) found in crude oil 30

alleles different versions of a gene on a pair of chromosomes 5–7, 9, 19

alpha particles (α) radioactive particles which are helium nuclei – helium atoms without the electrons (they have a positive charge) 107–108

alternating current (a.c.) an electrical current in which the direction of the current changes at regular intervals 104–106

amino acids small molecules from which proteins are built 4, 59–60, 68, 70

ammeter a device that measures the amount of current running through a circuit in Ampères 102

Ampères (amps) the unit of measurement used for the flow of electrical current or charge 53, 101

amplitude the maximum disturbance of a wave motion from its undisturbed position 46–47, 51

anaerobic respiration respiration that does not need oxygen 63–64

analogue equipment that can display data with continuous values 51–52

analogue signal transmitted data that can have any value 51–52

anode positive electrode 39, 86

antibiotic therapeutic drug acting to kill bacteria taken into the body 11–12, 16, 19–20

antibody protein normally present in the body, or produced in an immune response, which neutralises an antigen 10, 16

antidepressant a prescribed drug that makes synapses in the brain more sensitive to certain types of transmitter substances 75

anti-diuretic hormone (ADH) hormone which controls re-absorption of water in kidneys (and so water levels in the blood) 15

antigen harmful substance that stimulates the production of antibodies in the body 10–11, 16

antimicrobial substance that acts to kill bacteria 11–12, 16

arteries blood vessels that carry blood from the heart to other parts of the body 13–14, 16

asexual reproduction reproduction (creation of offspring) involving only one parent; offspring are genetically identical to the parent 8, 10

asteroid small object in orbit in the solar system 42, 47

atmosphere thin layer of gas surrounding a planet 49–52

atom the basic 'building block' of an element which cannot be chemically broken down 18, 25–26, 28–29, 31–34, 43, 48, 77–90, 107–111

attractive a force that pulls two objects together 31, 32, 83, 87, 97

auxin a plant hormone that affects the rate of growth 66, 70

average speed distance travelled divided by the time taken 95, 100

axon a long projection from a nerve fibre that conducts impulses away from the body of a nerve cell 71, 76

B

background radiation low-level radiation that is found all around us 31, 32, 83, 87, 97, 107, 112

bacteria single-celled microorganisms, some of which may invade the body and cause disease 8, 10–12, 16–20, 23, 34–35, 37, 48, 60, 63–64, 110

balanced symbol equation a symbolic representation showing the kind and amount of the starting materials and products of a chemical reaction 38, 79, 89

base (1) solid alkali; any substance that neutralises an acid; (2) one of the three molecules that makes up a single unit of DNA 4, 38, 41, 68, 70

behaviour the way in which an organism reacts to changes in its environment 73–74, 76

beta blockers a prescribed drug that blocks the adrenaline receptors in the synapses and stops the transmission of impulses 75

beta particles (β) particles given off by some radioactive materials (they have a negative charge) 108–109, 112

Big Bang the theoretical beginning of the Universe, when energy and matter expanded outwards from a point 43

binary digit a number that can only take the values 0 or 1 51

binary fission simple cell division 10

binding energy the energy that holds particles together in a nucleus 110, 112

biodegradable a material that can be broken down by microorganisms 21, 39

biodiversity the variety in terms of number and range of different life forms in an ecosystem 21–22

biofuel fuel such as wood, ethanol or biodiesel, obtained from living plants 27–28, 53, 55, 57–58

blood plasma yellow liquid in blood, in which the blood cells are carried 15

blood pressure the pressure of blood against the walls of the blood vessels 14, 16, 37

byte a measure of digital data consisting of 8 binary digits 51–52

C

capillaries small blood vessels that join arteries to veins 13, 16

carbon an element that combines with others, such as hydrogen and oxygen, to form many compounds in living organisms 13, 18, 21–28, 30–35, 38, 40–41, 50, 52–3, 57–64, 80–81, 83–86, 88–90, 108–109, 111

carbon cycle the way in which carbon atoms pass between living organisms and their environment 18, 22, 50

carbon dioxide gas whose molecules consist of one carbon and two oxygen atoms, CO_2; product of respiration and combustion; used in photosynthesis; a greenhouse gas 13, 18, 21, 23–28, 30, 38, 50, 52–53, 57–60, 63–64, 83–85, 88–89, 111

Glossary

carbon monoxide poisonous gas whose molecules consist of one carbon and one oxygen atom, CO 24, 26, 28

carrier someone who carries a gene but does not themselves have the characteristic 7, 9

carrier wave electromagnetic wave on which a signal is superimposed for transmission 50

catalyst chemical that speeds up a chemical reaction but is not itself used up 27, 33, 93–94

catalytic converter a device fitted to vehicle exhausts to reduce the level of nitrogen oxides and unburnt hydrocarbons emitted 27–28

cathode negative electrode 39, 86

cell body the part of a nerve cell that contains the nucleus 71, 73

cell membrane layer around a cell which helps to control substances entering and leaving the cell 60–61, 64

cell sampling removal of a small number of fetal cells, e.g. from the placenta or amniotic fluid, for testing 7

cellulose large polysaccharides made by plants for cell walls 59, 64

central nervous system (CNS) collectively the brain and spinal cord 70, 76

ceramics non-metallic solids made by heating and cooling a material, such as clay to make pottery 30, 32

cerebral cortex the outer layer of the brain 74, 76

chain reaction a fission reaction that is maintained because the neutrons produced in the fission of one nucleus are available to initiate fission in other nuclei causing a rapid production of energy 111–112

chemical synthesis combining simple substances to make a new compound 89–94

chlorophyll the green chemical in plants that absorbs light energy 60, 64

chloroplasts structures characteristic of plant cells and the cells of algae where photosynthesis takes place 60, 64

cholesterol chemical needed by the body for the formation of cell membranes, but too much in the blood increases the risk of heart disease 13

chromosomes thread-like structures in the cell nucleus that carry genetic information – each chromosome consists of DNA wound around a core of protein 4–6, 8–9, 67–68, 70

circulatory system a transport system in the body that carries oxygen and food molecules 13, 16, 65

classify put things into groups according to their properties 22

clinical trials scientific testing of drugs, vaccines and medical processes 12

clone organism genetically identical to another 7–9, 66, 70

combustion process in which substances react with oxygen releasing heat 18, 22, 25–26

comet lump of rock and ice in a highly elongated orbit around the Sun 42, 47

competition result of more than one organism needing the same resource, which may be in short supply 19, 22, 62

components devices such as lamps and motors on an electrical circuit to which energy is transferred 54, 101–103, 106

composite material consisting of a mixture of other materials 34

compound substance composed of two or more elements which are chemically joined together, for example H_2O 18, 22, 25–26, 30, 38, 79–86, 88–90, 94

compressive strength a measure of resistance to squeezing or crushing forces 29

condense to turn from a gas into a liquid, as in steam (water vapour) which condenses to liquid water 28, 31

conditioned (response) a learned response that occurs when animals link two or more stimuli that are not connected 73

conditioned reflex *see* conditioned response 73

conductor a substance in which electric current can flow freely 30, 87, 101–103, 105

conservation of energy when energy can not be created or destroyed 99

contaminated having mixed with something harmful such as a pollutant or radioactive substance 56, 111

contamination (radioactivity) something that comes into contact with radioactive material 56–57, 58, 108, 112

continental drift slow movement of continents (land masses) relative to each other 44, 47

continuous variation variation in organisms of features that can take any value, for example height 4

control group in a drugs trial, the group that receives the placebo allowing researchers to assess whether the drug has an effect in the experimental group 12, 16

control rods absorb excess neutrons in order to control a chain reaction 111–112

convection heat transfer in a liquid or gas, when particles in a warmer region gain energy and move into cooler regions, carrying this energy with them 47, 50, 55

coolant gas or liquid that circulates around a reactor to keep it cool 111–112

coronary arteries blood vessels that carry blood away from the heart 13

coronary heart disease (CHD) when arteries that supply the heart muscle gradually become blocked by fatty deposits, preventing the heart from working properly 13

correlation a link between two factors that shows they are related, but one does not necessarily cause the other; a positive correlation shows that as one variable increases, the other also increases; a negative correlation shows that as one variable increases, the other decreases 13, 24, 49–50, 52, 62

corrosive a substance that can destroy or eat away other substances by a chemical reaction, e.g. it will burn skin 79–80

covalent bonds these join together the atoms inside a molecule 83, 85, 88

cross-links bonds that link one polymer chain to another 32, 68,

crude oil black substance extracted from the Earth, from which petrol and many other products are made 30–31, 35

crust surface layer of Earth, made up of tectonic plates 30, 34, 36, 41, 44–45, 47, 85,

crystal lattice crystals formed by ionic compounds, such as sodium chloride, which have a regular repeating pattern and shape 81, 84

crystalline a solid material with atoms, molecules or ions arranged in a regular repeating pattern 32

crystallise when a liquid undergoes evaporation the product left behind cools and starts to form crystals 92

crystals the solid residue left after salts evaporate, they have regular shapes and flat sides 84, 88, 92, 94

current flow of electrons in an electric circuit 39, 47, 53–56, 58, 80, 86, 88, 100–106

cytoplasm a jelly-like substance within a cell where most of the chemical reactions take place 60, 64, 68, 70–71, 76

D

data information, often in the form of numbers obtained from surveys or experiments 11, 13–14, 18–21, 24–26, 29, 40, 44, 46–47, 50, 59, 61–62, 66, 77, 83, 87, 90

daughter cells in mitosis a cell splits to form two daughter cells which are identical to each other 65, 67, 70

daughter product the name given to the radioactive element formed from the decayed initial radioactive element 109

decay chain a series of radioactive decays of an unstable nucleus to the nucleus of a different element, until a stable nucleus is formed 109

decode to extract information from a code 51

decompose in chemistry, separation of a chemical compound into simpler compounds 21, 50, 86

decomposer in a food chain, an organism such as a fungus that uses materials from dead or decaying matter 17, 22

Glossary

decomposition the action of bacteria and fungi to break down previously living material 17, 22

deforestation the large-scale removal of trees from forested areas for building or farming 50

denaturing when an active site is destroyed and the enzyme molecules are broken apart 64

dendrite a short thread of cytoplasm on a neuron, carrying an impulse towards the cell body 71

denitrifying bacteria bacteria vital to the nitrogen cycle, which change nitrates in the soil to nitrogen 18

density the mass of a substance per unit volume 29–31, 34, 79, 83

detritivore in a food chain, an organism such as an earthworm that breaks down dead or decaying matter into smaller particles 21

diatomic molecules atoms that are joined together in pairs 80, 82

differentiation the change of an unspecialised body cell into a particular type of cell 8, 65, 69

diffusion the movement of molecules or particles from regions of high concentration to low concentration 61, 64

digital signal transmitted information that can take only a small number of discrete values, usually just 0 and 1 51, 52

direct current (d.c.) an electric current that flows in one direction only 104, 106

displacement the distance moved in a specific direction 80, 82, 95, 100

displacement reactions the difference in the reactivity of halogens. Where one halogen will take the place of another in its compounds 80, 82

displacement–time graph a visual way of showing the displacement (distance and direction from a starting point) of an object against time 95

dissolve to be soluble in water 23, 25–26, 28, 36–37, 84–86, 88–89, 91–92

dissolving the act of a solid mixing into a liquid to form a solution 41, 92

distance–time graph a visual way of showing the time taken for a journey and the distance travelled 95, 100

DNA large (polymer) molecule found in the nucleus of all body cells – its sequence determines genetic characteristics, such as eye colour, and gives each one of us a unique genetic code 4, 5, 8–10, 19–20, 22, 60, 64, 67–68, 70, 108

dominant (allele) the allele that is always expressed, irrespective of the other allele in the pair 5–7, 9

double helix two strands of the DNA molecule face each other in a way that looks like a ladder, these are then twisted around each other to form a double helix – like a spiral staircase 68, 70

drag *see* air resistance 98, 100

dry air air that has had all water vapour removed 23, 83, 88

dwarf planet spherical object orbiting the Sun, smaller than a planet and larger than an asteroid 42, 47

E

Ecstasy an illegal drug that affects the working of the chemical transmitter substance in nerve synapses in a similar way to antidepressants 14–16, 75

effector part of the body that responds to a stimulus 15–16, 71–73, 76

effervescence the fizzing and bubbling effect that occurs e.g. when an acid reacts with a carbonate ion 84

efficiency a measure of how effectively an appliance transfers the input energy into useful energy 17, 54, 56–58

egg female sex cell of an animal 4–6, 8–9, 17, 19, 65, 67, 69–70

elastic a material that returns to its original shape and size after a deforming force is removed 12, 29

electric current a negative flow of electrical charge through a medium, carried by electrons in a conductor 39, 53, 55, 58, 86, 88, 101, 104–106

electrodes solid electrical conductors through which the current passes into and out of the liquid during electrolysis – and at which the electrolysis reactions take place 74, 81

electrolysis decomposing an ionic compound by passing an electric current through it while molten or in solution 39, 41, 86, 88

electrolyte the liquid in which electrolysis takes place 86

electromagnetic induction a term used by Faraday to explain induced voltage 104, 106

electromagnetic radiation energy transferred as electromagnetic waves 48, 50, 52

electromagnetic spectrum electromagnetic waves ordered according to wavelength and frequency – ranging from radio waves to gamma rays 48

electron tiny negatively charged particle within an atom that orbits the nucleus – responsible for current in electrical circuits 48, 77–79, 81–83, 85–88, 101–103, 106–108, 111–112

electron arrangement the configuration of electrons in shells, or energy levels, in an atom 78, 81–82

electrostatic force a force caused by positive and negative charges 101

element substance made out of only one type of atom 18, 25–26, 28, 30, 38, 40, 43, 77–83, 85–86, 90, 107, 109, 112

embryo an organism in the earliest stages of development which began as a zygote and will become a foetus 6–9, 65, 69–70

embryonic stem cells cells in or from an embryo with the potential to become any other type of cell in the body 8–9, 65, 69–70

endothermic a reaction that reduces the temperature of the surroundings. The temperature falls in endothermic reactions 91, 94

end-point the sudden change of colour of an indicator, e.g. in titration 90

energy input the energy transferred into a device or appliance from elsewhere 54, 58

energy level describes the arrangement of electrons in an atom in shells 91, 94

energy level diagram visual way of showing the change in energy level during a chemical reaction 91, 94

energy output the energy transferred away from a device or appliance, which may be either useful or wasted 54

environment an organism's surroundings 4–5, 8–9, 15–22, 24, 34, 37, 39–41, 54, 56–58, 66, 70–71, 73, 76, 83–88, 93

enzymes proteins found in living things that speed up or catalyse reactions 4, 9, 59–60, 64, 66, 72

epidemiological studies studies of the patterns of health and illness in the population 13–14

equivalent dose a measure of radiation dose to biological tissue 108

erosion the wearing away of rock or other surface matter such as soil 41, 44–45, 47

error uncertainty in scientific data 29, 90

evaporate turn from a liquid to a gas, such as when water evaporates to form water vapour 31, 36, 63, 84, 92

evolution change in a species over a long period of time 19–20, 22

excrete to get rid of waste substances from the body 15, 18

exothermic a reaction that gives out heat to the surroundings. The temperature rises in exothermic reactions 91, 94

extinct a species that no longer survives 17, 19, 21–22

extinction the process or event that causes a species to die out 17, 19, 21–22

F

family tree diagram chart showing relationships between members of different generations of a family, which can be used to show inheritance of genetic characteristics 6

feral (children) children who have been isolated during their development are said to be 'feral'. Feral means wild or untamed 74

Glossary

fermentation the conversion of carbohydrates to alcohol and carbon dioxide by yeast or bacteria 63

fertilisation the moment when the nucleus of a sperm fuses with nucleus of an egg 5, 7, 67, 69–70

fetus a later–stage embryo of an animal; the body parts are recognisable 7

fibre a long thin thread or filament 29–32, 33, 35, 51–52

field in physics, a space in which a particular force acts 36, 44, 55, 104–106, 108

filtrate the insoluble products that remain trapped in a filter 92

filtration a method of separating one substance out from others. Filtering separates solids from liquids 92, 94

flue gas desulfurisation industrial process whereby sulfur is removed from waste gases 27–28

fold mountain a mountain caused by folding of the Earth's crust when two tectonic plates push against one another 45

food web flow chart showing how a number of living things in an environment depend on one another for their food 17, 22

force the push or pull that acts between two objects 29, 31–32, 35, 83–85, 88, 95, 97–101, 104–105–107, 112

formula (for a chemical compound) group of chemical symbols and numbers, showing elements, and how many atoms of each, a compound is made up of 26, 30, 79–86, 88–91, 94

fossil fuel fuel such as coal, oil or natural gas, formed millions of years ago from dead plants and animals 26–28

fossil record the information obtained over the years from fossil collections 19–22

fraction group of substances with similar boiling points, produced by fractional distillation 31, 35

fractional distillation process that separates the hydrocarbons in crude oil according to size of their molecules 31, 35

free electrons the outer electrons of atoms of materials that are good conductors which are loosely held and can break free easily so they can move freely 87, 101–103, 106

frequency the number of waves passing a set point, or emitted by a source, per second 46–48, 50–52, 108, 112

fuel rod long narrow tube in a nuclear reactor which contains nuclear fuel in pellet form 55, 111

functional protein a protein such as an enzyme that speeds up a chemical reaction 4

G

galaxy group of billions of stars 42–43, 47

gametes the male and female sex cells (sperm and eggs) 67, 70

gamma rays (γ) ionising high-energy electromagnetic radiation from radioactive substances, harmful to human health 48–49, 52, 107–110

gas state of matter in which atoms or molecules are spaced far apart and spread out to fill the available space 23–28

gene a section of DNA that codes for a particular characteristic by controlling the production of a particular protein or part of a protein by cells 4–10, 16, 19, 21, 59–60, 64–65, 68–70

gene pool the complete set of alleles in a population; a larger gene pool results in greater genetic variation 19

generator equipment for producing electricity 55, 58, 104, 106, 111

genetic code the information contained in a gene which determines the type of protein produced by cells 60, 68, 70

genetic diversity the differences between individuals (because we all have slight variations in our genes) 21

genetic screening testing large numbers of individuals for a gene, such as a gene for a genetic disorder 7, 9

genetic testing testing an individual for the gene for a genetic disorder 7, 9

genotype an individual's genetic make up, such as whether they are homozygous or heterozygous for a particular gene 4, 7, 9

geologist scientist who studies rocks and the changes in the Earth 36, 41, 44

giant covalent structures an element made with very strong covalent bonds between atoms in which a large number of carbon atoms are linked together in a regular pattern 85, 88

gland organ that secretes a useful substance 15, 71–72

global warming gradual increase in the average temperature of Earth's surface 24, 50, 52–53, 57

glucose a simple sugar 59–60, 63–64, 71

gradient the degree of slope of a line 93, 95, 99–100, 102

gram formula mass the number of grams of an element or compound represented by its RAM or RFM 86, 88

gravitational potential energy the energy an object gains due to its height 99–100

greenhouse effect the trapping of infra-red radiation by the Earth's atmosphere 50, 53

greenhouse gas a gas such as carbon dioxide that reduces the amount of infrared radiation escaping from Earth into space, thereby contributing to global warming 50, 52–53, 57

groups within the Periodic Table, the vertical columns are called groups 77

H

habitat the physical surroundings of an organism 17, 21–22, 37, 87

half-life the time taken for half of the atoms in a radioactive element to decay 109–112

halides compounds formed when halogens react with alkali metals and other metals 80

hardness a measure of resistance to change in shape of a solid, for example by scratching or by impact 29, 35

hazard something that is likely to cause harm, e.g. a radioactive substance 37, 56, 79, 89, 108–110

heart rate the number of heartbeats every minute 14, 16, 74–75

hertz unit for measuring wave frequency; 1 hertz (Hz) = 1 wave per second 46

heterozygous an individual who has two different alleles for an inherited characteristic 5, 9

high blood pressure blood pressure that is consistently abnormally high 14, 16, 37

high level waste for example, (radioactive) spent fuel rods, with a long half-life, which need to be disposed of carefully 111–112

homeostasis the way the body keeps a constant internal environment 15–16

homozygous an individual who has two identical alleles for an inherited characteristic 5, 9

hormones substances produced by animals and plants that regulate activities; in animals, hormones are produced by and released from endocrine tissue into the blood to act on target organs, and help coordinate the body's response to stimuli 4, 66, 70–72

hydrocarbon compound containing only carbon and hydrogen 25, 30–31, 35

hydroelectric description of power station generating electricity from the energy of moving water 55–58

hydrosphere made up of the water, ice and snow on the Earth's surface and the water vapour in the atmosphere 84

I

induced a term used to mean 'created' 104–106

inert an element that does not react with any other elements 78

identification key a way to find a scientific name for an organism by answering yes/no questions 62, 64

igneous rock rock formed by the solidification of molten magma or lava 36

immune when a person has resistance to a particular disease 10

Glossary

immune system a body system which acts as a defence against pathogens, such as viruses and bacteria 10–11, 16, 69

impulse an electrical signal that travels along an axon 69–76

in parallel when components are connected across each other in a circuit 102–103, 106

in series when components are connected end-to-end in a circuit 102–103, 106

indicator in chemistry, a substance that shows the presence of an acid or an alkali by a change in colour; in biology, a measure of the quality of a natural environment, for example, the number of sensitive species present in an aquatic environment, or the level of pollutants in the air 18, 22, 38, 79, 89–90, 94

insoluble not soluble in water (forms a precipitate) 23, 25, 37–38, 41, 84

instantaneous speed the speed at a particular moment in time 95–96

instinctive response behaviour that comes from reflex responses and does not have to be learned 74

insulator a substance in which electric current cannot flow freely 101, 106

insulin hormone produced by the pancreas that promotes the conversion of glucose to glucagon 71–72

intensity a measure of the power of a beam of radiation 48–50, 52, 106, 108

interdependence relationship between several organisms that depend on one another 17, 22

intermediate level waste for example, (radioactive) chemical sludge and reactor components, with short or longer half-lives that have to be disposed of with care 111–112

inversely proportional when there is an increase in one variable and a proportionate decrease in another variable 46, 48, 103

ion atom (or groups of atoms) with a positive or negative charge, caused by losing or gaining electrons 79, 81–82, 84–89, 91, 94, 107–108, 112

ionic bond chemical bond between two ions of opposite charges 84

ionic compounds salts made up of particles called ions which have a positive or negative electrical charge 81–82, 85, 87–89

ionic equation a chemical equation that describes changes that occur in aqueous solutions 84, 86, 88, 91, 94

ionisation the removal of electrons from atoms or molecules 48, 108, 112

ionising radiation electromagnetic radiation that has sufficient energy to ionise the material it is absorbed by 48–49, 52, 56, 58, 107–108, 110, 112

irradiation exposure to waves of radiation 56, 58, 110, 112

isolated separated, as in a strain of bacteria that can be separated from others, or as in an island that is remote from other land masses 20

isotopes atoms that have the same number of protons, but different numbers of neutrons. Different forms of the same element 110–111

J

joule unit of energy 53, 99, 102, 111

K

kidney the organ in the body that controls water balance 15–16, 18

kilowatt unit of power equal to 1000 watts or joules per second 53, 58

kilowatt-hour (kWh) the energy transferred in 1 hour by an appliance with a power rating of 1 kW (sometimes called a 'unit' of electricity) 53, 58

kinetic energy the energy an object has due to its motion 97, 99–100, 111

L

lattice a repeating pattern formed by the regular 3-D arrangement of ions 81, 84–85, 88

lava molten rock (magma) from beneath the Earth's surface when it erupts from a volcano 23, 36, 44

lichen small organism that consists of both a fungus and an alga 18

Life Cycle Assessment an analysis of the environmental impact of a product, including the production of raw materials, its manufacture, packing, transport, use and disposal 21, 40–41

light dependent resistor (LDR) a semiconductor device, where resistance changes with the amount of light 103

light pollution excessive artificial light that prevents us from *seeing* the stars at night and can disrupt ecosystems 42

light-year the distance travelled by light in 1 year 42

limiting factor a lack of something that prevents a reaction from increasing any further 61

limiting friction the maximum amount of force that can be applied to an object before it will move 97

line spectrum a set of different coloured lines produced when the light from a burning element is passed through a prism 77, 82

lithosphere the rocky outer section of Earth, consisting of the crust and upper part of the mantle 85

living indicator a species, the presence of which gives a measure of the quality of an environment; some species, such as the mayfly, are sensitive to pollutants and others are tolerant 18, 22

longitudinal a wave such as a sound wave in which the disturbances are parallel to the direction of energy transfer 45, 50

long-term memory information from our earliest memories onwards, which is stored for a long period of time 75–76

low blood pressure blood pressure that is consistently abnormally low 14

low level waste for example, contaminated (radioactive) paper and clothing that is not very dangerous, with a short half-life, but still needs to be disposed of carefully 111–112

M

magma molten (liquid) rock 45

magnetic field a space in which a magnetic material exerts a force 36, 44, 55, 104–106

malleable able to be beaten into a thin sheet; a common property of metals 30, 87

mantle semi-liquid layer of the Earth beneath the crust 44–45, 47, 85

mass extinction event the extinction of a large number of species at the same time 21

MDMA 3,4-methylenedioxymethamphetamine, the scientific name for Ecstasy 14–16, 75

mean an average of a set of data 26, 29, 31, 35, 59, 90

meiosis cell division that results in the formation of gametes 67, 70

melting point temperature at which a solid changes to a liquid 29, 32, 35, 78–79, 80–81, 83, 87–88

membrane (of a cell) the layer around a cell which helps to control substances entering and leaving the cell 39, 60–61, 72, 76

membrane cell electrolysis cell that uses a semi-permeable membrane to separate the reactions at the two electrodes, as in the electrolysis of brine 39, 60

memory the storage and retrieval (bringing back or remembering) of information 4, 10–11, 16, 74–76

memory cells white blood cells that form antibodies in response to a particular antigen and retain the ability to make that antibody should re-exposure to the antigen occur later in life 10–11, 16

meristems special regions in a plant where cells are able to divide 65–66, 70

messenger RNA (mRNA) a molecule that copies the base sequence of the DNA and carries it out of the nucleus of the ribosomes 68, 70

metal a group of materials (elements or mixtures of elements) with broadly similar properties, such as being hard and shiny, able to conduct heat and electricity, and able to form thin sheets (malleable) and wires (ductile) 23, 25, 28, 30, 34–35, 38, 49, 57, 62, 78–89, 94, 101, 103, 106

metallic bond the force in metals that attracts atoms together 87

methane a gas with molecules composed of carbon and hydrogen; a greenhouse gas 23, 25, 30, 50, 52, 56, 83

microwave electromagnetic wave similar to radio waves but with higher energy 49, 51–52, 54

microorganism very small organism (living thing) which can only be viewed through a microscope 10–12, 16–18, 21–22, 34, 39, 60, 63–64

Milky Way the galaxy in which our Sun is one of billions of stars 42, 47

minerals solid metallic or non-metallic substances found naturally in the Earth's crust 61, 85, 88

mitochondria found in the cytoplasm, where respiration takes place 60, 67

mitosis cell division that takes place in normal body cells and produces identical daughter cells 65, 67, 70

mixture one or more elements or compounds mixed together but not chemically joined, so they can be separated out fairly easily 23, 26, 30–31

molecular ion a charged ion composed of two or more atoms joined together by covalent bonds 84

molecule two or more atoms held together by strong chemical bonds 4, 9–10, 18, 23, 25–26, 28, 30–33, 35, 40–41, 48–49, 59–60, 63–64, 67–68, 70, 72, 75–76, 80, 82–83, 85, 88, 91, 101, 108

momentum the product of mass and velocity; momentum (kg m/s) = mass (kg) $\times$ velocity (m/s) 98, 100

monoculture when a single crop is grown 21, 25

monomers small molecules that become chemically bonded to one another to form a polymer chain 31, 35

moon a large natural satellite that orbits a planet 42, 47

motor an electric motor converts electrical energy into mechanical energy 102, 105–106

motor effect a term used to describe the force experienced when a current flows through a wire in a region where there is a magnetic field. If it is free to move this is known as the motor effect 105–106

motor neuron nerve carrying information from the central nervous system to muscles and glands 71, 73, 76

multicellular consisting of many cells 65, 70–71

multi-store model a type of model used by scientists to help explain how we remember and retrieve information 75–76

mutation a change in the DNA in a cell 5, 12, 19, 22, 108

myelin a fatty sheath that surrounds an axon, it acts as an insulator and makes an impulse travel faster 71, 76

N

nanometre (nm) unit used to measure very small things (one-billionth of a metre, or 10–9m) 33

nanoparticles very small particles on an atomic scale 33–35

nanotechnology technology making use of nanoparticles 34–34

nanotube a carbon molecule in the form of a cylinder 33–34

National Grid the network that distributes electricity from power stations across the country, using cables, transformers and pylons 56, 58

natural materials materials made from plant and animal products 30, 33, 35

natural selection process by which characteristics that can be passed on in genes become more common in a population over many generations (which are likely to give the organism an advantage that makes it more likely to survive) 19–20, 22

negative feedback information that causes a reversal in a control system, for example when we get too hot our body responds to bring our temperature back to normal through sweating and vasodilation 15, 19

nerve a group of nerve fibres 70–72, 75–76

nervous system sends messages between body cells using neurons; includes the central nervous system (brain and spinal cord) and the peripheral nervous system (network of neurons) 71, 73, 75–76

neuron a nerve cell that carries nerve impulses 71–76

neuron pathway neurons linked to pass nerve impulses 74, 76

neuroscientists scientists who study the nervous system 74

neutral (1) in chemistry the term neutral means 'between acid and alkali' (2) in physics, an atom with no overall charge 27–28, 38, 86, 89–91, 101, 108

neutralisation reaction between an acid and a base (H$^+$ ions and OH$^-$ ions), to make a salt and water 38, 90–91, 94

neutron small particle that does not have a charge – found in the nucleus of an atom 77, 82, 101, 107–109, 111–112

nitrates salts containing the nitrate ion (consisting of one nitrogen atom and three oxygen atoms); may be used as fertilisers, sometimes causing pollution of waterways 18, 22, 60–61, 64

nitrogen cycle the way in which nitrogen and nitrates pass between living organisms and the environment 18, 22

nitrogen-fixing bacteria bacteria vital to the nitrogen cycle, which change nitrogen from the air to nitrates in the soil, needed by plants 18

nitrogen oxides gaseous molecules containing nitrogen and oxygen atoms according to the formula NO_x, where $x = 1, 2$, etc.; these pollutants are formed due to the high temperatures created by the combustion of fossil fuels 24, 26, 28

noise random alteration to a communication signal, possibly due to interference 51, 57

non-living indicator a non-living measure of the quality of an ecosystem, such as water temperature 18, 22

nuclear fission a chain reaction employed in nuclear power reactors in which atoms are split, releasing huge amounts of energy 43, 111–112

nuclear fuel radioactive fuel, such as uranium or plutonium, used in nuclear power stations 53, 111–112

nuclear fusion nuclear reaction in which two small atomic nuclei combine to make a larger nucleus, with a large amount of energy released 43, 111–112

nucleus (1) the central core of an atom, which contains protons and neutrons and has a positive charge; (2) a distinct structure in the cytoplasm of cells that contains the genetic material 4, 8–9, 60, 64, 67–71, 77–78, 82–83, 101, 107–109, 111–112

O

oceanic ridge undersea mountain range formed by seafloor spreading and caused by the escape and solidification of magma where tectonic plates meet 44

oestrogen female hormone secreted by the ovary and involved in the menstrual cycle 71

Ohm's law law that states that the current through a metallic conductor is directly proportional to the voltage across its ends, if the conditions are constant 102, 106

optical fibre glass fibre that is used to transfer communication signals as light or infrared radiation 51, 55

orbit near-circular path of an astronomical body around a larger body 42, 44

orbits electrons are arranged in orbits (or shells) around the nucleus of an atom 77, 101, 107, 112

ores rocks that contain minerals, including metals, e.g. iron ore 85, 88

organ a part of the body made up of different tissues that work together to do a particular job 34, 65, 70–73

Glossary

oscilloscope laboratory equipment for displaying waveforms 46

osmosis the diffusion of water molecules through a partially permeable membrane 62, 64

outlier a measurement that does not follow the trend of other measurements 26, 29, 35, 61, 90

oxidation chemical process that increases the amount of oxygen in a compound; the opposite of reduction 25, 38, 85, 88

oxidised a substance that has undergone oxidation 27, 31, 85

ozone gas found high in the atmosphere which absorbs ultraviolet rays from the Sun 39, 49, 52

P

parallax angle between two imaginary lines from two different observation points on Earth to an object such as a star or planet, used to measure the distance to that object 42, 47

pancreas organ that produces hormones insulin and glucagon (from endocrine tissue) and digestive enzymes (from exocrine tissue) 71

partially permeable membrane a cell membrane that lets small molecules pass through but not large ones 61, 64

particulates pollution in the form of particles in the air, such as soot 24, 27, 33

pathogen harmful organism which invades the body and causes disease 10–11

period a horizontal row in the periodic table 78, 82

Periodic Table a table of all the chemical elements based on their atomic number 77–82, 86, 90

peripheral nervous system (PNS) network of neurons leading to and from the brain and spinal cord 71, 76

pH a measure of the acidity or alkalinity of a substance 15, 25, 38, 59, 64, 79, 89, 91, 94

phenotype the physical expression of a gene; different genotypes can give the same phenotype 4, 9

phloem plant cells that carry dissolved substances to every part of the plant 65, 70

photon a 'packet' of electromagnetic energy, the amount of energy depending on the frequency of the electromagnetic wave 48, 52

photosynthesis process carried out by green plants in which sunlight, carbon dioxide and water are used to produce glucose and oxygen 17–18, 22–23, 26, 28, 50, 59, 60–61, 63–66

phototropism a plant's growth towards or away from the stimulus of light 66, 70

phytoplankton microscopic plant life, often forming the basis of aquatic food chains 18, 60

pixel a tiny area (for example a dot or square) on a screen which conveys the data relating to a small part of a picture 51

placebo 'dummy' treatment given to some patients in a drug trial, that does not contain the drug being tested 12, 16

planet large sphere of gas or rock orbiting a star 19, 42, 47

plastic a compound produced by polymerisation, capable of being moulded into various shapes or drawn into filaments and used as textile fibres 21, 29–32, 34–35, 39–41, 54, 101, 106

plasticiser small molecules which fit between polymer chains and allow them to slide over each other 32, 35, 40–41

plate boundary where two adjacent tectonic plates of the Earth's crust meet or are moving apart 45

plate tectonics the theory that explains how changes to the Earth's surface occur at tectonic plate boundaries 23

pollutant harmful substance in the environment 23–28, 87

polymer large molecule made up of a chain of smaller molecules (monomers) 30–32, 35

polymerisation chemical process that combines monomers to form a polymer 31

positively phototropic plant shoots that grow towards a light source 66

potential difference (p.d.) another term for voltage, a measure of the energy carried by the electrical charge 102–103, 106

power amount of energy that something transfers each second, measured in watts (or joules per second) 53–54, 56–58, 101–107

precipitate insoluble solid formed in a solution during a chemical reaction 84, 88

pre-implantation genetic diagnosis (PGD) genetic testing of embryos created by in vitro fertilisation for a genetic disorder, so that healthy embryos can be transferred into the mother's uterus 7, 9

primary coil the input coil of a transformer 105–106

primary energy source a source of energy before conversion to useful energy; examples include fossil fuels, wind, biomass and solar energy 53, 55, 58

principal frequency the main frequency of electromagnetic radiation emitted by an object; hotter objects have higher principal frequencies 50

processing centre a centre of control that acts in response to information, for example the hypothalamus in the brain which responds to changes in body temperature 15–16

products chemicals produced at the end of a chemical reaction 25–30, 38–41, 59–61, 65, 86, 90–92, 94

protein a type of chemical with important functions in living organisms 4, 9–10, 18, 22, 59–61, 64, 68–70

proton small positively charged particle found in the nucleus of an atom 77–78, 101, 107–109, 112

proton number the number of protons in an atom 77–78, 82

pulse in the body, a beat of the heart that may be felt in an artery close to the skin; in data transmission, a short on-phase 14, 16, 51, 71

pulse rate a measure of the number of times per minute the heart is beating 14, 16

Punnett square a diagram that can be used to work out the probability of outcomes resulting from a genetic cross 6, 9

pure a substance that has nothing else mixed with it 25, 83, 87, 89, 92

PVC a type of polymer (short for polyvinyl chloride) 32, 39–41

P-waves longitudinal shock waves following an earthquake that can travel through the molten core of the Earth; they change direction at the boundary between different layers of the Earth 45, 47

Q

quadrat a frame, usually square, of wood or metal, that ecologists put on the ground and count the number of plants within it 18, 62, 64

R

radiation energy transfer by electromagnetic waves or fast-moving particles 42–43, 48–50, 52, 56, 58, 107–110, 112

radioactive a material that randomly emits ionising radiation from its atomic nuclei 36, 49, 56–58, 107–112

radioactive decay the disintegration of a radioactive substance, the process by which an atomic nucleus loses energy 36, 109

radioactive tracer a radioactive isotope with a short half-life that can be ingested and traced through a patient's body or to monitor the movement of waste products in industry 110

radiographer medical worker who takes and processes body images 49, 110

radiotherapy a technique that uses ionising radiation to kill cancer cells in the body 110

range in a series of data, the spread from the highest number to the lowest number 90

rate a measure of speed; the number of times something happens in a set amount of time 12, 14, 53, 55, 58–59, 61, 64, 79, 93–94, 96, 98, 101–102, 104, 107, 109

rate of flow of charge the current, or charge, that flows per second, measured in Ampères 101

rate of reaction the speed with which a chemical reaction takes place 79, 93–94

rating an assessment or classification according to a scale, as in electrical appliances that are rated in terms of power or energy efficiency 53

reactants chemicals that react together in a chemical reaction 25, 30, 86, 89–94

reaction force an equal force that acts in the opposite direction to the action force 97, 100

reactor the part of a nuclear power station where energy is released from nuclear fuel 55, 111

real brightness a measure of the light emitted by a star compared to the Sun, taking into account how far away it is 42

receptor part of a neuron that detects stimuli and converts them into nerve impulses 15–16, 71–73, 76

receptor molecules found on the membrane of cells, they allow transmitter substances to bind to them 72, 76

recessive (allele) an allele that is only expressed if the other allele in the pair is also recessive; it is hidden if the other allele in the pair is dominant 5–7, 9

redox reaction the reaction for extracting metals from their ores, involving both oxidation and reduction 85

redshift the shift of lines in a spectrum towards the red (longer wavelength) end, due to the motion of the source away from us 43, 47

reduces when the atoms of a substance are oxidised resulting in the reduction of another substance 85

reduction process that reduces the amount of oxygen in a compound, or removes all the oxygen from it – the opposite of oxidation 25, 85, 88

reflect in the case of light, re-direction of the light wave, usually back to the point of origin from a shiny surface 48–50

reflex a muscular action that we make without thinking 72–73, 76

reflex arc pathway taken by nerve impulse from receptor, through the nervous system to effector, bringing about a reflex response 73, 76

relative atomic mass (RAM) the mass of an atom compared to the mass of an atom of carbon (which has a value of 12) 77, 86, 88, 90, 94

relative brightness the apparent brightness of a star as *seen* from Earth; a dim star close to Earth may appear brighter than a bright one that is further away 42

relative formula mass (RFM) the sum of the RAMs of all the atoms or ions in a compound 86, 88, 90, 94

relay neurons found in the CNS they connect sensory neurons to motor neurons and so co-ordinate the body's response to stimuli 73, 76

reproducibility the ability of the results of an experiment to be reproduced by another experimenter 29

repulsive a force that pushes two objects apart 97, 107

resistance (in an electric circuit) a measure of how hard it is for an electric current to flow through a material; (in biology) ability of an organism to resist death/disease/harm, for example resistance may develop in some microorganisms against antimicrobials 11–12, 16, 19, 98–99, 102–103, 106

respiration process occurring in living things where oxygen is used to release the energy in foods 59–61, 63–64, 15–18, 22

response the action taken as the result of a stimulus 15, 73–74, 76

resting heart rate a person's heart rate when inactive 14

resultant (force) the overall forces acting on an object added together 97–98, 100

ribosomes small structures in the cytoplasm that make proteins 68

risk the likelihood of a hazard causing harm 34, 36–41, 49, 52, 56–57, 89, 108, 110–111

S

salt generically, the dietary additive sodium chloride; in chemistry, an ionic compound formed when an acid neutralises a base 13, 15, 23, 33, 36–38, 41, 65, 77, 81, 84, 88–89, 91, 94

Sankey diagram diagram showing how the energy supplied to something is transferred into 'useful' or 'wasted' energy 54, 56

saturated fat a component of the diet that, when eaten in excess, can contribute to coronary heart disease and other health problems 13

seafloor spreading an extension of the seafloor caused by tectonic plate movement and the extrusion of magma between two plates which solidifies to form rock 23, 44

secondary coil the output coil of a transformer 105, 107

secondary energy source more convenient form of energy, such as electricity and refined fuels, produced from primary energy sources 53, 58

sediment particles of rock etc. in water that settle to the bottom 23, 28, 36, 44, 46

sedimentary rock rock formed when sediments are laid down and compacted together 23, 28, 36, 41, 44

sedimentation the settling of particles in water to the bottom 41, 44

seismic waves vibrations that pass through the Earth following an earthquake 45

selective breeding choosing organisms with desired characteristics to breed with one another 19

sensory neuron nerve cell carrying information from receptors to the central nervous system 71

serotonin a transmitter substance found at the synapses in the brain 72

sex cells the male and female gametes (sperm and eggs) 5

sex-determining gene a gene carried on the Y sex chromosome that causes a fetus to develop into a male 6

sexual reproduction reproduction of an organism that involves two parents 5

shells electrons are arranged in shells (or orbits) around the nucleus of an atom 77–78

short-term memory information from our most recent experiences, which is only stored for a short period of time 75

side effects unwanted effects produced by medicines 11–12

sievert (Sv) the SI unit of equivalent dose 108

signal information that is transmitted by, for example, an electrical current or an electromagnetic wave 51, 72

smog air pollution that is caused, for example by vehicle emissions and industrial fumes 26

solar system the planetary system around the Sun, of which the Earth is part 42

soluble able to dissolve (usually in water) 38, 81, 84, 92

specialised (cells) a cell that has a particular function 65

species basic category of biological classification, composed of individuals that resemble one another, can breed among themselves, but cannot breed with members of another species 17, 19–21

speed how fast an object travels, calculated using the equation: speed (metres per second) = distance/time 95

speed–time graph a visual way of showing how an object's speed changes over a period of time 96

sperm male sex cell of an animal 5

state symbols symbols used in equations to show whether something is solid (s), liquid (l), gas (g) or in solution in water (aq) 84

stem cells unspecialised body cells that can develop into other, specialised cells 8, 69

step down transformer device used to change the voltage of an a.c. supply to a lower voltage 105

step up transformer device used to change the voltage of an a.c. supply to a higher voltage 105

Glossary

sterile containing no living organisms 108

stiffness a measure of the resistance of a solid to bending forces 29

stimulus (*pl* stimuli) a change in the environment that causes a response by stimulating receptor nerve cells, e.g. a hot surface 72–73

stores the name given to different types of memory storage in the multi-store model 75

strong nuclear force a force that holds all the particles together in a nucleus of an atom 107

structural protein a protein, such as collagen, whose function is to build tissues 4

substrates the chemicals that enzymes work on 59

sulfur dioxide pollutant gas released from burning sulfur-containing fuels, which causes acid rain 24–27, 87

superbug harmful microorganism that has become resistant to antimicrobials 12

supercontinent very large land mass 36

supernova explosion of a large star at the end of its life 43

sustainability measure of whether a resource or process we use now will still be able to be used by future generations 21

sustainable resource or process that will still be available to future generations 21

S-waves transverse shock waves following an earthquake that cannot travel through the molten core of the Earth 45

switched off genes in body cells that are not active 69

switched on genes that are active, or turned on and control how a cell behaves and looks 69

synapse the gap between two adjacent neurons 72

synthetic material material manufactured from chemicals 30

T

tarnish the reaction that occurs when a metal reacts with oxygen in the air 79

tectonic plate section of Earth's crust that slowly moves relative to other plates 45

tensile strength a measure of the resistance of a solid to a pulling or stretching force 29

terminal velocity the maximum speed achieved by any object falling through a gas or liquid 98

theory a creative idea that may explain an observation and that can be tested by experimentation 19–20, 23, 36, 43, 44

therapeutic cloning a procedure in which a nucleus is removed from an egg and is replaced with a nucleus from a body cell in order to produce new cells with identical genes 69

thermistor a semiconductor device in which resistance changes with temperature 103

thermoplastic plastic with a shape that can be changed by heating 32

thermosetting plastic with a shape that becomes permanent after heating and cooling 32

tissue a group of cells that work together and carry out a similar task, such as lung tissue 65

tissue culture (in plants) a method of cloning by taking small pieces of plant tissue from the root or stem and treating it with enzymes to separate the cells, which then grow into separate identical plants 66

titration common laboratory method used to determine the unknown concentration of a known reactant 90

toxic a poison or hazardous substance that can cause serious medical conditions or death 80, 87

toxins poisons or hazardous substances 10, 75

transect a line of quadrats 108

transmitted radiation that passes through a material 108

transmitter substance a chemical that passes across a synapse 72

transparent a material that allows light to pass through 34

transverse a wave in which the disturbances are at right angles to the direction of energy transfer 45

trend the changes in a property across a period of the Periodic Table 78, 80

true value a theoretically accurate value that could be found if measurements could be made without errors 90

turbine device which makes a generator spin to generate electricity 55

U

Universe the whole of space and all the objects and energy within it 42–43

urea waste product excreted by the kidneys 15

V

vaccination medical procedure, usually an injection, that provides immunity to a particular disease 11

vaccine weakened microorganisms that are given to a person to produce immunity to a particular disease 11

vacuum a space where there are no particles of any kind 42, 48, 111

variable resistor a device that allows the amount of resistance in a circuit to be varied 102

variation differences between individuals belonging to the same species 4–5, 19

vasodilate increase in diameter of small blood vessels near the surface of the body to increase the flow of blood 15

veins blood vessels that carry blood from parts of the body back to the heart 13

velocity the speed of an object in a certain direction 95

velocity–time graph a visual way of showing direction of travel and acceleration 96

volcano landform from which molten rock erupts onto the surface 23, 44–45

volt the unit of voltage 53

voltage a measure of the energy carried by an electric current (*see* potential difference) 102, 104–105

voltmeter a device used to measure the voltage across a component 102

W

watt (W) unit of power, or rate of transfer of energy, equal to a joule per second 53

wave a periodic disturbance that transfers energy 45–46, 48–49, 51

wave equation the speed of a wave is equal to its frequency multiplied by its wavelength 46

wave technology in renewable energy, equipment that allows us to harness the power of ocean waves 56

wavelength distance between two successive wave peaks (or troughs, or any other point of equal disturbance) 43, 46–47

white blood cell blood cell that defends the body against disease 10

wind turbine device that uses the energy in moving air to turn an electricity generator 56, 58

work work is done when a force moves an object 99–100

X

X-rays ionising electromagnetic radiation 48–49, 52

xylem cells plant cells that carry water and mineral salts to where they are needed 65, 70

Z

zygote the cell formed by the fusion of a male and female gamete at fertilisation 5, 65, 70

Data sheet

Fundamental physical quantity	Unit(s)
length	metre (m); kilometre (km); centimetre (cm); millimetre (mm); nanometre (nm)
mass	kilogram (kg); gram (g); milligram (mg)
time	second (s); millisecond (ms); year (a); million years (Ma); billion years (Ga)
temperature	degree Celsius (°C); kelvin (K)
current	ampere (A); milliampere (mA)

Prefixes for units			
nano (n)	one thousand millionth	0.000 000 001	$\times 10^{-9}$
micro (μ)	one millionth	0.000 001	$\times 10^{-6}$
milli (m)	one thousandth	0.001	$\times 10^{-3}$
kilo (k)	$\times$ one thousand	1 000	$\times 10^{3}$
mega (M)	$\times$ one million	1 000 000	$\times 10^{6}$
giga (G)	$\times$ one thousand million	1 000 000 000	$\times 10^{9}$
tera (T)	$\times$ one million million	1 000 000 000 000	$\times 10^{12}$

Useful equations
speed = distance travelled ÷ time taken
acceleration = change in velocity ÷ time taken
momentum = mass $\times$ velocity
change of momentum = resultant force $\times$ time it acts
change in gravitational potential energy = weight $\times$ height difference
kinetic energy = ½ $\times$ mass $\times$ [velocity]^2
resistance = voltage $\times$ current
energy transferred = power $\times$ time
electrical power = voltage $\times$ current
efficiency = (energy usefully transferred ÷ total energy supplied) $\times$ 100%

Data sheet

Tests for negatively charged ions

Ion	Test	Observation
carbonate CO_3^{2-}	add dilute acid	effervesces, and carbon dioxide gas produced (the gas turns lime water milky)
chloride (in solution) Cl^-	acidify with dilute nitric acid, then add silver nitrate solution	white precipitate
bromide (in solution) Br^-	acidify with dilute nitric acid, then add silver nitrate solution	cream precipitate
iodide (in solution) I^-	acidify with dilute nitric acid, then add silver nitrate solution	yellow precipitate
sulfate (in solution) SO_4^{2-}	acidify, then add barium chloride solution or barium nitrate solution	white precipitate

Tests for positively charged ions

Ion	Test	Observation
calcium Ca^{2+}	add sodium hydroxide solution	white precipitate (insoluble in excess)
copper Cu^{2+}	add sodium hydroxide solution	light blue precipitate (insoluble in excess)
iron(II) Fe^{2+}	add sodium hydroxide solution	green precipitate (insoluble in excess)
iron(III) Fe^{3+}	add sodium hydroxide solution	red-brown precipitate (insoluble in excess)
zinc Zn^{2+}	add sodium hydroxide solution	white precipitate (soluble in excess, giving a colourless solution)

Formulae of some common molecules and compounds*

H_2	hydrogen gas	CH_4	methane	KCl	potassium chloride
O_2	oxygen gas	NH_3	ammonia	MgO	magnesium oxide
N_2	nitrogen gas	H_2SO_4	sulfuric acid	$Mg(OH)_2$	magnesium hydroxide
H_2O	water	HCl	hydrochloric acid	$MgCO_3$	magnesium carbonate
Cl_2	chlorine gas	HNO_3	nitric acid	$MgCl_2$	magnesium chloride
CO_2	carbon dioxide	NaCl	sodium chloride	$MgSO_4$	magnesium sulfate
CO	carbon monoxide	NaOH	sodium hydroxide	$CaCO_3$	calcium carbonate
NO	nitrogen monoxide	Na_2CO_3	sodium carbonate	$CaCl_2$	calcium chloride
NO_2	nitrogen dioxide	$NaNO_3$	sodium nitrate	$CaSO_4$	calcium sulfate
SO_2	sulfur dioxide	Na_2SO_4	sodium sulfate		

* These will not be provided in your exam. You need to learn them.

The Periodic Table

Key

relative atomic mass
1
H
hydrogen
1

atomic symbol
name
atomic (proton) number

Group 1	Group 2											Group 3	Group 4	Group 5	Group 6	Group 7	Group 0
																	4 **He** helium 2
7 **Li** lithium 3	9 **Be** beryllium 4											11 **B** boron 5	12 **C** carbon 6	14 **N** nitrogen 7	16 **O** oxygen 8	19 **F** fluorine 9	20 **Ne** neon 10
23 **Na** sodium 11	24 **Mg** magnesium 12											27 **Al** aluminium 13	28 **Si** silicon 14	31 **P** phosphorus 15	32 **S** sulfur 16	35.5 **Cl** chlorine 17	40 **Ar** argon 18
39 **K** potassium 19	40 **Ca** calcium 20	45 **Sc** scandium 21	48 **Ti** titanium 22	51 **V** vanadium 23	52 **Cr** chromium 24	55 **Mn** manganese 25	56 **Fe** iron 26	59 **Co** cobalt 27	59 **Ni** nickel 28	63.5 **Cu** copper 29	65 **Zn** zinc 30	70 **Ga** gallium 31	73 **Ge** germanium 32	75 **As** arsenic 33	79 **Se** selenium 34	80 **Br** bromine 35	84 **Kr** krypton 36
85 **Rb** rubidium 37	88 **Sr** strontium 38	89 **Y** yttrium 39	91 **Zr** zirconium 40	93 **Nb** niobium 41	96 **Mo** molybdenum 42	[98] **Tc** technetium 43	101 **Ru** ruthenium 44	103 **Rh** rhodium 45	106 **Pd** palladium 46	108 **Ag** silver 47	112 **Cd** cadmium 48	115 **In** indium 49	119 **Sn** tin 50	122 **Sb** antimony 51	128 **Te** tellurium 52	127 **I** iodine 53	131 **Xe** xenon 54
133 **Cs** caesium 55	137 **Ba** barium 56	139 **La*** lanthanum 57	178 **Hf** hafnium 72	181 **Ta** tantalum 73	184 **W** tungsten 74	186 **Re** rhenium 75	190 **Os** osmium 76	192 **Ir** iridium 77	195 **Pt** platinum 78	197 **Au** gold 79	201 **Hg** mercury 80	204 **Tl** thallium 81	207 **Pb** lead 82	209 **Bi** bismuth 83	[209] **Po** polonium 84	[210] **At** astatine 85	[222] **Rn** radon 86
[223] **Fr** francium 87	[226] **Ra** radium 88	[227] **Ac*** actinium 89	[261] **Rf** rutherfordium 104	[262] **Db** dubnium 105	[266] **Sg** seaborgium 106	[264] **Bh** bohrium 107	[277] **Hs** hassium 108	[268] **Mt** meitnerium 109	[271] **Ds** darmstadtium 110	[272] **Rg** roentgenium 111							

Elements with atomic numbers 112–116 have been reported but not fully authenticated.

* The Lanthanides (atomic numbers 58–71) and the Actinides (atomic numbers 90–103) have been omitted.
Cu and Cl have not been rounded to the nearest whole number.

Exam tips

The key to successful revision is finding the method that suits you best. There is no right or wrong way to do it.

Before you begin, it is important to plan your revision carefully. If you have allocated enough time in advance, you can walk into the exam with confidence, knowing that you are fully prepared.

Start well before the date of the exam, not the day before!

It is worth preparing a revision timetable and trying to stick to it. Use it during the lead up to the exams and between each exam. Make sure you plan some time off too.

Different people revise in different ways, and you will soon discover what works best for you.

Remember!

There is a difference between *learning* and *revising*.

When you revise, you are looking again at something you have already learned. Revising is a process that helps you to remember this information more clearly.

Learning is about finding out and understanding new information.

Some general points to think about when revising

- Find a quiet and comfortable space at home where you won't be disturbed. You will find you achieve more if the room is ventilated and has plenty of light.

- Take regular breaks. Some evidence suggests that revision is most effective when tackled in 30 to 40 minute slots. If you get bogged down at any point, take a break and go back to it later when you are feeling fresh. Try not to revise when you're feeling tired. If you do feel tired, take a break.

- Use your school notes, textbook and this Revision guide.

- Spend some time working through past papers to familiarise yourself with the exam format.

- Produce your own summaries of each module and then look at the summaries in this Revision guide at the end of each module.

- Draw mind maps covering the key information on each topic or module.

- Review the **Grade booster checklists** on page 255–272.

- Set up revision cards containing condensed versions of your notes.

- Prioritise your revision of topics. You may want to leave more time to revise the topics you find most difficult.

Workbook

The **Workbook** (pages 145–254) allows you to work at your own pace on some typical exam-style questions. These are graded to show the level you are working to (G–E, D–C or B–A*). You will find that the actual GCSE questions are more likely to test knowledge and understanding across topics. However, the aim of the Revision guide and Workbook is to guide you through each topic so that you can identify your areas of strength and weakness.

The Workbook also contains example questions that require longer answers (**Extended response questions**). You will find one question that is similar to these in each section of your written exam papers. The quality of your written communication will be assessed when you answer these questions in the exam, so practise writing longer answers, using sentences. The **Answers** to all the questions in the Workbook are detachable for flexible practice and can be found on pages 273–288.

Collins
Workbook

NEW GCSE SCIENCE

Science A and Additional Science A

OCR

Twenty First Century Science

Authors: John Beeby
Michael Brimicombe
Brian Cowie
Sarah Mansel
Ann Tiernan

Revision Guide +
Exam Practice Workbook

What genes do

G–E

1 Genes help to determine your characteristics and make you who you are.

   a Write down the name of the part of the cell in which genes are found. .... [1 mark]

   b Write down the name of the chemical that makes up your genes. .... [1 mark]

D–C

2 Write down one example of:

   a a structural protein: .... [1 mark]

   b a functional protein: .... [1 mark]

B–A*

3 The Human Genome Project researched the genes found on human chromosomes.

   a Explain **one benefit** of the Human Genome Project.

   ....

   .... [2 marks]

   b Explain **one ethical consideration** of the Human Genome Project.

   ....

   ....

   .... [2 marks]

G–E

4 Look at the drawing of Jodie.

pink hair — hair is straight — pale skin colour has been changed by tanning — blue eyes — scar — decayed tooth — pierced ear — dimple

   a Write down **two** characteristics controlled by Jodie's **genes**.

   ....

   .... [2 marks]

   b Write down **two** characteristics that have been caused by Jodie's **environment**.

   ....

   .... [2 marks]

   c Write down one characteristic that is a result of Jodie's genes being affected by her environment. Explain your answer.

   .... [2 marks]

D–C

5 Write down a characteristic controlled by more than one pair of genes. Describe how this will affect the appearance of the characteristic in the population.

   ....

   .... [2 marks]

B–A*

6 Explain the difference between the terms genotype and phenotype for a named characteristic in humans.

   ....

   ....

   ....

   .... [3 marks]

B–A*

7 Here are some statements about identical twins. Put a tick (✓) in the box next to the correct statement.

   a They have slightly different combination of genes in their cells. ☐

   b If separated at birth, they will still look identical as they grow. ☐

   c Their appearance can be affected by their environment. ☐

   d Their hair colour is a feature always dependent on their genes. ☐   [1 mark]

Genes working together and variation

1 Here are some statements made by students about chromosomes. Some are correct, while some are incorrect. Put ticks (✓) in the boxes next to the **two** correct statements.

a There are 46 chromosomes in human cells. ☐

b Egg cells contain 23 pairs of chromosomes. ☐

c Sex cells contain one chromosome from each pair. ☐

d The genes for a characteristic are in different places on each chromosome of the pair. ☐ **[2 marks]**

G–E

2 Mutations can occur in humans that result in abnormalities in a baby.

a Describe how a chromosome mutation can occur.

....

.... **[2 marks]**

D–C

b Write down one example of a chromosome mutation. Genetically, how is this person different from an individual without the chromosome mutation?

....

.... **[2 marks]**

3 Explain why you are similar to, but different from, your parents.

You are similar because:

....

....

.... **[3 marks]**

B–A*

You are different because:

....

....

.... **[4 marks]**

4 Write down the definition of an allele.

....

.... **[1 mark]**

D–C

5 The following diagrams of chromosomes show the combinations possible for the alleles, with dimples, D, and without dimples, d.

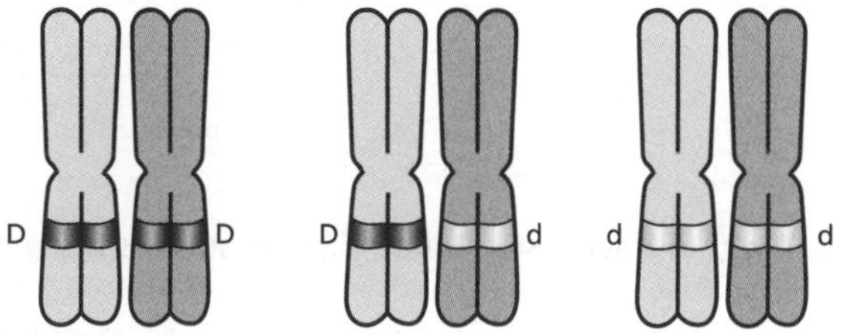

D   D   D   d   d   d

B–A*

With reference to the diagrams, explain the terms **homozygous** and **heterozygous** for the inheritance of dimples.

....

....

....

.... **[4 marks]**

Genetics crosses and sex determination

1 Use the words provided to complete the sentences. You can use the words once, or not at all.

alleles   bases   chromosomes   different positions

DNA   each chromosome   identical positions   one chromosome

Genes for a particular trait are located at ....

on .... of the chromosome pair. Different versions of

genes are called ....   **[3 marks]**

2 People are either able to roll their tongue into a U-shape, or unable to roll their tongue. The diagram shows a pair of chromosomes.

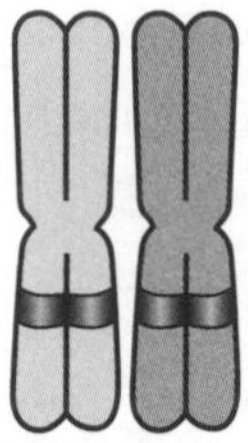

a Is the allele for tongue rolling dominant or recessive? How do you know?

.... **[2 marks]**

b Write down the other possible genotypes related to tongue rolling.

.... **[2 marks]**

c Write down the phenotypes for these genotypes.

.... **[2 marks]**

Genotype TT: can roll tongue

3 A man and woman who can both roll their tongues have children. Use the information in the genotype diagram above to answer the questions.

a Write down the possible genotypes of the couple. .... **[2 marks]**

b The couple have children. Their first child cannot roll his tongue; the second one can.

  i What does this tell you about the genotypes of the couple? Explain your answer.

....

.... **[3 marks]**

  ii Show the genetic cross involved. Use a Punnett square to illustrate your answer.

                                                  [3 marks]

4 Explain why some embryos develop into male babies.

....

....

....

.... **[3 marks]**

5 Explain why males are more likely to have a sex-linked disease such as haemophilia or colour blindness.

....

....

....

....

.... **[5 marks]**

Gene disorders, carriers and genetic testing

1 Some diseases are caused by defective genes. Use the words provided to complete the sentences.

concentrating   **cystic fibrosis**   **digesting food**

Huntingdon's disease   **recessive**   **dominant**

a For some genetic disorders a single allele is sufficient to cause the disorder. These are

called .... disorders. One example of this is ...., where symptoms

include tremors, memory loss and difficulty in ....   **[3 marks]**

b In other types of genetic disorder, two alleles are needed for the disease to occur. These are

called .... disorders. One example is ...., where symptoms include

difficulty in breathing and difficulty in ....   **[3 marks]**

G–E

2 About 1 in every 10 000 babies born will have inherited the allele for Huntington's disease.

In the UK, around 700 000 babies are born every year. How many of these would you expect
to have inherited the allele for Huntington's disease? Show your working.

....

....

....   **[2 marks]**

D–C

3 Genetic testing is carried out on adults, children and unborn fetuses.

a Write down two reasons for carrying out genetic testing on **children** or **adults**.

....

....   **[2 marks]**

b Write down two genetic testing techniques used during pregnancy.

....

....   **[2 marks]**

D–C

4 Genetic screening programmes have advantages, but they also have ethical implications.

a Explain the implications of genetic screening programmes carried out:

i on week-old babies in the heal prick blood test:

....

....   **[2 marks]**

ii by employers and insurance companies:

....

....   **[2 marks]**

b Explain the advantages of testing embryos (by pre-implantation genetic diagnosis) before
implanting the embryos following in vitro fertilisation.

....

....

....   **[2 marks]**

B–A*

Cloning and stem cells

1 There are advantages and disadvantages to asexual reproduction.

   a Explain **one** advantage. .... **[1 mark]**

   b Explain **one** disadvantage. ....

   .... **[2 marks]**

D–C

2 Describe the technique used to clone Dolly the sheep artificially.
Annotate the diagram to illustrate your answer.

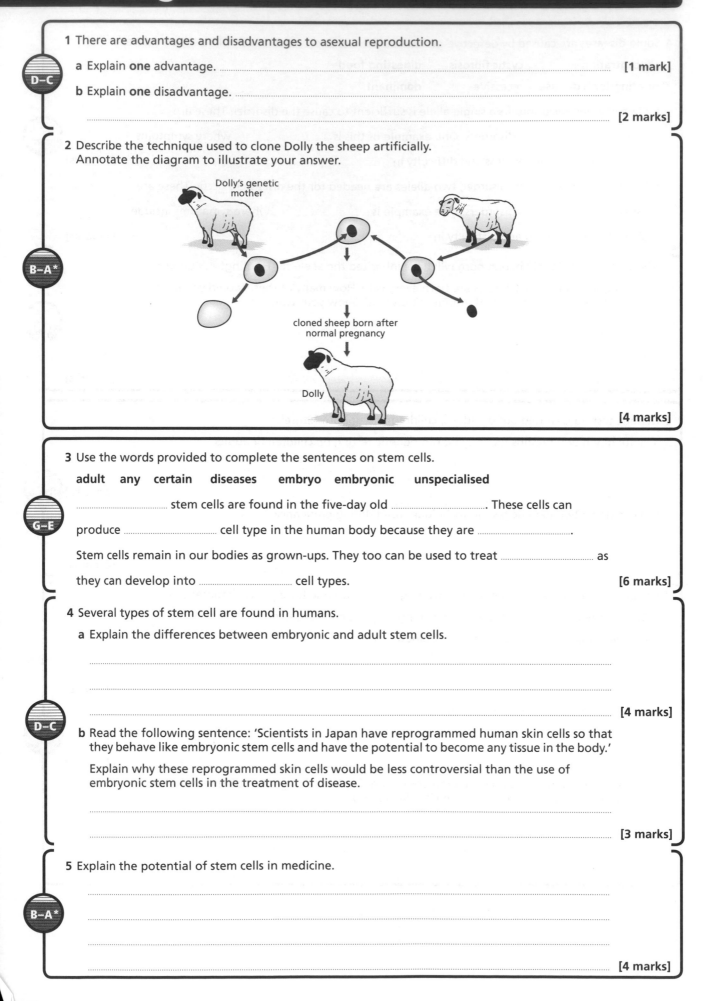

Dolly's genetic mother

cloned sheep born after normal pregnancy

Dolly

[4 marks]

B–A*

3 Use the words provided to complete the sentences on stem cells.

adult   any   certain   diseases   embryo   embryonic   unspecialised

.... stem cells are found in the five-day old .... These cells can

produce .... cell type in the human body because they are ....

Stem cells remain in our bodies as grown-ups. They too can be used to treat .... as

they can develop into .... cell types. **[6 marks]**

G–E

4 Several types of stem cell are found in humans.

   a Explain the differences between embryonic and adult stem cells.

   ....

   ....

   .... **[4 marks]**

   b Read the following sentence: 'Scientists in Japan have reprogrammed human skin cells so that
they behave like embryonic stem cells and have the potential to become any tissue in the body.'

   Explain why these reprogrammed skin cells would be less controversial than the use of
embryonic stem cells in the treatment of disease.

   ....

   .... **[3 marks]**

D–C

5 Explain the potential of stem cells in medicine.

....

....

....

.... **[4 marks]**

B–A*

B1 Extended response question

A couple visit their doctor. Their first child was born with a single gene disorder called phenylketonuria (PKU). The phenylketonuria allele is recessive. The couple are both healthy, but the disorder has been seen before in both families.

The doctor first explains to the family how, as healthy individuals, they could have produced a child with phenylketonuria. She draws a Punnett square to help with this explanation.

The couple see the doctor's diagram and think because they've had one child with PKU, there will be less chance, next time, of having a child with the disorder. The doctor explains why this isn't true.

Use your knowledge and understanding of genetics to write out the explanations that the doctor gives to the couple.

The quality of written communication will be assessed in your answer to this question.

[6 marks]

Microbes and disease

1 Here are some statements about microorganisms and disease. Some statements are correct, while some are incorrect. Put ticks (✓) in the boxes next to the **two** correct statements.

G–E

Bacteria make us feel ill by releasing toxins. ☐

Microorganisms that cause disease are called pathogens. ☐

The only microorganisms that cause disease are bacteria and viruses. ☐

Diseases caused by viruses include colds and TB. ☐ **[2 marks]**

2 A person has an accident and 10 cells of the flesh-eating bacterium *Streptococcus pyogenes* enter the wound. At human body temperature, the flesh-eating bacterium can divide into two every 10 minutes. For every million bacteria, 1 cm² of flesh is eaten away. What area of flesh would be eaten 3 hours after infection?

D–C

....

....

.... **[2 marks]**

3 The diagram below shows the growth curve of a bacterium.

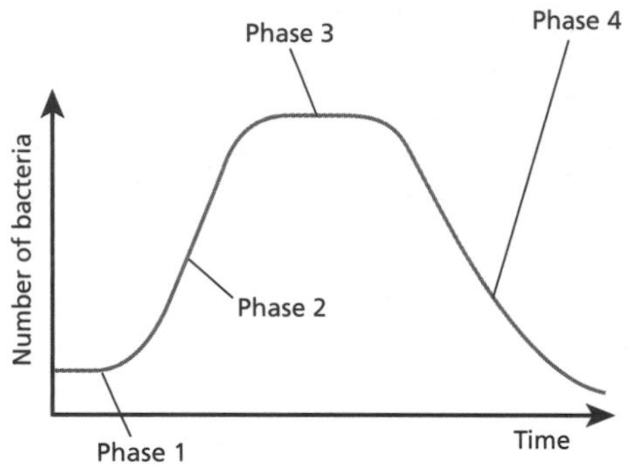

B–A*

a On the diagram, label the four phases of bacterial growth. **[1 mark]**

b Describe what is happening in each phase.

Phase 1: ....

Phase 2: ....

Phase 3: ....

Phase 4: .... **[4 marks]**

4 Here are some statements about ways in which the body has external barriers to microorganisms. Choose from the words provided to complete the sentences.

acid   alkali   intestines   oil

saliva   skin   stomach   tears

D–C

Our .... is able to heal itself, and produces ...., which can stop

bacteria from reproducing. Chemicals in our .... and .... also have

antibacterial properties. Our .... produces ...., which destroys many

bacteria in our digestive system. **[6 marks]**

B2 Keeping healthy

Vaccination

1 Here are some statements made by students about vaccines. Put a tick (✓) in the box next to the **one** correct statement.

Vaccines are always made from dead microorganisms. ☐

The microorganisms in vaccines have antibodies on their surface. ☐

Memory cells are left in the blood after vaccination. ☐

An infecting microorganism is destroyed very quickly by antigens in the blood. ☐

G–E

[1 mark]

2 Courses of vaccinations are given to babies and children.

a Write down what a vaccination against a microorganism does.

.... [1 mark]

b Explain why a course of vaccinations is given to babies and children.

.... [1 mark]

D–C

c Explain why a single flu vaccine would not give permanent protection against the disease, and new ones have to be developed.

....

....

.... [3 marks]

3 After a science lesson about vaccinations, some friends are discussing the use of vaccines to eradicate disease. Here are some quotes:

William: 'Using vaccines, we've already eradicated one disease from the world.'

Xavier: 'We could do this for others if we vaccinated the whole population.'

Yvonne: 'But vaccines are very often unsafe.'

Zak: 'And we can't vaccinate against all diseases.'

Which statements are correct? Use these ideas to discuss whether vaccinations could eradicate infectious diseases globally.

B–A*

....

....

....

....

....

....

.... [5 marks]

4 Some people get unwanted reactions when they are given a vaccination or a prescription drug.

a What is this type of unwanted reaction called?

.... [1 mark]

G–E

b Why is this type of reaction more severe in some people than others?

.... [1 mark]

5 Write down two effects of antimicrobials that are effective on microorganisms.

....

B–A*

.... [2 marks]

Safe protection from disease

D–C

1 a Some bacteria develop resistance to antibiotics. Explain how this resistance develops.

....

....

.... [5 marks]

b Write down three recommendations for the use of antibiotics.

....

....

.... [3 marks]

2 The resistance that some bacteria show to antibiotics was investigated in Belgium between 1985 and 2008.

- The graph below left shows the resistance of the bacterium that causes pneumonia to four antibiotics.
- The graph below right shows the total antibiotic use between 1997 and 2007, in packages per 100 000 inhabitants per day.

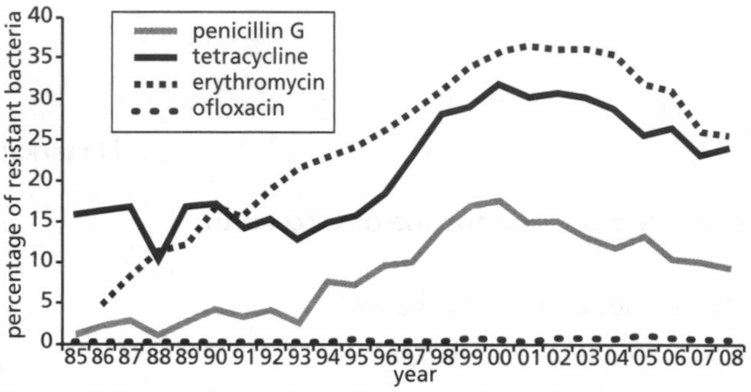

B–A*

(Source: National Reference Centre S. *pneumoniae*, University *Leuven*)

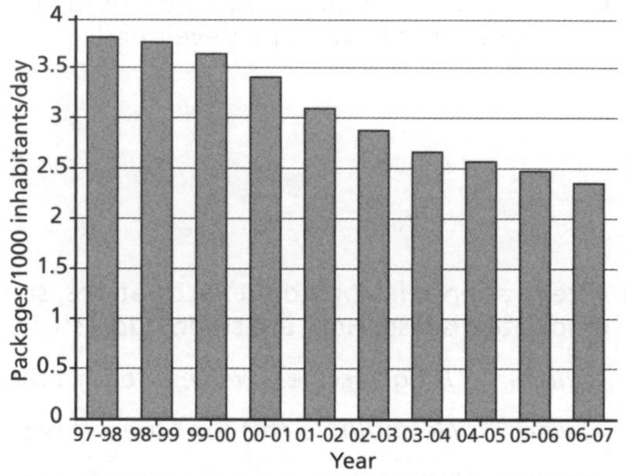

(Source: Goossens *et al*. (2008). Achievements of the Belgian Antibiotic Policy Coordination Committee. *Eurosurveillance* **13**, 10–13)

a Describe the trends shown in resistance of the bacterium to the antibiotics in the left-hand graph.

....

....

.... [5 marks]

b What can you conclude from the graph of total antibiotic use?

.... [1 mark]

c What can you conclude from the two graphs together?

....

.... [2 marks]

G–E

3 Describe **two** types of clinical trials.

....

.... [2 marks]

B–A*

4 Three types of clinical trial used are open-label, blind study and double-blind study. Explain why the double-blind test is considered to generate the most reliable results.

....

....

.... [3 marks]

The heart

1 Use the words provided to complete the sentences about the heart and circulatory system. You can use the words once, or not at all.

blood   double pump   half   heart   nutrients   oxygen

quarter   single pump   veins   wastes   lungs   blood vessels

a The blood carries .... and .... to the body's cells.

It removes .... from the body's cells.   **[3 marks]** G–E

b The heart acts as a .... As one .... of the heart is carrying blood

to the body, the other is carrying blood to the ....   **[3 marks]**

c The circulatory system is made up of the ...., ... and

....   **[3 marks]**

2 Complete the table to explain the structure and function of arteries, capillaries and veins.

	Function	How their structure is related to function
Arteries		
Capillaries		
Veins		

[6 marks] D–C

3 Heart disease kills more people in the UK than any other disease. Explain what coronary heart disease is.

....

....   **[2 marks]** G–E

4 Trends in heart disease are studied across the world. The following graph shows male deaths from CHD for different age groups, from 1968 to 2008.

(Source: The British Heart Foundation)

a Describe the trend in heart disease in men, from 1968 to 2008.

....

....   **[2 marks]**

b In which group has the change been greatest? ....   **[1 mark]**

c Suggest three factors that could have led to this change.

....

....   **[3 marks]**

d Trends in data suggest that the more deprived the socio-economic group, the greater the risk of death from CHD. How might scientists look for evidence of factors that may be involved?

....   **[1 mark]**

B–A*

1 Write down **two** possible consequences of leaving high blood pressure untreated.

....

....

[2 marks]

2 Write down **three** lifestyle factors that can lead to heart disease.

....

....

[3 marks]

3 Here are some statements about the measurements scientists make of our cardiovascular fitness.

Some statements are correct, while some are incorrect. Put ticks (✓) in the boxes next to the **two** correct statements.

Pulse rate is measured in beats per minute. ☐

Blood pressure is measured using a sphygmomanometer. ☐

The resting heart rate in teenagers is 50–70. ☐

Blood pressure measurements indicate the pressure in our veins. ☐

The fitter we are, the faster our pulse rate during exercise. ☐ **[2 marks]**

4 One epidemiological study carried out in the USA looked at the effects of passive smoking (being exposed to and inhaling someone else's cigarette smoke) on groups of female nurses. The results are shown below.

Type of coronary heart disease (CHD)	Number of cases		
	Never exposed to cigarette smoke	Occasional exposure to cigarette smoke	Regular exposure to cigarette smoke
Non-fatal heart attack	14	63	50
Fatal heart attack	3	11	11

(Source: Kawachi I., Colditz G.A., Speizer F.E, Manson JoAnn, Stampfer M.J. Willett W.C, and Hennekens C.H. (1997). A Prospective Study of Passive Smoking and Coronary Heart Disease. *Circulation* 95 2374-2379)

a What was the total number of CHD cases in the study:

i in nurses not exposed to cigarette smoke? .... **[1 mark]**

ii in nurses exposed to any amount of cigarette smoke? .... **[1 mark]**

b How many times more likely to get CHD were the nurses who were exposed to smoke, compared with those exposed to no smoke?

....

[1 mark]

c What conclusions can be drawn from the study about the effect of passive smoking on non-fatal heart attacks?

....

....

....

[2 marks]

d Why are epidemiological studies of this type difficult to carry out?

....

....

....

....

[3 marks]

Keeping things constant

1 Write down a definition of **homeostasis**.

.... [1 mark]

2 A person develops hypothermia. Which factor in the body has been allowed to change? Put a tick (✓) in the box next to the correct answer.

Blood pressure ☐   Water ☐

Blood sugar ☐   Temperature ☐   Salt ☐   [1 mark]

G–E

3 When running, Jane's body produces more carbon dioxide. Complete the flow chart below to show the control systems involved in homeostasis.

[ ] ➡ [ ] ➡ [ ]   [3 marks]

D–C

4 a Atiq's body temperature rises during strenuous exercise. On the diagram below, complete the labels to show how the changes in his body return his temperature to normal.

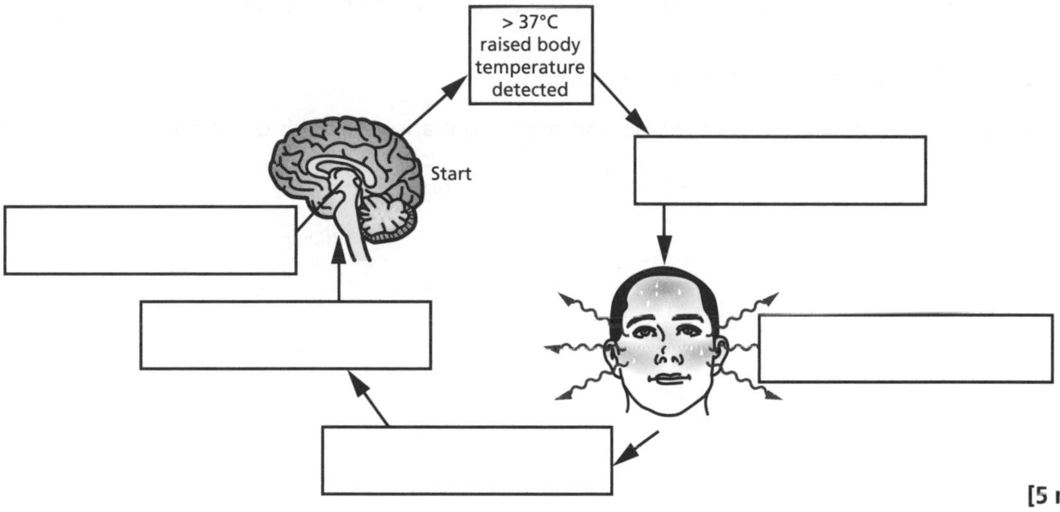

> 37°C raised body temperature detected

Start

[ ]

[ ]

[ ]

[ ]

[ ]

[5 marks]

B–A*

b What happens when Atiq's body temperature returns to normal?

....

.... [2 marks]

c Name the control mechanism that reverses the changes that have put Atiq's body system off balance?

.... [1 mark]

5 Explain what happens to our blood plasma and urine if we:

a drink a lot of water: ....

.... [3 marks]

G–E

b eat a lot of salt: ....

.... [3 marks]

6 a Anti-diuretic hormone (ADH) helps us to control the water balance of our bodies. How does ADH work?

.... [2 marks]

B–A*

b Explain the effect that alcohol has on the water balance of the body.

....

.... [3 marks]

B2 Extended response question

Gary sees this article in a magazine.

In Europe prior to 1940, the numbers of deaths from many diseases, including diphtheria, were negligible. But after mass vaccination programmes against diseases began, wave after wave of epidemics followed in fully vaccinated people, with many deaths. Vaccinations are harmful!

Gary does not think that this is correct. He does some research and finds this table in the Department of Health's Vaccination 'Green Book'.

Disease	Total deaths in 2003	Deaths in the UK before the vaccination was introduced (year of introduction)
Diphtheria	0	2133 (1939)
Hib	14	24 (1991)
Measles	0	99 (1967)
Meningococcal C	19	101 (1998)
Polio	0	241 (1955)
Whooping cough	2	92 (1956)

(Source: The Department of Health (2006). Immunisation against infectious disease. ('The Green Book'). The Department of Health.)

Use the data to discuss the claims made in the book.

The quality of written communication will be assessed in your answer to this question.

[6 marks]

Species adaptation, changes, chains of life

1 Species of plants and animals are adapted to their environment.

   a Write down a definition of a species.

   .... **[1 mark]**

   b A cactus is adapted to live in hot, dry conditions. Write down three adaptations and explain how they enable the cactus to live there.

     i ....

     ii ....

     iii .... **[3 marks]**

G–E

2 The diagram below shows a typical food chain of a moorland.

A high concentration of insecticide sprayed onto farmland near an area of moorland ends up on the moorland and kills most of the insects and spiders. Describe what will happen to:

   i the foxes ....

   .... **[2 marks]**

   ii the insect eating birds ....

   .... **[2 marks]**

   iii the lizards ....

   .... **[3 marks]**

   iv the owls ....

   .... **[2 marks]**

B–A*

3 Write down one reason why species become extinct.

.... **[1 mark]**

G–E

4 a In a grassland food web, grasshoppers transfer 111 MJ of energy into their bodies during the summer. Of that energy, 94 MJ is lost during respiration and in faeces. What is the efficiency of energy transfer?

.... **[1 mark]**

D–C

   b When the grasshoppers die, their remains store 16 MJ of energy. Explain what happens to this energy on their death.

.... **[1 mark]**

B–A*

1 Carbon is recycled through the environment in the carbon cycle. Choose from the words provided to complete the sentences.

air   bacteria   carbon dioxide   combustion   decomposition

fixed   fungi   photosynthesis   respiration   soil

G–E

a Carbon enters the carbon cycle as .... from the .... Plants use this

to produce food by the process of .... Carbon is said to be .... by

this process.   **[4 marks]**

b When obtaining energy from food, organisms produce carbon dioxide by the process

of .... As organisms die and are broken down, this gas is also returned to the

air by a process of .... This involves microorganisms such as ....

and .... that live in the ....   **[5 marks]**

2 Explain the processes of nitrogen-fixation and denitrification.

B–A*

....

....

....   **[2 marks]**

3 Here are some statements about living and non-living indicators to monitor environmental change. Some are correct, while some are incorrect. Put ticks (✓) in the boxes next to the two correct statements.

D–C

Lichens are biological indicators that give information on air temperature.   ☐

Mayfly larvae are good biological indicators because they can only live in clean water.   ☐

Nitrates in the water and air are non-biological indicators that are monitored.   ☐

Monitoring carbon dioxide gives information on climate change.   ☐   **[2 marks]**

4 The following graph shows the carbon dioxide concentration in the air measured at the monitoring station of Mauna Loa, in Hawaii.

Weekly average CO_2 concentrations derived from continuous data at Mauna Loa, Hawaii

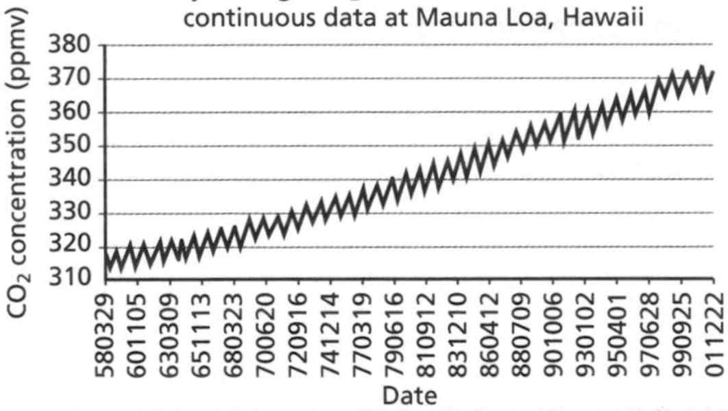

(year, month and day, e.g. 760610 indicates June 10th, 1976)

(Source: Keeling, C.D. and T.P. Whorf. (2004) Atmospheric CO_2 concentrations derived from flask air samples at sites in the SIO network. *In Trends: A Compendium of Data on Global Change.* Carbon Dioxide Information Analysis Center, Oak Ridge, USA.)

B–A*

a Describe how levels in carbon dioxide in the air change from 1958 to 2001. Suggest why these changes have occurred.

....

....

....

....   **[4 marks]**

b Scientists also measure levels of carbon dioxide from other sources. Write down two of these sources.

....

....   **[2 marks]**

Variation and selection

1 Choose from the words provided to complete the sentences:

complex   evolution   genetics   million   simple   thousand   variation

Life began around 3500 .... years ago with .... forms of life.

More .... forms of life developed later. The process by which new species emerge

over a very long period of time is called .... The changes involved in this process

have come about by .... between individuals. **[5 marks]**

G–E

2 Here are some statements about fossils. Put a tick (✓) in the box next to the **one** correct statement.

It is not usually possible to find out the age of fossils. ☐

The fossils we find are mostly of animals that are alive today; few are extinct. ☐

Fossils are the remains of dead plants and animals. ☐

Fossils of eggs that have been found are where the eggs have been turned into rock. ☐ **[1 mark]**

3 Explain what a mutation is, and how mutations can lead to changes in a population of organisms over a period of time.

....

....

.... **[5 marks]**

D–C

4 Explain how changes in the gene pool lead to evolution.

....

....

....

.... **[5 marks]**

B–A*

5 An animal breeder is trying to develop a new breed of cattle that produces large volumes of lower fat milk. Explain how the cattle breeder would use selective breeding to achieve this.

....

....

.... **[4 marks]**

D–C

6 Scientists in Germany investigated the effectiveness of the poisons warfarin and bromadiolone on rats. Some of their results are shown in the table opposite.

a Explain how the data are evidence for natural selection.

Town	Not resistant to either poison (%)	Resistant to warfarin (%)	Resistant to both poisons (%)
Dorsten	44	56	0
Drensteinfurt	90	5	5
Ludwigshafen	100	0	0
Olfen	21	21	58
Stadtlohn	5	8	87

(Source: Kohn M.H, Pelz H-J, and Wayne R.K (2000). Natural selection mapping of the warfarin-resistance gene. *Proc. Natl. Acad. Sci. USA*. **97** 7911–7915)

B–A*

....

....

.... **[4 marks]**

b Suggest which town had used the rat poison the most prior to the study. Explain your answer.

....

.... **[2 marks]**

Evolution, fossils and DNA

1 The following statements are the stages in the development of a new species. They are not in the correct order. Put numbers in the boxes to place the sequence into order.

D–C

a The environments are different, so natural selection operates differently on the populations. ☐

b Two populations of a species become isolated by a physical barrier. ☐

c Individuals from the different populations become unable to reproduce with each other. ☐

d Different alleles become more frequent in different populations. ☐

e A new species has been produced. ☐ **[5 marks]**

2 Explain how Darwin's observations of the animal and plant life on the Galapagos Islands, volcanic islands off the coast of South America, led to his 'theory of evolution by natural selection'.

B–A*

....

....

....

.... **[4 marks]**

3 Choose from the words provided to complete the sentences about classifying organisms.

ancestor   appearance   classified   DNA   evolved   groups

G–E

Organisms are .... by putting them into .... Previously, many

scientists grouped organisms mainly on their ...., but now they look at the

organisms' .... Using this type of grouping helps us to understand how

organisms .... over thousands or millions of years. **[5 marks]**

4 Explain how evidence from the fossil record and DNA provides evidence for evolution.

D–C

....

....

....

.... **[3 marks]**

5 A French scientist called Lamarck thought that animals changed during their lifetime, and these characteristics would be passed on to their young. For example, as a giraffe stretched to reach leaves on a tree, its neck would get longer. The giraffe would pass on its longer neck to its offspring.

Explain why Darwin's description of the origin of new species is a better explanation.

B–A*

....

....

....

....

.... **[5 marks]**

Biodiversity and sustainability

1 Here are some statements about biodiversity. Put a tick (✓) in the box next to the **one** correct statement.

A habitat with a narrow range of species of organisms has a high biodiversity. ☐

Biodiversity includes a wide range of genetic variation. ☐

A rainforest has very high biodiversity. ☐

A plantation of tropical oil palm trees has very high biodiversity. ☐

The term biodiversity refers to the range of different groups of organisms. ☐ **[1 mark]**

(G–E)

2 Put a tick (✓) in the box next to the **two** correct examples of sustainability.

Meeting our current needs without depriving future generations ☐

Reducing the planet's biodiversity ☐

Recycling and re-using items instead of throwing them away ☐

Removing rainforest for housing ☐

Increasing the amount of land used to grow crops ☐ **[1 mark]**

(G–E)

3 Explain why the removal of one species will have a big impact on the ecosystem.

....

....

....

.... **[4 marks]**

(D–C)

4 Intensive crop production has involved large-scale planting of one type of crop. This practice is called monoculture.

  a Explain why monoculture has negative effects on biodiversity and is non-sustainable.

....

....

.... **[3 marks]**

  b Describe the methods scientists, working with farmers, have used to improve biodiversity.

....

....

.... **[3 marks]**

(B–A*)

5 Write down three ways in which sustainability can be improved in the manufacture of a product.

....

....

.... **[3 marks]**

(D–C)

6 Explain why it is preferable to decrease the amount of packaging, even if it is biodegradable.

....

....

.... **[2 marks]**

(B–A*)

Competition between plants is important to crop growers.

In one study, scientists investigated the growth of banana plants grown at different densities in three different plots. Each plot had an area of three hectares. The following results were obtained

Plot	Number of plants grown in plot	Average plant height in m	Time to reach harvest in months	Average mass of a bunch of bananas in kg	Total yield of bananas in tonnes per hectare
1	5 000	3.5	15	15.3	23
2	10 000	4.2	18	14.3	40
3	15 000	4.3	20	13.3	52

Analyse the data and suggest reasons for the differences between the plants grown in the three plots.

The quality of written communication will be assessed in your answer to this question.

B–A*

[6 marks]

The changing air around us

1 a Underline the **two** main gases that make up the air.

helium   hydrogen   oxygen   nitrogen   **[2 marks]**

b Finish the sentence by choosing the best word from this list:

compound   element   mixture   pressure

Air in the atmosphere is a gas .... **[1 mark]**

c Why are clouds are not parts of the air?

.... **[1 mark]**

2 This diagram represents a molecule of oxygen. ◯◯

This diagram represents a molecule of nitrogen. ●●

Use the diagrams to draw a representation of a sample of air in the box below.

[2 marks]

3 Describe an experiment that can be used to find the percentage of oxygen in the air.

....

....

....

.... **[3 marks]**

4 a Tick **two** boxes showing the gases released by volcanoes.

carbon dioxide   ☐

chlorine   ☐

hydrogen   ☐

water vapour   ☐   **[2 marks]**

b What did these two gases make about 4 billion years ago?

.... **[1 mark]**

5 These sentences describe how our present atmosphere formed. They are not in the correct order.

i Plants produced oxygen.

ii Simple bacteria removed carbon dioxide by photosynthesis.

iii The Earth cooled, allowing water vapour to condense.

iv Four billion years ago the atmosphere was very hot.

v Fossil fuels formed from buried organisms.

The correct order of the sentences is: .... **[4 marks]**

6 Most scientists now agree on the explanation of how our atmosphere formed.

What is needed for a theory to become accepted?

....

....

....

.... **[3 marks]**

Humans, air quality and health

G–E

1 a Name **two** uses of fuels.

.... [2 marks]

b Tick **one** box that shows a gas that is not an air pollutant.

carbon monoxide ☐

carbon dioxide ☐

hydrogen oxide ☐

sulfur dioxide ☐ [1 mark]

c What is mean by 'good' air quality?

.... [1 mark]

D–C

2 This chart shows how carbon dioxide levels have increased over the last 50 years.

Year	1960	1970	1980	1990	2000	2010
Carbon dioxide levels (ppm)	310	325	340	350	370	395

a Describe the pattern shown by the data.

....

....

.... [3 marks]

b Suggest how solid particulates get into the air by:

i a natural process: .... [1 mark]

ii human activity: .... [1 mark]

c Name the **two** pollutants that contribute to acid rain.

.... [2 marks]

d Name **two** health problems made worse by poor air quality.

.... [2 marks]

B–A*

3 a The table in Question 2 shows units of ppm. What is ppm and what does it mean?

....

.... [2 marks]

b In the UK air quality is regularly monitored. Suggest why.

....

....

.... [3 marks]

B–A*

4 Airborne carbon particulates have been linked to lung disease. Lung disease is also linked to smoking.

a Describe why the facts in both sentences are correlations.

....

.... [1 mark]

b Explain why a correlation between smoking and lung disease does not necessarily mean this factor is the cause.

....

.... [2 marks]

Burning fuels

1 a Name the **two** elements in hydrocarbons. .... [2 marks]

 b When coal burns, carbon joins with oxygen to form carbon dioxide.

   Use the information to write the word equation for this reaction.

   .... [2 marks]

G–E

2 a Complete this word equation for combustion:

   hydrocarbon fuel + ... ⟶ .... + ... [2 marks]

 b Underline a word from the list below to describe this reaction: sulfur + oxygen ⟶ sulfur dioxide.

   acidification   combustion   neutralisation   oxidation   reduction   [1 mark]

D–C

3 Gas welding uses ethyne gas (C_2H_2) and oxygen (O_2).

 Draw a visual representation to show this reaction.

 .... [3 marks]

B–A*

4 Use the best words from this list to complete the sentences.

 joined   complex   molecules   new   old   rearranged

 a Atoms of non metal elements join to make .... [1 mark]

 b In chemical reactions, atoms are .... to make

   .... substances. [2 marks]

G–E

5 Complete these sentences.

 a The substances on the left side of a chemical equation are the ....

   and the right side shows the .... [2 marks]

 b The total mass on each side of a chemical equation is .... [1 mark]

D–C

6 All the atoms in all the fossil fuels ever burnt are still present on Earth. Explain why.

 ....

 .... [2 marks]

B–A*

7 a Burning sulfur in air is a chemical reaction. State two things you see that shows this.

   .... [2 marks]

 b Give two differences between sulfur and sulfur dioxide.

   .... [2 marks]

G–E

8 a Explain why fossil fuels contain sulfur.

   ....

   .... [2 marks]

 b Write a symbol equation for burning sulfur.

   .... [2 marks]

D–C

9 a Describe how acid rain forms, and how it damages the environment.

   ....

   ....

   .... [3 marks]

 b Why is acid rain described as a indirect pollutant?

   .... [1 mark]

B–A*

Pollution

1 a Name two processes that burn large amounts of fuel.

.... **[2 marks]**

b When carbon burns in a good air supply, carbon dioxide is formed. What is made if carbon is

burnt in a limited oxygen supply? .... **[1 mark]**

c Which element in fuels becomes sulfur dioxide when the fuel burns? .... **[1 mark]**

2 Name the two gases in the air responsible for nitrogen oxide formation.

.... **[2 marks]**

3 a During high temperature combustion, nitrogen monoxide is released.
Draw a visual representation to show this reaction.

.... **[3 marks]**

b Describe, giving examples, why nitrogen oxides are classified as pollutants.

....

....

....

.... **[3 marks]**

4 Name these compounds.

a .... **b** .... **c** .... **[3 marks]**

5 Suggest natural processes that remove the following pollutants from the air.

a Particulate carbon: .... **[1 mark]**

b Sulfur and nitrogen oxides: .... **[1 mark]**

c Carbon dioxide: .... **[1 mark]**

6 This data shows the levels of nitrogen oxides in air on one street.

All the data was taken over a 5-minute period.

Reading	1	2	3	4	5	6
Value (units)	15	12	14	8	14	12

a Suggest a reading that is likely to be an outlier. .... **[1 mark]**

b Write down the range of the results. .... **[1 mark]**

c Calculate the mean value. Show your working out.

....

.... **[2 marks]**

7 Air quality measurements occasionally produce unexpected results.

Explain why unexpected results are not necessarily outliers.

....

....

....

.... **[2 marks]**

Improving power stations and transport

1 Over the last decade, many new gas-fired power stations have been built.

   Suggest **two** advantages of producing electricity from gas rather than coal.

   ....

   .... [2 marks]

   D–C

2 Suggest **two** methods of removing sulfur dioxide at power stations.

   ....

   .... [2 marks]

   B–A*

3 Suggest **two** ways of saving electricity in the home.

   ....

   .... [2 marks]

   G–E

4 Why is it important to save fossil fuels?

   .... [1 mark]

5 Some power stations burn biofuels. Give **two** examples of biofuels.

   .... [2 marks]

   D–C

6 a Explain how biofuels are made, and why they are classed as 'carbon neutral'.

   ....

   .... [3 marks]

   b Describe **two** disadvantages of replacing fossil fuels with biofuels.

   ....

   .... [2 marks]

   B–A*

7 Cars and lorries release pollutants into the air.

   Suggest **two** ways the amount of pollutants could be reduced.

   .... [2 marks]

   G–E

8 Catalytic convertors contain a platinum catalyst than helps pollutant gases react.

   a Where are catalytic convertors found in cars?

   .... [1 mark]

   b Complete this word equation to show how the pollutant gases are removed.

   Carbon monoxide + ... ⟶ .... + ... [2 marks]

   D–C

   c Describe why the above reaction is both oxidation and reduction.

   ....

   .... [2 marks]

9 Many motor companies are developing electric cars.

   a Suggest problems of using electric cars at present.

   ....

   .... [3 marks]

   B–A*

   b Explain why using electric cars is not 'carbon neutral'.

   ....

   .... [2 marks]

Air quality varies from day to day.

This chart shows the carbon particulate level (PM10) and nitrogen oxide levels (NO_x) in Birmingham City Centre over a week in January.

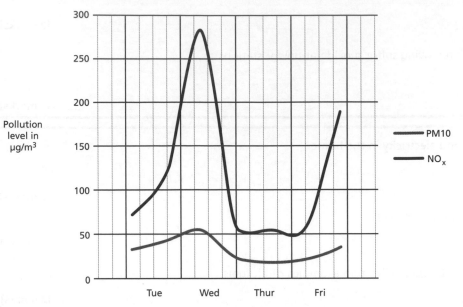

Describe what the graph shows. Include ideas about how the pollutants change, whether a correlation exists, and possible causes.

The quality of written communication will be assessed in your answer to this question.

B–A*

[6 marks]

Using and choosing materials

1 a Draw lines to match each material to its best property and to its use.

Material	Property	Use
Rubber	Can be moulded	Washing-up bowls
Plastic	Hard and elastic	Making clothes
Fibres	Can be woven	Car tyres

[2 marks]

G–E

b What does the word 'property' mean when talking about materials?

....

.... [1 mark]

2 Write in the missing property to complete each sentence.

a The temperature at which a solid turns to a liquid: .... [1 mark]

b The force needed to crush a material: .... [1 mark]

c How well a material stands up to wear: .... [1 mark]

d The mass of a given volume of material: .... [1 mark]

D–C

3 Climbers need to consider a number of properties when buying a rope.

a Suggest why climbing ropes need to be:

  i dynamic (stretchy)

  .... [1 mark]

  ii dry-treated (waterproof)

  .... [1 mark]

B–A*

b Single ropes need to be a minimum of 9 mm. Why are 11-mm ropes stronger?

....

.... [2 marks]

4 Bill and Joy measured how far a fibre stretched when 200 g was added. Here are the results.

Result	1	2	3	4	5	6
Stretch (cm)	5.8	5.7	5.3	5.6	5.8	5.6

a Identify the outlier in the results. .... [1 mark]

b State the range. .... [1 mark]

c Why do many results need to be taken?

.... [1 mark]

D–C

d What is the true value of these results? .... [1 mark]

e Suggest **two** reasons for an outlier being discarded.

....

....

.... [2 marks]

B–A*

Natural and synthetic materials

1 Underline **three** properties which most metals have.

   brittle   good conductor   hard   insulator   malleable   soft   **[3 marks]**

2 Give **three** examples of ceramics.

   ....

   .... **[3 marks]**

3 Rubber, plastic and fibres are all made of large molecules called .... **[1 mark]**

4 Give one example of a natural material which comes from:

   a plants ....

   b animals ....

   c the Earth's crust .... **[3 marks]**

5 What are synthetic materials?

   ....

   ....

   .... **[2 marks]**

6 Give **three** reasons why synthetic materials have replaced some natural materials.

   ....

   ....

   .... **[3 marks]**

7 Name the two elements present in hydrocarbon molecules.

   .... **[2 marks]**

8 Finish the sentences.

   a Most crude oil is used for making .... **[1 mark]**

   b When oil burns the number of atoms that react are .... to
   the number of atoms of products made. **[1 mark]**

9 The general formula for an alkane is C_nH_{2n+2}

   a What would be the formula for butane, which contains 4 carbon atoms?

   .... **[1 mark]**

   b What would be the formula for hexane, which contains 6 carbon atoms?

   .... **[1 mark]**

10 Suggest why the composition of crude oil varies from place to place.

   ....

   .... **[1 mark]**

11 When hydrocarbon fuels burn in oxygen, carbon dioxide and water are made.
One hydrocarbon fuel is methane, CH_4.

   Draw a visual representation to show the complete combustion of methane.

   .... **[3 marks]**

G–E

D–C

B–A*

G–E

D–C

B–A*

Separating and using crude oil

1 Finish the sentences. Choose the best words from this list:

condenses   distillation   evaporates   fractions   molecules

a Crude oil is separated by fractional .... [1 mark]

b The oil is heated up so it .... [1 mark]

c When the gas rises up and cools it .... into a liquid. [1 mark]

d Gases with similar boiling points collect together into .... [1 mark]

2 Describe the link between the size of hydrocarbon molecules and the boiling point.

.... [1 mark]

3 Explain why petrol boils at a lower temperature than diesel.

....

....

....

.... [3 marks]

4 a Draw and label paper clips to show the difference between monomers and polymers.

[1 mark]

b Polymers have replaced many natural materials. Name a polymer that has replaced:

i aluminium for tennis rackets: .... [1 mark]

ii metal for buckets: .... [1 mark]

5 The expanded formula for the monomer ethene is shown below.

Next to it, draw the expanded formula for polyethene.

$$\begin{array}{c} H \\ \\ H \end{array}\!\!\!\!\!\!C = C\!\!\!\!\!\!\begin{array}{c} H \\ \\ H \end{array}$$

[2 marks]

6 Describe why PET polymer is superior to glass for making bottles.

....

....

.... [3 marks]

7 Describe how the properties of polymer chains can be changed.

....

....

.... [2 marks]

Polymers: properties and improvements

G–E

1 Look at these polymer molecules.

i
ii
iii

Which molecule will have:

a the strongest forces between molecules? .... **[1 mark]**

b the highest melting point? .... **[1 mark]**

D–C

2 Low density polyethene (LDPE) has long molecules with many branches.

a Suggest what the properties of LDPE are likely to be.

....

.... **[3 marks]**

b Describe the structure of HDPE and why it is likely to have a higher melting point than LDPE.

....

.... **[3 marks]**

B–A*

3 Explain how the degree of crystallinity affects a polymer.

....

....

.... **[3 marks]**

G–E

4 Finish the sentences. Choose the best words from this list:

higher   less   lower   more   stronger   weaker

a The longer the chain length, the .... the polymer. **[1 mark]**

b Longer chains need .... force to separate them. **[1 mark]**

c Longer chains have a .... melting point. **[1 mark]**

5 Describe plasticisers and how they work.

....

....

.... **[4 marks]**

D–C

6 How are thermoplastics different to thermosetting plastics?

....

.... **[3 marks]**

7 Natural rubber is too soft for using as a car tyre. Suggest **two** ways the rubber can be made harder.

.... **[2 marks]**

B–A*

8 a Describe how crystallinity can be increased.

....

.... **[1 mark]**

b Suggest a disadvantage of increased crystallinity.

.... **[1 mark]**

Nanotechnology and nanoparticles

1 Underline the instrument used to view very small objects.

bathyscope   milliscope   microscope   telescope   [1 mark]

2 The width of a human hair is about 0.1 mm. How many would fit into 1 mm?

....   [1 mark]

3 a Name **one** naturally occurring nanoparticle.

....   [1 mark]

b What is the name of the nanoparticles released when a fuel burns?

....   [1 mark]

c Nanoparticles can occur naturally, or by accident. Name another way they occur.

....   [1 mark]

4 Underline the correct answer. Nanoparticles are about the same size as a:

human hair   large molecule   oxygen molecule   salt crystal   [1 mark]

5 How many nanometres are in a metre? ....   [1 mark]

6 Explain the difference between 'buckyballs' and nanotubes.

Buckyball   Nanotubes

....
....
....
....   [4 marks]

7 Explain why nanoparticles are effective catalysts.

....
....   [2 marks]

8 a Explain why a carbon nanotube can have a diameter of just a few nanometres but a length of many millimetres.

....
....   [2 marks]

b Explain why cutting a 1 cm³ cube into four means the volume stays the same but the surface area increases.

....
....
....
....   [4 marks]

G–E

D–C

B–A*

The use and safety of nanoparticles

G–E

1 Suggest why nanoparticles are added to socks.

.... [1 mark]

2 Nanoparticles can be added to tennis balls to make them bouncy for longer. Look at these results.

Test / date	Bounce height in cm from 100 cm height			
	New	3 months old	6 months old	9 months old
1	79	75	78	77
2	81	80	77	76
3	82	79	78	77
4	80	79	79	76

D–C

a Calculate the mean result when the ball is new. .... [1 mark]

b State the range shown by the 6-months-old results. .... [1 mark]

c Identify one outlier contained in the data. .... [1 mark]

d Describe if the data shows whether or not the ball has stayed bouncy.

....

....

.... [3 marks]

e What name is given to a mixture of materials? .... [1 mark]

3 What is meant by the term nanotechnology?

.... [1 mark]

4 Carbon nanotubes are being incorporated into body armour. Suggest why this makes the armour strong, low density and flexible.

B–A*

....

....

....

.... [3 marks]

G–E

5 Silver nanoparticles can be washed out of clothing. Why can this be a problem?

....

.... [2 marks]

6 Nanoparticles can be incorporated into sunscreen to prevent damage by ultraviolet rays. Suggest why some people are concerned about the safety of this type of sunscreen.

D–C

....

....

.... [3 marks]

7 Some environmental and health groups are campaigning to have nanoparticles banned. Outline arguments **against** this point of view.

B–A*

....

....

....

.... [3 marks]

C2 Extended response question

Badminton rackets can be made using different materials.

The table shows information about badminton racket design for players of different abilities.

Rackets for...	Stem material	Mass (g)	Cost (£)
Beginner	Titanium and steel	120	15
Intermediate	Graphite and titanium	95	30
Advanced	Graphite	80	45
Elite	Nano-filled graphite	85	70

With reference to the table, describe the properties that are needed in a badminton racket, and explain why the materials used result in these different properties.

The quality of written communication will be assessed in your answer to this question.

[6 marks]

Moving continents and useful rocks

G–E

1 Underline the name that describes scientists who study rocks.

astronomers   geologists   meteorologists   oceanologists

[1 mark]

2 Tectonic plates can move by sliding past each other. State **two** other ways they can move.

....

[2 marks]

D–C

3 A supercontinent called Pangea existed about 225 million years ago. Describe how this happened, and what formed as a result.

....

....

[2 marks]

4 Why has the climate in Britain changed over the last 600 million years?

....

....

[2 marks]

B–A*

5 Describe how magnetism can help date igneous rock.

....

....

....

....

[4 marks]

G–E

6 a Why did the Industrial Revolution start in the north-west of Britain?

....

[1 mark]

b Name **three** raw materials needed for the Industrial Revolution.

....

[3 marks]

D–C

7 Describe the conditions needed for limestone to form.

....

....

[3 marks]

8 Describe how rock salt deposits form.

....

....

....

[4 marks]

B–A*

9 a What evidence does this picture show about this sedimentary rock?

....

....

....

....

[2 marks]

b Describe **one** other form of evidence which can be used to show the conditions needed for a sedimentary rock.

....

....

[2 marks]

C3 Chemicals in our lives – risks and benefits

Salt

1 Name **two** different ways used to obtain salt.

....

.... [2 marks] G–E

2 State why salt is put onto icy roads and explain how it works.

....

....

....

.... [3 marks] D–C

3 Suggest why salt is not obtained from sea water in the UK.

.... [1 mark]

4 Describe **two** problems with salt extraction and their environmental impact.

....

....

....

.... B–A*

....

.... [4 marks]

5 Give **two** uses of salt in the food industry.

....

.... [2 marks]

6 The table shows how salt affects bacterial growth.

% salt concentration	0	12	24	36	48	60
Number of bacteria	58	54	50	46	32	5

G–E

Describe the pattern shown by the data.

....

.... [3 marks]

7 a Explain why salt is classified as a possible hazard, and how the risks can be estimated.

....

....

....

.... [3 marks] D–C

b Describe how people can reduce the risk from salt.

....

....

.... [2 marks]

8 Describe how the UK government regulates food safety.

....

....

....

.... B–A*

....

.... [4 marks]

Reacting and making alkalis

1 Tick **two** boxes showing properties of alkalis.

a pH of between 1 and 6 ☐ c Turn litmus paper blue ☐

b Convert stale urine to potassium hydroxide ☐ d Neutralise acid soil and are used to make glass ☐

[2 marks]

2 Alkalis react with acids to make salts. If nitric acid reacts with calcium hydroxide, a salt called calcium nitrate is made. What salt is made when:

a sulfuric acid reacts with potassium hydroxide? .... [1 mark]

b hydrochloric acid reacts with sodium hydroxide? .... [1 mark]

3 Name **two** traditional sources of alkalis.

.... [2 marks]

4 Before industrialisation, alkalis were in short supply. Explain why.

....

....

.... [3 marks]

5 Lime is made by heating limestone. Limestone has the formula $CaCO_3$. Carbon dioxide gas (CO_2) is also produced.

a Draw a visual representation to show the reaction for making lime.

.... [1 mark]

b Lime is used to neutralise acidic soil. Describe why this is preferable to adding an alkali.

....

....

....

.... [4 marks]

6 Describe the role of a mordant in the dying process.

.... [1 mark]

7 The Leblanc process manufactured sodium carbonate by reacting limestone and salt.

a Name two toxic by-products of this process.

.... [2 marks]

b Explain how the pollution problems arising from this process can be solved.

....

.... [2 marks]

8 a Predict the product of this reaction:

sodium carbonate + sulfuric acid ⟶ .... [1 mark]

b Predict the product of this reaction:

ammonium hydroxide + nitric acid ⟶ .... [1 mark]

9 All metal hydroxides are classed as bases. Why are some metal hydroxides alkalis?

....

.... [2 marks]

G–E

G–E

D–C

B–A*

D–C

B–A*

Uses of chlorine and its electrolysis

1 This bar chart shows deaths from typhoid in different countries.

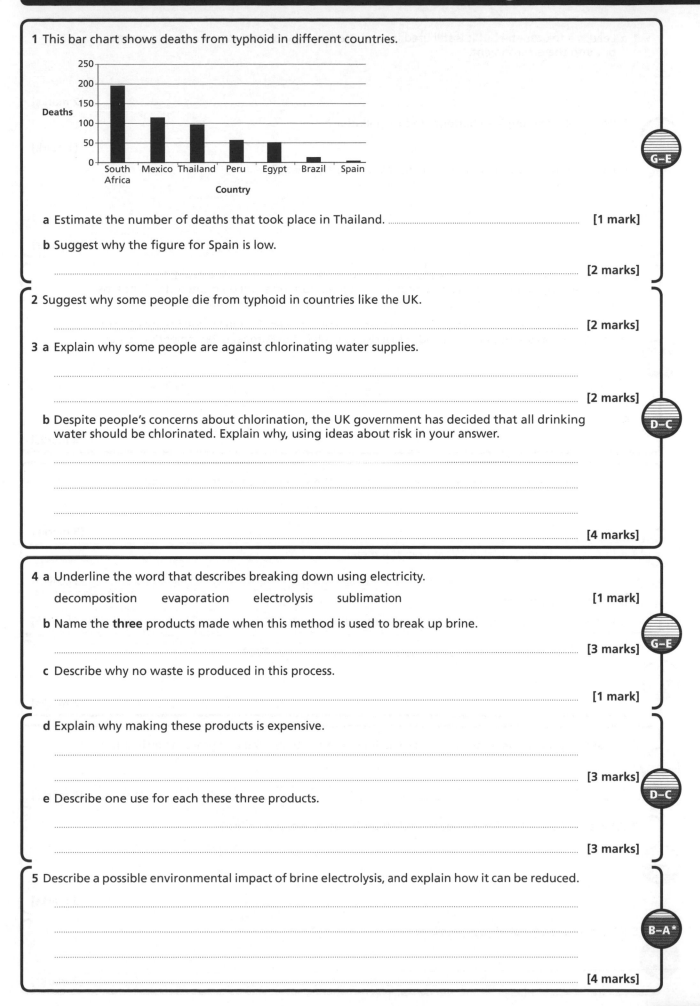

a Estimate the number of deaths that took place in Thailand. .... [1 mark]

b Suggest why the figure for Spain is low.

.... [2 marks]

2 Suggest why some people die from typhoid in countries like the UK.

.... [2 marks]

3 **a** Explain why some people are against chlorinating water supplies.

....

.... [2 marks]

b Despite people's concerns about chlorination, the UK government has decided that all drinking water should be chlorinated. Explain why, using ideas about risk in your answer.

....

....

....

....

.... [4 marks]

4 **a** Underline the word that describes breaking down using electricity.

decomposition   evaporation   electrolysis   sublimation   [1 mark]

b Name the **three** products made when this method is used to break up brine.

.... [3 marks]

c Describe why no waste is produced in this process.

.... [1 mark]

d Explain why making these products is expensive.

....

.... [3 marks]

e Describe one use for each these three products.

....

.... [3 marks]

5 Describe a possible environmental impact of brine electrolysis, and explain how it can be reduced.

....

....

.... [4 marks]

G–E

D–C

G–E

D–C

B–A*

Industrial chemicals and LCA

1 a Lead is a toxic metal but it is still used in car batteries. Suggest **two** ways lead could spread out into the environment.

....

.... [2 marks]

b Why does recycling lead batteries reduce pollution?

.... [1 mark]

c Lead is a cumulative poison in humans. What does this mean?

....

....

.... [3 marks]

D–C

2 New chemicals need to pass a risk assessment before they can be used. One example is bromine-based flame-retardant chemicals used in foam furniture and on printed circuit boards for games and phones. Some people believe these chemicals cause behaviour problems.

The European Parliament has voted to ban these chemicals, even though little evidence of harm exists. Use ideas about perceived and actual risk to explain the European Parliament's decision.

....

....

....

....

.... [4 marks]

B–A*

3 Name the **three** different chemical elements in PVC. The start of each word is provided.

a C is ....   **b** Cl is ....

c H is ....   [3 marks]

4 a PVC is non-biodegradable. What does this mean?

.... [1 mark]

b Finish this sentence:

PVC can be made softer by adding small molecules called .... [1 mark]

G–E

5 Why are some people worried about adding small molecules to PVC?

....

.... [2 marks]

D–C

6 Explain why a Life Cycle Assessment is sometimes referred to as 'cradle to grave' assessment.

....

.... [3 marks]

G–E

7 Why is the LCA better for a wooden table than for using wood on a fire?

....

....

.... [3 marks]

D–C

8 Why is a full LCA not always possible?

....

.... [2 marks]

B–A*

C3 Chemicals in our lives – risks and benefits

Many local councils collect plastic bottles for recycling into new plastics. Plastics can also be made directly from crude oil.

The two flow charts show the different processes.

Collecting and recycling	Making from crude oil
Collection by lorry from homes	Drill for oil and distil it
Sort into types and clean	'Crack' chains to make monomers
Crush and shred	Join monomers to make polymers
Make into pellets	Make into pellets

Describe what is meant by the term Life Cycle Assessment, and use the charts to compare the Life Cycle Assessment of the two methods.

The quality of written communication will be assessed in your answer to this question.

B–A*

[6 marks]

Our solar system and the stars

G–E 1 Our solar system contains asteroids. Name five other types of object in the solar system.

.... [5 marks]

2 Here are four planets in the solar system: Earth, Jupiter, Mercury, Uranus.

D–C
 a Put these planets in order of increasing mass.

.... [3 marks]

 b Put these planets in order of increasing size of orbit.

.... [3 marks]

B–A* 3 Explain why a light-year is about 1×10^{13} km.

....

....

.... [3 marks]

4 Complete the sentences by choosing words from this list.

G–E
galaxies   stars   planets   moons

The Milky Way is one of many .... in the Universe. Each galaxy is made of

many .... [2 marks]

5 A scientist has a hypothesis that the Sun is just another star. Suggest how they could attempt to disprove this hypothesis.

D–C
....

....

.... [3 marks]

6 Explain why observing very distant galaxies is equivalent to observing the Universe as it was a long time ago.

B–A*
....

....

....

.... [3 marks]

7 Here are some objects in the Universe: Earth, Milky Way, solar system, Sun. Put them in order of size, starting with the smallest.

G–E
.... [3 marks]

8 Explain how the brightness of a star can be used to estimate its distance.

D–C
....

....

....

.... [3 marks]

9 Explain why the distance to most stars is not accurately known.

B–A*
....

.... [2 marks]

The fate of the stars

1 Where does the Sun get its energy from?

.... [1 mark] **G–E**

2 Shortly after the Big Bang, the only elements in the Universe were hydrogen and helium.
Now the Universe contains many different elements. Explain where they came from.

....

....

....

....

.... [4 marks] **D–C**

3 Explain why our solar system with solid planets must be much younger than the Universe.

....

....

.... [2 marks] **B–A***

4 Explain what we know about the motion of galaxies.

....

....

....

....

.... [4 marks] **D–C**

5 Draw lines to link the start of each sentence to its correct ending.

The Sun is ...	... 14 000 million years old.
The Earth is ...	... 5000 million years old.
The Universe is ...	... 4500 million years old.

[2 marks] **G–E**

6 Scientists believe that the Universe around us today came into being 14 000 million years ago.
Explain why they are much less certain about what will happen to the Universe in the future.

....

....

....

....

....

.... [4 marks] **B–A***

Earth's changing surface

1 Draw lines to link the start of each sentence to its correct ending.

G–E

Fossils are ...		... new mountains made from lava.
Mountains are ...		... materials from the erosion of mountains.
Volcanoes are ...		... the remains of plants and animals in rock.
Sediments are ...		... made by folding the rocks of the Earth's crust.

[3 marks]

2 Describe and explain the formation of sedimentary rocks.

D–C

....

....

....

....

[4 marks]

3 The Earth is about 4000 million years old. Explain why erosion has not reduced all the continents to sea level in that time.

B–A*

....

....

....

....

[4 marks]

4 Which of these observations provide evidence for Wegener's theory of continental drift? Put a tick (✓) next to your answer.

G–E

a There is life on all of the continents. ☐

b The continents fit together like a jigsaw. ☐

c Many continents are surrounded by the sea. ☐

d There are sedimentary rocks on all of the continents. ☐

e There are chains of mountains at the edges of some continents. ☐

[2 marks]

5 Give reasons why scientists didn't accept Wegener's theory of continental drift when it was first published.

D–C

....

....

....

....

....

[5 marks]

6 Describe and explain the pattern of magnetisation of rocks on the Atlantic seafloor.

B–A*

....

....

....

[4 marks]

Tectonic plates and seismic waves

1 State **three** things which happen at plate boundaries.

....

....

.... [3 marks] D–C

2 What is a plate boundary?

.... [2 marks]

3 Explain the presence of volcanoes at plate boundaries.

....

....

....

.... [4 marks]

4 Explain why earthquakes happen at plate boundaries.

....

.... [1 mark]

B–A*

5 Although the Earth was formed about 4000 million years ago, the majority of the rocks on its surface are much younger. Use the rock cycle to explain why.

....

....

....

.... [4 marks]

6 Draw lines to link each part of the Earth with its best description.

| core | | solid rock |

| crust | | liquid iron |

| mantle | | semi-liquid rock |

G–E

[2 marks]

7 Complete the sentences by choosing the correct words from the list below.

continent   **faster**   **slower**   **P-waves**

S-waves   **seismic**   **sound**   **tectonic**

Sudden movement of .... plates makes two types of .... waves.

The ones which can only move through the crust and mantle are called ....

The ones which can also move through the core are called ....

P-waves move .... than S-waves. [5 marks]

D–C

8 Explain the evidence from earthquakes of the structure of the Earth.

....

....

....

.... [4 marks]

B–A*

Waves and their properties

1 Draw lines to link each property of a wave to its correct description.

Speed	Maximum value of the disturbance in one wave
Frequency	How far the energy of the wave travels in a second
Amplitude	The number of vibrations of the wave source in one second
Wavelength	Distance along wave from one zero disturbance to the next

G–E

[3 marks]

2 What does a wave carry away from its source?

....

[1 mark]

3 Ben watches a firework display. He hears the bang from a firework 1.5 seconds after he sees the flash. How far away was the firework?

(Speed of sound = 300 m/s, speed of light = 300 000 000 m/s)

D–C

....

....

....

[2 marks]

4 Draw lines to link each quantity with its correct units.

Speed	m
Frequency	Hz
Wavelength	m/s

G–E

[2 marks]

5 Complete the sentence by choosing the correct word from the list below.

height   length   number   time

The frequency of a wave is the .... of vibrations per second produced by it.   [1 mark]

D–C

6 What is the speed of a wave which has a frequency of 20 kHz and a wavelength of 40 cm?

....

....

[2 marks]

7 a The speed of red light in glass is 2.0×10^8 m/s. The wavelength of red light in glass is 0.44 μm. Calculate its frequency.

....

....

[2 marks]

B–A*

b Blue light has a frequency of 6.7×10^{14} Hz. Estimate its wavelength in glass.

....

....

....

[3 marks]

Explain the evidence from distant galaxies for an expanding Universe.

The quality of written communication will be assessed in your answer to this question.

B–A*

[6 marks]

Waves which ionise

1 Sentences a–d explain how you see this page. Put them in the correct order (1–4).

    a The light travels away from the source.

    b A light source in the room emits some light.

    c The light is absorbed by detectors in your eyes.

    d The light bounces off the paper but is absorbed by the ink.   **[3 marks]**

G–E

2 How fast does light pass through a vacuum? .... **[1 mark]**

D–C

3 Arrange these parts of the electromagnetic spectrum in order of increasing energy: infrared; gamma rays; radio waves; ultraviolet. .... **[3 marks]**

4 Complete these sentences about the transfer of energy by X-rays. Choose words from this list.

    amplitude   energy   frequency   light   photons

    power   protons   speed   sound

B–A*

    X-rays transfer .... in packets called .... from one place to another

    at the speed of .... Increasing the energy of a photon requires an increase of

    its ....   **[4 marks]**

5 Tick (✓) to show which of these changes increases the intensity of an electromagnetic wave.

    a Increase the speed of the photons   c Increase the energy of each photon

    b Increase the frequency of the wave   d Increase the number of photons per second   **[3 marks]**

G–E

6 Explain why the intensity of a wave decreases as it moves away from its source.

    ....

    .... **[3 marks]**

D–C

7 A laser beam transfers 4.8 mJ of energy in 8.0 s to a black surface of area 2.0×10^{-6} m^2. Calculate the intensity of the laser beam.

B–A*

    .... **[2 marks]**

8 Explain what happens to an atom when one of its electrons is knocked out.

    ....

    .... **[3 marks]**

G–E

9 a Which parts of the electromagnetic spectrum are ionising radiations?

    .... **[3 marks]**

    b Explain how an electromagnetic wave can ionise an atom.

D–C

    ....

    .... **[2 marks]**

10 Explain why exposure to ionising radiation is harmful to people.

    ....

B–A*

    ....

    ....

    .... **[5 marks]**

Radiation and life

1 Here are four statements about ionising radiation.

   i They are absorbed by muscle but not by bone.   **iii** They pass easily through both muscle and bone.

   ii They are absorbed by bone but not by muscle.   **iv** They are produced by radioactive materials.

   a Which of the statements are **only** true for X-rays? .... [1 mark]

   b Which of the statements are **only** true for gamma rays? .... [2 marks]

(G–E)

2 Complete the sentences. Choose words from this list.

   absorbed   **clothes**   **goggles**   **increasing**   **reducing**

   reflected   **shields**   **stopping**   **transmitted**

   X-rays can be used to image people's luggage at airports. The X-rays are .... by

   high density items but are .... by the low density items. The people who operate

   these X-ray scanners can reduce their risk of cancer with .... made of lead

   and .... the time they spend near the scanner.   **[4 marks]**

(D–C)

3 Which of these actions **increase** the amount of thermal energy transferred from a light source to an object?

   a Leave the light switched on for longer   ☐

   b Increase the intensity of the light source   ☐

   c Increase the frequency of the light radiation   ☐

   d Put the light source and object in a darkened room   ☐

   e Increase the distance of the object from the light source   ☐   **[3 marks]**

(G–E)

4 Explain why dry food does not heat up in a microwave oven.

   ....

   .... **[2 marks]**

(D–C)

5 Describe how scientists study people to determine the risk of using mobile phones.

   ....

   ....

   ....

   .... **[5 marks]**

(B–A*)

6 a Name the electromagnetic radiation in sunlight which gives you sunburn. .... **[1 mark]**

   b In what other way can this radiation damage you? .... **[1 mark]**

   c What can you do to protect yourself when exposed to sunlight?

   .... **[2 marks]**

(G–E)

7 Sunbathing has both risks and benefits.

   a What are the risks? .... **[2 marks]**

   b What are the benefits? .... **[2 marks]**

(D–C)

8 What is the effect of ultraviolet radiation on the ozone layer?

   ....

   .... **[2 marks]**

(B–A*)

Climate and carbon control

1 Tick (✓) to show which of these statements about radiation from the Sun is correct.

a It contains all possible frequencies ☐ **d** All of it passes through our atmosphere ☐

b It contains a range of frequencies ☐ **e** It contains just one frequency – light ☐

c All of it is absorbed by our atmosphere ☐ **f** Some is absorbed by our atmosphere ☐ **[2 marks]**

G–E

2 Complete the sentences. Choose the best words from this list.

all   average   cold   decreasing   enduring

greatest   increasing   hot   least

Electromagnetic radiation is emitted by .... objects. The principal frequency

of that radiation is the one with the .... intensity, and it increases

with .... temperature. **[3 marks]**

D–C

3 The diagram represents the flow of some carbon on the Earth.

Match each arrow with what it represents.

i Animals eat plants for food ☐

ii Animals breathe out carbon dioxide ☐

iii Plants use carbon dioxide to build tissues ☐

iv Plants emit carbon dioxide from respiration ☐

[3 marks]

G–E

4 Explain why the level of carbon dioxide in the atmosphere has been steady for thousands of years.

....

.... **[3 marks]**

D–C

5 The temperature of both the atmosphere and the amount of carbon dioxide in the atmosphere has been rising for the past century. How do scientists determine which of these factors is the cause and which is the effect?

....

....

.... **[3 marks]**

B–A*

6 State three effects of global warming which could affect food grown in the UK.

....

.... **[3 marks]**

G–E

7 Explain how scientists can test and use a climate model to predict the effects of global warming.

....

....

....

.... **[4 marks]**

B–A*

Digital communication

1 The waves of the electromagnetic spectrum are: gamma rays; infrared; microwaves; radio; visible; ultraviolet; X-rays. Which four do we use for communication?

.... [4 marks]

G–E

2 Explain why radio waves are used to carry TV broadcasts.

.... [2 marks]

D–C

3 Explain why infrared is used for long-distance telephone communication.

.... [2 marks]

4 What effect does modulation have on a carrier wave?

....

B–A*

.... [2 marks]

5 Which one of these graphs shows a digital signal? .... [1 mark]

signal

time
A

signal

time
B

signal

time
C

G–E

6 Sound information is carried by a radio wave as a code using 1 and 0.

a What are 1 and 0 in this context?

.... [2 marks]

b Explain how the radio receiver is able to reproduce the sound information.

....

....

D–C

.... [4 marks]

7 Explain why digital radio receivers are often unaffected by background radio noise.

....

....

B–A*

.... [3 marks]

8 Complete the passage about digital information. Choose from this list: *0, 1, 2, 4, 8*.

Digital information for sound or pictures has only .... values. Sound information is coded

as a string of binary digits called .... or .... and stored in groups of ....

G–E

as bytes. [4 marks]

9 Explain how sounds can be created by an MP3 player from digital information.

....

....

D–C

.... [4 marks]

10 List four advantages of using digital signals to transfer information from one place to another.

....

B–A*

.... [4 marks]

P2 Extended response question

Explain how cutting down forests affects the amount of carbon dioxide in the atmosphere.

The quality of written communication will be assessed in your answer to this question.

D–C

[6 marks]

Energy sources and power

1 State the names of **three** different fossil fuels.

.... [3 marks]

2 State the names of **four** different renewable energy sources.

....

.... [4 marks]

3 Why is electricity called a secondary energy source?

.... [1 mark]

G–E

4 Explain why the way that we make our electricity will have to change in the future.

....

.... [2 marks]

D–C

5 Explain the environmental impact of building more gas-fired power stations in the UK.

....

....

.... [4 marks]

B–A*

6 Complete the sentences. Choose words from this list.

energy   force   joules   newtons   seconds   watts   weight

The power of a component is the .... which transfers to it in one second.

It is measured in .... when the energy is in .... and the time

in .... [4 marks]

G–E

7 A 250-W computer is left on for 24 hours. How much energy in kWh does it consume?

.... [2 marks]

8 A kettle is plugged into the mains electricity supply. Link the start of each sentence about the kettle to its correct ending.

The power of the kettle ...	... is energy transfer per second.
The energy in the circuit ...	... provides the current with energy.
The current in the circuit ...	... transfers energy from the supply.
The voltage of the supply ...	... flows from the supply to the kettle.

[3 marks]

D–C

9 Complete the equation for a microwave oven.

The .... of the oven (in watts) = the supply voltage (in ....) ×

the .... (in ....) [4 marks]

10 An appliance draws a current of 3.0 A from the 230-V mains supply. At what rate is energy transferred by the electric current?

.... [2 marks]

B–A*

11 What is the current drawn from the 230-V mains supply when an appliance of power 1.15 kW is plugged in?

.... [2 marks]

Efficient electricity

1 Draw lines to link the start of each sentence about measuring electricity to its correct ending.

One kilowatt-hour is ...	... a decimal code.
	... 3 600 000 joules.
	... units of kilowatt-hours.
The meter readings are in ...	... a unit of electrical power.
	... the energy transferred into a house.
	... the current transferred out of a house.
An electricity meter records ...	... the voltage of the mains supply of a house.

[3 marks]

2 A washing machine has an average power rating of 800 W when it is run off a 120-V mains supply. A single wash cycle takes 90 minutes. If a unit of electricity costs 15p, how much does the electricity for a wash cycle cost?

[3 marks]

3 Here is the Sankey diagram for a TV set.

   a What does the Sankey diagram show?

   14 J sound
   42 J input
   18 J heat

[2 marks]

   b One of the labels of the diagram is missing. What should it be? **[2 marks]**

4 Link the start of each sentence about a Sankey diagram to its correct end.

An arrow splits ...	... out to the right.
Electrical energy flows ...	... in from the left.
Waste heat energy flows ...	... out downwards.
The electrical energy input is ...	... proportional to its energy.
The thickness of each arrow is ...	... where there is an energy transfer.
Useful transferred energy flows ...	... equal to the sum of the energy outputs.

[5 marks]

5 An electric drill transfers every 50 J of electricity into 20 J of useful kinetic energy. The rest is wasted as heat and sound. Calculate the percentage efficiency of the drill.

[2 marks]

6 Suggest three ways in which the government could make all of us use less energy.

[3 marks]

7 Explain why global demand for energy is likely to rise in the future.

[4 marks]

Generating electricity

1 Complete the sentences. Choose words from this list.

coil   generators   length   magnet   motors   transformers

Power stations use .... to make electricity. Each one contains a ....

which spins around near a .... of wire.   **[3 marks]**   G–E

2 A magnet spins near a coil of wire. What appears across the wire? Circle the correct answer.

charge   electricity   power   voltage   **[1 mark]**

3 Sentences a–e describe how a power station makes electricity. Number the boxes and put them in the correct order (1–5).

a Water boils into steam. ☐

b A magnet is made to spin round. ☐

c A voltage appears across a coil of wire. ☐

d The high pressure gas spins the turbine round. ☐

e A primary fuel transfers energy to thermal energy. ☐   **[4 marks]**   D–C

4 a State **three** primary energy sources which boil water into steam in a power station.

....   **[3 marks]**

b State **three** primary energy sources which spin turbines directly.

....   **[3 marks]**

5 What is the job of a turbine in a power station?

....

....   **[2 marks]**   G–E

6 Many power stations use steam to spin the turbines. State **three** other substances used to spin turbines for electricity.

....   **[3 marks]**

7 Describe how fossil fuels are used to spin turbines in power stations.

....

....

....

....   **[4 marks]**   D–C

8 Explain why a gas-fired power station has two different types of turbine.

....

....

....

....   **[4 marks]**

9 Describe in detail how the fuel transfers energy to a turbine in a nuclear power station.

....

....

....   B–A*

....

....   **[5 marks]**

Electricity matters

1 Why is nuclear waste a health risk?

.... [1 mark]

2 People who work with nuclear waste can be affected when they are irradiated by it.

  a Why is irradiation a risk to health?

.... [2 marks]

  b Suggest what steps people should take to reduce the risk of irradiation from nuclear waste.

.... [2 marks]

  c Explain why people who work with nuclear waste need to worry more about contamination.

....

....

.... [3 marks]

3 Explain how governments assess the risks of different technologies for making electricity.

....

....

.... [3 marks]

4 Here are some energy sources which can be used to spin turbines directly. Circle the three renewable ones.

gas   **hydroelectric**   **tidal**   **wind**   [2 marks]

5 Hydroelectricity is renewable. Explain other advantages and disadvantages of using hydroelectricity.

....

....

....

.... [4 marks]

6 Explain how the use of renewable technology can reduce the environmental impact of generating electricity.

....

....

....

.... [4 marks]

7 What is the voltage of the mains supply in our homes?

.... [1 mark]

8 The efficiency of transfer of energy from coal in a power station to the mains supply in your house is less than 50%. State the sources of wasteful energy transfer in order of their importance.

....

....

.... [4 marks]

Electricity choices

1 Draw lines to link each energy source with its impact on the environment.

Energy source

| Fossil fuel |

| Wind power |

| Nuclear power |

| Hydroelectricity |

Impact on environment

| Noise and visual pollution |

| Floods large areas of land |

| Produces radioactive waste |

| Produces greenhouse gases |

[3 marks]

2 Name four different energy sources which will run out in the next thousand years.

.... **[4 marks]**

3 Which of these energy sources do not contribute to global warming?
Circle your answers.

coal   gas   geothermal   nuclear   oil   solar   wind   **[4 marks]**

4 The government decides to increase generating capacity in the UK by 20 000 MW. They shortlist these two ways of doing this: 1000-MW nuclear power stations; 500-MW wind farms.
Discuss what they should decide to do. Justify your recommendation.

....

....

....

....

....

[4 marks]

5 The USA produces about twice as much carbon dioxide per person as the UK. Why should this be of concern to people in both countries?

....

.... **[1 mark]**

6 Explain why people in the UK will have to use less energy in the future to keep global energy demand unchanged.

....

....

....

.... **[4 marks]**

7 In planning for the future, governments are always anxious to avoid the possibility of power cuts. Explain what they could do to avoid power cuts.

....

....

....

.... **[4 marks]**

P3 Extended response question

Some power stations in the UK are coming to the end of their working life. The government has to decide on replacements for them. One factor which needs to be considered is the cost of building the replacement. What **other** factors should the government consider?

The quality of written communication will be assessed in your answer to this question.

D–C

[6 marks]

The chemical reactions of living things

1 Read the statements below about the chemical reactions of living things. Some statements are correct, while some are incorrect. Put ticks (✓) in the boxes next to the **two** correct statements.

a Only animals carry out respiration; plants photosynthesise instead. ☐

b The end producers of photosynthesis are carbon dioxide and water. ☐

c Respiration is the process by which organisms release energy from food. ☐

d During photosynthesis, light energy is converted to chemical energy. ☐ **[2 marks]**

G–E

2 Write down the definition of an enzyme.

.... **[1 mark]** G–E

3 Here are some statements about enzymes. Use the words provided to complete the sentences.

acid   active site   amino acids   genes   lock and key

product   proteins   optimum   substrate   sugars

Enzymes are .... They are made up of long chains

of .... joined together. Enzymes are assembled according

the instructions in our ....

A molecule that is broken down by an enzyme is called the ....

The part of the enzyme that links with the molecule to be broken down is called

the ....

This model of enzyme action is called the .... model. **[6 marks]**

D–C

4 The graphs show the effect of pH on the activity of two proteases (enzymes that break down proteins).

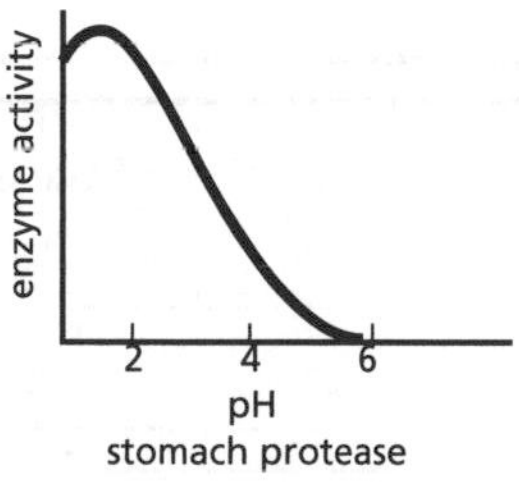

stomach protease

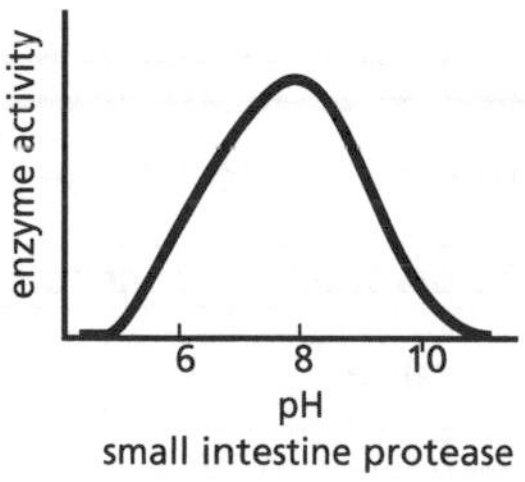

small intestine protease

a What are the optimum pHs of stomach and small intestine proteases?

.... **[2 marks]**

b Explain why enzyme activity is sensitive to pH.

....

....

.... **[3 marks]**

B–A*

5 Olivia finds a graph on the Internet showing how pH affects the activity of a protease called papain, from the papaya plant.

a Olivia says that the graph shows that papain is unaffected by pH. Is Olivia correct? Explain your answer.

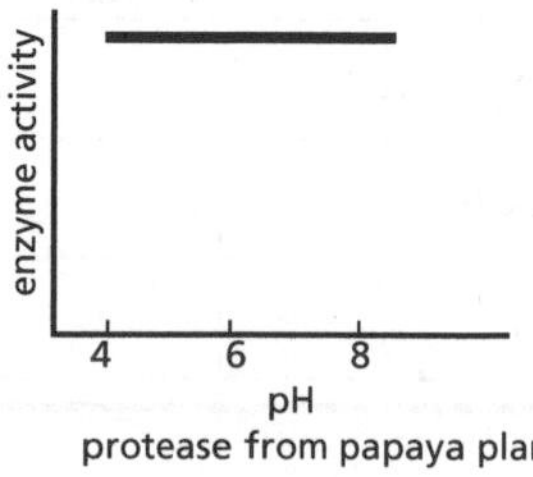

....

....

....

protease from papaya plant

.... **[2 marks]**

b In terms of enzyme structure, suggest what these results show.

.... **[1 mark]**

How do plants make food?

1 Fill in the spaces below to complete the word equation for photosynthesis.

[___] + [___] → [___] + [___]

[___]

[2 marks]

2 Here are some statements about photosynthesis. Use the words provided to complete the sentences below.

amino acids   cellulose   chlorophyll   chloroplasts   glucose

proteins   respiration   starch   sugar   the Sun

The energy for photosynthesis comes from .... The energy is

absorbed by the green pigment called .... This is found in

structures in the cells called ....

The main product of photosynthesis is a .... called

.... Some is used for .... ,

to release energy. Some is converted into large molecules the plant needs, such

as .... , ... and

....

[7 marks]

3 Write down the symbol equation for photosynthesis.

[___]

[___] + [___] → [___] + [___]

[2 marks]

4 Complete the table below to describe the function of parts of the plant cell.

Part of the plant cell	Function
a Cell membrane	
b Cell wall	
c Cytoplasm	
d Mitochondrion	
e Nucleus	

[5 marks]

5 Complete the table below comparing cell structure and function in microorganisms.

	Bacteria	Yeast
Outer layer of cell		
Genetic code		
Respiration		

[3 marks]

Providing the conditions for photosynthesis

1 Write down a definition of diffusion.

....

....

[2 marks]

G–E

2 Here are some statements about the movement of gases in and out of plants. Use the words provided to complete the sentences:

| active | carbon dioxide | decrease | increase | into | diffusion |
| nitrogen | osmosis | out of | oxygen | passive | |

.... gas, required by the plant for photosynthesis, moves

.... the plant by the process of ....

By the same process, ... gas, produced by photosynthesis,

moves .... the plant.

The process is ...., so if the temperature is increased, the rate at

which the process will take place will ....

[7 marks]

D–C

3 In the school lab, Adam cut and weighed two potato chips. He placed one in distilled water, and one in concentrated sucrose solution.

a After 30 minutes, Adam weighed the potato chips again. How did the mass of the chips change? Explain your answer.

....

....

....

[5 marks]

b Adam wanted to find the concentration of dissolved chemicals in the potato chips. Describe how he could extend his experiment to do this.

....

....

....

[5 marks]

4 Explain why nitrates are taken up into plant roots by active transport.

....

....

[3 marks]

B–A*

5 The graph opposite shows the effect of light intensity on photosynthesis in a single-celled plant.

a Describe and explain the effect of light intensity on the plant.

....

....

....

.... [4 marks]

Rate of photosynthesis vs Light intensity

D–C

b The investigation was also carried out in a high concentration of carbon dioxide.

i Sketch a graph of what you would expect on the axes above.

ii Explain its shape.

....

....

[2 marks]

Fieldwork to investigate plant growth

1 Ruby is carrying out a survey of the trees in a wood. Describe and explain how she should use a key to identify the trees from their leaves.

G–E

....

....

....

....

....

[4 marks]

2 Ecologists sample the plants in an area to find out their abundance.

  a Quadrats are used in several sizes. For instance:

    5 cm × 5 cm   20 cm × 20 cm   0.5 m × 0.5 m   0.5 km × 0.5 km

    Which size quadrat would be most suitable for sampling:

    i green algae growing on a tree trunk? .... [1 mark]

    ii trees growing in a wood? .... [1 mark]

    iii daisies growing on the school playing fields? .... [1 mark]

  b The diagram shows the results of a survey to estimate the distribution of dandelions in an area. The quadrats are 0.5 m × 0.5 m.

D–C

    Estimate the distribution of dandelions as plants per m².

.... [2 marks]

  c Ecologists sometimes place quadrats on transect lines to carry out their surveys. Write down **one** situation in which they would do this.

.... [2 marks]

3 In an investigation on the number of species of plant identified growing around the base of an oak tree, the following results were obtained.

Distance from oak tree (m)	0	5	10	15	20	25	30
Number of plant species	0	2	8	14	22	33	34

  a A student produced a hypothesis that this could be the result of the shade from the oak tree.

    Explain how light intensity might affect plant growth.

....

....

B–A*

....

....

[4 marks]

  b On measuring the light intensity at different distances from the tree, the increase in light intensity correlated with the increase in number of plant species present.

    What conclusions can be drawn about the hypothesis?

....

....

....

....

[4 marks]

How do living things obtain energy?

1 Here are some statements about **aerobic respiration**. Some statements are correct, while some are incorrect. Put a tick (✓) in the boxes next to the **two** correct statements.

a Glucose is a product of respiration in plants. ☐

b Some plants carry this out in the absence of oxygen. ☐

c It provides the energy essential for our cells to live. ☐

d One product is carbon dioxide. ☐

e Body cells use starch and protein for aerobic respiration. ☐

[2 marks]

G–E

2 Fill in the spaces below to complete the word equation for respiration.

☐ + ☐ → ☐ + ☐ + ☐

[2 marks]

D–C

3 Aerobic respiration is an essential process in cells.

a Write down the symbol equation for aerobic respiration.

.... **[2 marks]**

b Why does this equation not fully explain the process?

....

.... **[2 marks]**

B–A*

4 Write down **three** situations in living organisms in which anaerobic respiration takes place.

....

....

.... **[3 marks]**

G–E

5 Chemical products of anaerobic respiration in microorganisms are important in certain food products.

a Write down **three** chemical products produced by anaerobic respiration in microorganisms.

....

.... **[3 marks]**

D–C

b Write down **one** type of microorganism that respires anaerobically, and the food product it is used to make.

....

.... **[2 marks]**

6 Write down **three** differences between aerobic respiration and anaerobic respiration. In your answer, refer to:

– the use of oxygen

– the products of the reaction

– the amount of energy released.

....

....

.... **[3 marks]**

B–A*

B4 Extended response question

An enzyme called invertase is often used when brewing beer. It is added to the mixture of brewing ingredients at the beginning of the process.

Invertase breaks down a sugar called sucrose in the mixture, into the sugars glucose and fructose.

Explain:

- how the enzyme invertase works
- why invertase will not break down the sugar maltose, which is also found in the mixture of brewing ingredients
- why pH 4.5 is the optimum for brewing beer.

Use diagrams to illustrate your answer.

The quality of written communication will be assessed in your answer to this question.

D–C

[6 marks]

How organisms develop

1 Here are some statements about cell specialisation in animals. Use the words provided to complete the sentences.

brain   multicellular   nerve cells   nervous tissue   organs

single-celled   specialised   tissue   tissues   unspecialised

a In organisms that are ...., cells are

.... to do different jobs. **[1 mark]**

b Cells of the same type are grouped into a ....

One example of this is ...., which are grouped together

to form .... **[1 mark]**

c Different .... are grouped together, and work together,

in structures called .... One example of this is

the .... **[1 mark]**

> G–E

2 Explain how cells in a human embryo become specialised to do a particular job.

....

....

....

.... **[4 marks]**

> B–A*

3 Here are the names of some plant tissues and organs:

flower   phloem   root   stem   xylem

Put these in the correct columns in the table below.

Plant tissue	Plant organ

[5 marks]

> G–E

4 Here are some statements about cell specialisation and growth in plants. Some statements are correct, while some are incorrect.

Put ticks (✓) in the boxes next to the **two** correct statements.

a Plants keep growing throughout their lives. ☐

b Plant cell division occurs in meristems. ☐

c Plant cell division occurs throughout the shoot and root. ☐

d Meristems produce growth of plant height only. ☐ **[2 marks]**

> D–C

5 Write down **four** areas where meristems are found in plants. For each of these areas, describe how the meristem produces growth.

....

....

....

....

.... **[4 marks]**

> B–A*

Plant development

1 Describe what happens when a shoot is cut off and placed in water. Why is this technique useful?

G–E

....

....

....

[3 marks]

2 Describe **one** method used to produce a plant clone.

D–C

....

....

....

[3 marks]

3 Write down **two** functions of auxins in producing plant clones.

B–A*

....

....

[2 marks]

4 Daniel notices that a plant on the kitchen window sill is growing towards the light. Explain how this is of benefit to the plant.

G–E

....

....

....

[3 marks]

5 Here are some statements about the effects of plant hormones on plant growth. Use the words provided to complete the sentences below.

D–C

all directions   **away from**   **one direction**   **photosynthesis**

phototropism   **respiration**   **towards**

When plants are exposed to light from ...., they

grow .... the light. This response of plants to the direction

of light is called .... Responding to light in this way increases

a plant's chances of survival, as light is required for ....   **[4 marks]**

6 A scientist investigating the response of plants to light placed one group of plants in the dark, one group exposed to light from one direction, and one group in even illumination. The plants were in an atmosphere of radioactive carbon dioxide, and after five hours, the amount of radioactive auxin in the area below the shoot tip was measured. The scientist's results are shown below.

| | Plants in the dark | Plants in the light | Plants exposed to light from one side | |
			Dark side	Lighted side
Total radioactive auxin in counts per minute	3004	2985	2173	878

B–A*

Explain fully what these results tell the scientist about the effect of light on auxin in the plants.

....

....

....

....

....

[4 marks]

Cell division

1 Mitosis is a process that occurs in humans and other organisms.

   a Write down a definition of mitosis.

   ....

   .... [3 marks]

   b When does mitosis occur in human cells?

   ....

   .... [2 marks]

2 One way of estimating the percentage of time spent during and between mitosis is using the formula below:

   $$\text{Percentage of cells in the stage} = \frac{\text{number of cells in the stage}}{\text{total number of cells}} \times 100$$

   A student observed 1000 cells taken from different parts of the human gut. Carry out calculations to complete the table below.

Region of human gut	Number of cells in mitosis	Percentage of total time spent in mitosis
Stomach	22	
Small intestine	39	
Large intestine	13	

   ....

   .... [3 marks]

3 The cell cycle is the series of events a) between and leading up to cell division, and b) cell division itself.

   The following table gives timings of these two phases.

Cell type	Events leading up to cell division, minutes	Cell division (mitosis)
Meristem of pea plant	1300	150
Chick skin cells	700	25
Rat intestine	2000	30
Developing fruit fly egg	3	7

   a What processes are occurring in the events leading up to cell division?

   ....

   .... [3 marks]

   b In which of the cells above does mitosis take up the shortest proportion of the cell cycle?

   .... [1 mark]

   c In which cell type is mitosis longer than the events leading up to it? Suggest why this is.

   ....

   .... [2 marks]

4 Here are some statements about meiosis in humans. Some statements are correct, while some are incorrect.

   Put ticks (✓) in the boxes next to the **two** correct statements.

   a Meiosis is used to produce eggs and sperm. ☐

   b Two daughter cells are produced by meiosis. ☐

   c The number of chromosomes in a cell halves during meiosis. ☐

   d Meiosis takes place during growth and is used to repair tissues. ☐ [2 marks]

Chromosomes, genes, DNA and proteins

1 Here are some statements about chromosomes, genes and DNA. Use the words provided to complete the sentences.

amino acids   chloroplasts   chromosomes   cytoplasm   DNA   genetic   nucleus

.... are thread-like structures made from a chemical

called ....

This chemical carries the .... code, which codes for the production

of .... in the .... of the cell.   **[5 marks]**

2 The diagram below is a simplified diagram of a DNA molecule.

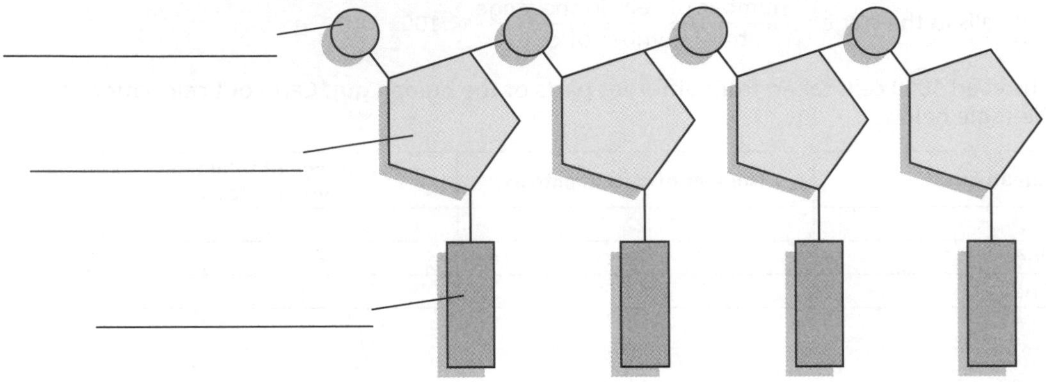

a Label the parts of the molecule shown.   **[3 marks]**

b The DNA molecule is shown flat. Describe how this would appear in three dimensions.

....   **[1 mark]**

3 Here are some statements about protein synthesis. One statement is correct, while the rest are incorrect.

Put a tick (✓) in the box next to the correct statement.

a The genetic code carries the instructions for protein synthesis.   ☐

b In the first stages of protein synthesis, RNA is copied in the nucleus.   ☐

c Protein synthesis takes place in the nucleus.   ☐

d The order of bases in a gene codes for the synthesis of several proteins.   ☐   **[1 mark]**

4 The statements below describe how a protein is produced. They are in the wrong order. Put numbers in the boxes to show the correct sequence. The first one has been done for you.

a In the nucleus, the strands of the DNA molecule separate.   ☐ 1

b The mRNA passes from the nucleus to the cytoplasm.   ☐

c A molecule of mRNA is synthesised, using the DNA as a template.   ☐

d Amino acids are bonded together on a ribosome.   ☐

e Amino acids are ferried to the ribosome.   ☐   **[3 marks]**

G–E

D–C

B–A*

Cell specialisation

1 Here are some statements about the switching on and off of genes in cells.

Some statements are correct, while some are incorrect. Put ticks (✓) in the boxes next to the **two** correct statements.

a A red blood cell makes only the proteins it needs to function. ☐

b In most cells, all the genes are switched on. ☐

c In most cells in the human body, genes are frequently switched on and off. ☐

d In embryonic stem cells, any gene can be switched on. ☐

e Because of their genes, adult stem cells can replace any cell in the body. ☐

[2 marks]

D–C

2 Here are some statements about stem cell research and therapy. Use the words listed below to complete the sentences.

adult   damaged   diseased   embryonic   embryo   limited   many   new

Stem cells have the potential to grow cells to replace .... or

.... cells.

.... stem cells are found at various locations in the body, e.g. the

bone marrow. These cells can be used to produce .... types.   **[5 marks]**

G–E

3 After a science lesson about stem cells, some friends are discussing the use of stem cells in medicine. Here are some quotes:

Michael: 'Collecting embryonic stem cells from an embryo destroys the embryo in the process. I think using them is therefore unethical.'

Ahmed: 'But most embryonic stem cells come from surplus embryos from IVF treatments. So no harm is being done.'

Beatrice: 'If stem cells can treat conditions that are currently incurable, I think it's OK to produce embryos specifically for therapy.'

Maia: 'I think adult stem cells are just as good, so actually, we don't need to worry about killing any embryos.'

Use these ideas to discuss the use of stem cells in therapy.

....

....

....

....

....

....   **[4 marks]**

D–C

4 Complete the table below on the use of stem cells from different sources to replace damaged cells of a patient.

Source of stem cells	One advantage	One disadvantage
Embryo		
Adult		
Therapeutic cloning		
Transformed body cells		

[4 marks]

B–A*

B5 Extended response question

Explain what happens when a plant is exposed to light from one side, and why this helps in the plant's survival.

The quality of written communication will be assessed in your answer to this question.

B–A*

[6 marks]

The nervous system

1 Humans have two communication systems.

  a Write down the names of the two communication systems.

  .... **[2 marks]**

  b Write down **two** differences between the systems.

  ....

  ....

  ....

  .... **[2 marks]**

2 Here are some statements about neurons. Use the words provided to complete the sentences.

central	effectors	eyes	motor	muscles	peripheral

receptors	sensory	stimuli

  .... neurons connect ....,

  which detect ...., with the ....

  nervous system. .... neurons connect the

  .... nervous system to ....,

  e.g. ...., which produce a response.   **[4 marks]**

3 Write down the names of **two** hormones. For each hormone, state:

  – where in the body that it is made

  – the effects it has on the body.

  ....

  ....

  ....

  ....

  .... **[4 marks]**

4 The diagram below shows the structure of a neuron.

  a Label the diagram.   **[5 marks]**

  b On the diagram, draw an arrow showing the direction of the nerve impulse.   **[1 mark]**

5 Write down **three** factors that affect the speed of transmission of a nerve impulse. Describe how each factor affects the speed.

  ....

  ....

  ....

  .... **[3 marks]**

Linking nerves together

1 Here are some statements about the way nerves link with other. Some statements are correct, while some are incorrect.

Tick (✓) the boxes next to the **two** correct statements.

a One nerve can connect physically with many others. ☐

b As a nerve impulse reaches the end of the nerve, a chemical signal is released. ☐

c The junction between one nerve and another is called a synapse. ☐

d Few nerves in the body pass messages to other nerves. ☐ **[2 marks]**

2 The human body is thought to use around 50 different neurotransmitters.

a Give **three** reasons why we need different neurotransmitters.

....

....

.... **[3 marks]**

b How are nerves adapted to work with different neurotransmitters?

....

.... **[1 mark]**

3 Here are some statements about nervous co-ordination. Use the words provided to complete the sentences.

axon	brain	ear	effectors	eye
receptors	spinal cord	stimulus	muscle	organ

The nervous system responds to a change in the environment called a ....

These are detected by special cells called ....

Sometimes these special cells are grouped together or form part of an .... ,

e.g. the .... and the .... **[5 marks]**

4 Here are some statements about nervous co-ordination. Use the words provided to complete the sentences.

central	contraction	enzymes	expansion	heartbeat
hormones	limb movement	peripheral	stimuli	transmitters

a The part of the nervous system that co-ordinates responses is called

the .... nervous system.

b Glands make and release chemicals such as ....

and ....

c Muscles are used for movement. Their .... helps

the body to move away from or towards .... Muscles are

also used for movement we're not conscious of, e.g. our .... **[3 marks]**

Reflexes and behaviour

1 If a bright is shone into your eyes, muscles in the iris of your eye contract, reducing the amount of light that enters your eye.

 a Complete the flow chart below to show this process.

| Light (stimulus) | → | | → | | → | | → | | → | Muscles in iris (effector) |

[4 marks] **G–E**

 b Explain how this reflex is useful.

....

....

[1 mark]

2 You pick up a dinner plate that is hot. The dinner plate is very expensive, and you do not drop it.

 Explain how you have prevented yourself from dropping the plate.

....

....

....

....

[3 marks] **B–A***

3 A scientist carried out an experiment on the behaviour of woodlice. Twenty woodlice were placed in a choice chamber (see diagram), where four different environmental conditions had been produced.

 After one hour, the following results were obtained:

Conditions	Dry		Wet	
	Light	Dark	Light	Dark
Number of woodlice	1	5	3	11

(diagram: circle divided into quarters labelled dry light, wet light, dry dark, wet dark)

 a What percentage of woodlice are found in the light; dark; wet; dry?

....

....

[4 marks] **G–E**

 b What conclusion can be drawn from the experiment?

....
[2 marks]

 c What type of behaviour are the woodlice showing?

....
[1 mark]

 d Suggest why this type of behaviour is essential to simple animals.

....
[2 marks]

4 The doorbell rings and a person's dog starts to bark loudly. Explain how this is an example of a conditioned reflex.

....

....

....

[3 marks] **D–C**

5 The hoverfly is a harmless insect that has black and yellow stripes resembling those of a wasp. Explain how a conditioned reflex that develops in predatory birds increases a hoverfly's chances of survival.

....

....

....

....

[3 marks] **B–A***

The brain and learning

G–E

1 The brain co-ordinates the activities of the body.

  a Label the diagram of the brain below.

  b Write down **four** traits, most developed in humans, that the cerebral cortex is most involved with.

  ....

  ....

  ....

  [4 marks]

D–C

2 When investigating how the brain works, explain the advantages of using techniques such as magnetic resonance imaging (MRI) over invasive techniques.

  ....

  ....

  ....

  ....

  [5 marks]

B–A*

3 List **four** traits linked with the highly developed structure of our brains that make us human.

  ....

  ....

  [4 marks]

G–E

4 Here are some statements about how we learn things. Use the words provided to complete the sentences.

axons	drugs	gaps	impulses	neuron pathway
neurons	preventing	repeating	links	stimuli

  Transmitting impulses between .... in the brain leads

  to .... forming between the neurons. This is called

  a .... These are strengthened by ....

  the experience, so more and more .... follow the same route.

  Another way of strengthening these is using strong .... **[3 marks]**

D–C

5 Explain why children find it easier to learn new skills than adults.

  ....

  ....

  [1 mark]

B–A*

6 Describe and explain what happens if a child is not given the appropriate stimuli early in life.

  ....

  ....

  ....

  ....

  [5 marks]

Memory and drugs

1 Jodie is trying to remember a list of things for her science exam next week.

a Write down a definition of the term **memory**.

.... **[2 marks]**

b Which type of memory will Ruby need to use?.. **[1 mark]**

c Write down **two** ways that might help Ruby to remember items in the list.

.... **[2 marks]**

G–E

2 Complete the diagram of the multi-store model of memory opposite.

[5 marks]

D–C

3 Some friends are discussing how they are revising for their science exam. Here are some quotes:

Amir: 'When I'm preparing, I condense my science notes into key points.'

Justine: 'When I'm revising a list of points, I use the initial letter of each word, and arrange them into a word or list that I can remember easily. It's called a mnemonic.'

Lucas: 'When I've finished reading through my science notes, I write down as much as possible of what I've read.'

Bethany: 'If I listen to loud rock music while I'm revising, it helps things to sink in.'

a Which of the friends has used a stimulus to help them to memorise their science?

.... **[2 marks]**

b Which of the friends is using both processes involved in memory? Explain your answer.

....

.... **[2 marks]**

B–A*

4 Write down the names of **two** groups of chemicals that interfere with nerve impulses moving between a nerve and another nerve, and a nerve and a muscle.

.... **[2 marks]**

G–E

5 When a transmitter substance called acetylcholine crosses a synapse between a nerve and a muscle, it causes the muscle to contract.

Bungarotoxin, a venom produced by the banded krait snake, blocks acetylcholine receptors. Explain what happens to the muscles of someone bitten by a banded krait.

....

.... **[2 marks]**

D–C

6 One of the effects of the drug MDMA (Ecstasy) is to block the re-uptake of a chemical called serotonin into a neuron at a synapse. Serotonin is a chemical transmitter, which in the brain, is important in regulating mood. Explain the science involved when a nerve impulse is transmitted, and the effect of Ecstasy on this.

....

....

....

....

.... **[5 marks]**

B–A*

B6 Extended response question

Use the work of Pavlov to explain how animals can develop and learn a reflex response to a stimulus by introducing a new, unrelated stimulus.

The quality of written communication will be assessed in your answer to this question.

B–A*

[6 marks]

Atoms, elements and the Periodic Table

1 Mendeleev used data about the elements to arrange them into the Periodic Table.

Give **two** types of data that Mendeleev used.

.... **[2 marks]**

G–E

2 The table shows three elements that Döbereiner put into a 'triad'.

Elements	Relative atomic mass
Top element: lithium	7
Middle element: sodium	23
Bottom element: potassium	39

a Work out the mean of the relative atomic mass of the top and bottom element.

.... **[1 mark]**

b Explain why Döbereiner thought sodium fitted as the middle element.

.... **[1 mark]**

D–C

3 Other scientists, such as Newlands and Mendeleev, had different ideas about how to organise the elements.

a Both Newlands and Mendeleev also put lithium, sodium and potassium together in their arrangement of the elements.

Explain why they both thought that these three elements belong together.

.... **[1 mark]**

B–A*

b Give **two** reasons why Mendeleev's arrangement of elements was an improvement on Newlands' arrangement.

....

.... **[2 marks]**

4 The diagram shows the line spectrum of helium and hydrogen.

Scientists used line spectra of elements to find out that hydrogen and helium are in our Sun.

Explain how they did this.

Helium

Hydrogen

....

....

.... **[3 marks]**

D–C

5 Complete the sentences about the structure of an atom. Choose words from the list below.

electrons   groups   ions   molecules   neutrons   protons   shells

An atom contains a tiny nucleus, which contains particles called ....

and .... travel around the outside

of the atom. They are arranged in .... **[2 marks]**

G–E

6 The table shows information about some atoms. Complete the table.

Proton number	Relative atomic mass	No. of protons	No. of neutrons	No. of electrons
9	19			
	27			13
			4	3

[3 marks]

B–A*

1 Lithium and sodium are both elements in Group 1 of the Periodic Table.

The table shows some information about a lithium and a sodium atom.

	No. of protons	No. of neutrons	No. of electrons
lithium	3	4	3
sodium	11	12	11

a The diagram below left shows the electron arrangement in a lithium atom.

Complete the diagram and labels to show the electron arrangement in a sodium atom.

Lithium

Sodium

Electron arrangement: 2.1   Electron arrangement: .... **[2 marks]**

b Potassium is another element in Group 1. Potassium has a proton number of 19.

Describe the **similarities** and **differences** between the electron arrangements of potassium and sodium.

....

....

.... **[3 marks]**

---

2 a The table shows some information about some elements.

Name of element	Symbol	Proton number
neon		10
fluorine		
	Pb	82

Complete the table. Use the Periodic Table to help you.   **[3 marks]**

b Which element is **not** in Period 2 of the Periodic Table? Put a ring around your choice.

Lithium   **Beryllium**   **Carbon**   **Aluminium**   **Oxygen**   **[1 mark]**

---

3 The following statements are about trends across a period of the Periodic Table.
Put a tick (✓) in the boxes next to the **two** correct statements.

From left to right across a period in the Periodic Table...

a the number of electrons in the outer shell increases. ☐

b the elements are more likely to be non-metals. ☐

c proton number decreases. ☐

d the elements are more likely to be solids. ☐   **[2 marks]**

---

4 A textbook gives this statement: *The number of electrons in the outer shell gives information about the group number and the metal or non-metal character of the element.*

Explain what this statement means.

....

....

....

.... **[3 marks]**

Reactions of Group 1

1 The table shows some data about Group 1 elements.

Element	Melting point in °C	Boiling point in °C
lithium	180.0	1330
sodium	97.8	
potassium	63.7	774

a What is the trend in melting point down Group 1?

.... [1 mark]

b Predict the boiling point of sodium. Explain your reasoning.

.... [1 mark]

2 a Joe adds a piece of sodium to some water that contains pH indicator. The boxes show his observations during the reaction.

Use straight lines to connect each observation with each reason.

Observation   **Reason**

| **a** The sodium goes into a ball shape. | | **i** Hydrogen is made. |

| **b** The sodium fizzes and bubbles form. | | **ii** Sodium hydroxide is made. |

| **c** The pH indicator turns blue. | | **iii** The metal melts because the reaction gives out energy. |

[2 marks]

b Joe repeats the experiment. This time he uses potassium rather than sodium. Give **two** similarities and **two** differences between the reactions of potassium and sodium with water.

....

....

.... [4 marks]

3 Eve investigates the reactions of Group 1 metals with chlorine.

a She reacts a hot piece of lithium with chlorine.

Complete and balance the symbol equation for the reaction.

Li   +   ⟶ LiCl   [2 marks]

b Eve reacts sodium and potassium with chlorine. She notices that the reactions get faster when she uses a metal further down the group.

She writes this conclusion in her notes: *I think that for Group 1 elements, the more electron shells there are in the atom, the more reactive the element.*

Do you agree with Eve? Use ideas about electron arrangement to explain your reasoning.

....

....

....

.... [4 marks]

Group 7 – The halogens

1 Draw straight lines to connect each **halogen** to its **state** and **colour** at room temperature and pressure.

State	Halogen	Colour
		red-brown
solid	chlorine	
		dark grey
liquid	bromine	
		purple
gas	iodine	
		pale green

[3 marks]

2 The halogens all contain **diatomic** molecules.

Give the formula for a chlorine molecule and explain why it is diatomic.

.... **[2 marks]**

3 The states of the halogens show a trend down the group.

a Describe the trend in the states of the halogens down the group.

.... **[1 mark]**

b Describe **one** other trend shown by the halogens down the group.

....

.... **[2 marks]**

4 Rose heats some iron wool and puts it into a gas jar that contains chlorine.

The iron wool glows very hot and a brown solid is formed.

a What is the name of the brown solid that is made in the reaction?

b What safety precautions should Rose take when she uses chlorine?

Explain your reasoning.

....

.... **[2 marks]**

c Rose does the experiment again. This time she reacts iron with bromine gas.

How does the rate of the reaction change when Rose uses bromine instead of chlorine?
Explain your answer.

....

.... **[2 marks]**

5 Astatine is a halogen. It is below iodine in Group 7.

a Jane adds some chlorine water to a solution of potassium astatide.

Complete and balance the symbol equation for the reaction that happens.

$$Cl_2 \quad + \quad KAt \quad \longrightarrow \quad KCl \quad + \qquad$$ **[2 marks]**

b Chlorine is much more reactive than astatine. Use ideas about electron shells to explain why.

....

.... **[2 marks]**

Ionic compounds

1 Which of the statements about ionic compounds are true? Put a tick (✓) in the boxes next to the **two** correct answers.

a Ionic compounds are usually gases. ☐

b Ions in a solid ionic compound are arranged in a regular pattern. ☐

c Compounds of Group 1 and Group 7 elements are ionic. ☐

d Ionic compounds have low melting points. ☐

e Ionic compounds conduct electricity when they dissolve in water. ☐

[2 marks]

2 The table shows information about some chemicals.

Chemical	Does it conduct electricity when solid?	Does it conduct electricity when melted?	Melting point
A	no	yes	high
B	no	no	low
C	yes	yes	high

a Which chemical, A, B or C, is a metal?

.... [1 mark]

b Which chemical, A, B or C, is an ionic compound?

.... [1 mark]

3 a How does the electrical conductivity of a solid ionic compound change when its temperature increases above its melting point?

....

.... [2 marks]

b Use ideas about ions to explain why the electrical conductivity changes.

....

.... [2 marks]

4 The table gives some information about a sodium atom and a chlorine atom.

	No. of protons	No. of neutrons	No. of electrons
sodium atom	11	12	11
chlorine atom	9	10	9

The sodium atom reacts to become a sodium ion.

a Give **two** similarities and **one** difference between a sodium atom and a sodium ion. Use the table to help you.

....

.... [3 marks]

b Give **one** difference between what happens when a sodium atom becomes an ion and when a chlorine atom becomes an ion.

.... [2 marks]

5 The formula for sodium chloride is NaCl. The formula for calcium chloride is $CaCl_2$.

Use ideas about charges on the ions to explain why the formulae are different.

....

....

....

.... [4 marks]

G–E

G–E

D–C

D–C

B–A*

C4 Extended response question

Lithium is at the top of Group 1 in the Periodic Table. Fluorine is at the top of Group 7 in the Periodic Table.

The position of an element in the Periodic Table gives information about:

• the arrangement of the outer-shell electrons.
• whether the element is a metal or a non-metal
• the reactivity of the element.

Describe the differences between lithium and fluorine based on their positions in the Periodic Table.

The quality of written communication will be assessed in your answer to this question.

[6 marks]

Molecules in the air

1 Air contains a mixture of gases

Draw straight lines to connect each **gas** with its correct **formula** and **percentage** in the air.

Formula	Gas	Percentage
O_2	nitrogen	21%
Ar	oxygen	78%
CO_2	argon	0.04%
N_2	carbon dioxide	about 1%

[2 marks]

2 The atoms in the molecules of gases in the air are held together by covalent bonds.

Which of the following statements about covalent bonds are true? Put a tick (✓) in the boxes next to the **two** correct answers.

a When a covalent bond forms, atoms lose or gain electrons. ☐

b The nuclei of both atoms are attracted to the electrons in the bond. ☐

c Atoms share electrons to form a bond. ☐

d Very little energy is needed to overcome the attraction between atoms. ☐

e The bonds are always arranged in a 2-D arrangement. ☐

[2 marks]

3 The table shows data about some elements in the air.

Substance	Melting point (°C)	Boiling point (°C)
nitrogen	−210	−196
oxygen	−218	−183
argon	−189	−186

a Which element is a liquid over the largest temperature range?

.... **[1 mark]**

b Use ideas about forces between molecules to explain why the melting points and boiling points of the elements are all very low.

....

.... **[2 marks]**

c Ben looks at the table. He writes this note in his book: *Nitrogen has the lowest boiling point. There is a link because gases with very low melting points have very low boiling points.*

Do you agree with Ben? Explain your reasoning.

....

....

....

....

.... **[4 marks]**

Ionic compounds: crystals and tests

1 The table shows some information about solid sodium chloride.

Melting point	Solubility in water	Electrical conductivity
801 °C	very soluble	does not conduct when solid

a Explain why solid sodium chloride has a high melting point.

....

....

.... **[3 marks]**

b Explain why sodium chloride does not conduct electricity when it is solid.

.... **[1 mark]**

c Describe what happens to the arrangement and movement of the ions when sodium chloride dissolves in water.

....

....

.... **[2 marks]**

2 The table shows the ions and formulae for some ionic compounds. Complete the table.

Name	Positive ion	Negative ion	Formula
potassium chloride	K^+	Cl^-	KCl
calcium chloride	Ca^{2+}	Cl^-	
potassium sulfate			
calcium sulfate	Ca^{2+}	SO_4^{2-}	$CaSO_4$

[3 marks]

3 The table below shows the results for tests in which dilute sodium hydroxide was added to known cations.

Ion	Observation
calcium Ca^{2+}	a white precipitate forms; the precipitate does not dissolve in excess sodium hydroxide
copper Cu^{2+}	a light blue precipitate forms; the precipitate does not dissolve in excess sodium hydroxide
iron(II) Fe^{2+}	a green precipitate forms; the precipitate does not dissolve in excess sodium hydroxide
iron(III) Fe^{3+}	a red-brown precipitate forms; the precipitate does not dissolve in excess sodium hydroxide
zinc Zn^{2+}	a white precipitate forms; the precipitate dissolves in excess sodium hydroxide

Katy does some tests on a salt. These are her results.

Test	Observation
Add dilute sodium hydroxide	White precipitate that does not dissolve in excess
Add dilute acid	Fizzing

a Use the table to identify the salt that Katy tests. .... **[2 marks]**

b Katy adds dilute sodium hydroxide to a zinc salt. She compares her results to the results in the table.

Give **one** similarity and **one** difference Katy will see when she tests a zinc salt compared to the results in the table.

....

.... **[2 marks]**

c The precipitate does not dissolve in excess sodium hydroxide. What word can be used to describe a precipitate that does not dissolve? Put a ring around the correct answer.

ionic   insulator   insoluble   soluble   solution   **[1 mark]**

d Write an ionic equation, with state symbols, for the reaction that happens when dilute sodium hydroxide is added to the salt.

.... **[2 marks]**

C5 Chemicals of the natural environment

Giant molecules and metals

1 The table shows information about the percentages of the elements in the Earth's crust.

Element	oxygen	silicon	aluminium	iron	other elements
Percentage in the Earth's crust	47	28	8	5	

a What percentage of the Earth's crust is other elements? .... **[1 mark]**

b Quartz is a common mineral. It contains the compound silicon dioxide.

Which elements does silicon dioxide contain?

.... **[1 mark]**

c Iron and aluminium are both extracted from their oxides.

What type of reaction happens when a metal is extracted from its oxide? Put a ring around the correct answer.

conduction   **neutralisation**   **oxidation**   **reduction**   **[1 mark]**

2 Copper extraction causes environmental problems due to large amounts of waste rock.

Explain why copper mining produces so much waste rock.

....

....

.... **[2 marks]**

3 The diagrams show the structures of diamond and carbon dioxide.

Carbon dioxide

Diamond

a Carbon dioxide has a simple molecular structure and diamond has a giant covalent structure.

Describe the main differences between these two types of structure.

....

....

.... **[3 marks]**

b Tick (✓) to show which of the following statements about carbon dioxide and diamond are true and which are false.

	True	False
i Both carbon dioxide and diamond conduct electricity.	☐	☐
ii Carbon dioxide has a lower melting point and boiling point than diamond.	☐	☐
iii The atoms in both substances are held together by shared electrons.	☐	☐
iv Each carbon atom in both substances are bonded to four other atoms.	☐	☐

[4 marks]

Equations, masses and electrolysis

1 Iron is extracted from iron ore in a blast furnace. In the furnace, carbon monoxide reacts with iron oxide to make iron: iron oxide + carbon monoxide $\longrightarrow$ iron + carbon dioxide.

Tick (✓) to show which substances are reactants and which are products in the reaction.

	iron oxide	carbon monoxide	iron	carbon dioxide
reactant				
product				

[1 mark]

2 Copper can be extracted from copper oxide by heating with carbon.

Complete the word and symbol equations for this reaction by filling in the missing names and formulae.

copper oxide   +   carbon $\longrightarrow$ +

     $2CuO$ + C $\longrightarrow$ $2Cu$ +   **[2 marks]**

3 a The table shows the relative atomic masses of some elements and the relative formula masses of some compounds.

Complete the table by filling in the missing information. Use the Periodic Table to help you.

Name	Formula	Relative atomic masses of each element	Relative formula mass
sodium chloride	NaCl	Na: 23 Cl: 35.5	
magnesium chloride	$MgCl_2$	Mg: Cl: 35.5	95
calcium sulfate	$CaSO_4$	Ca: 40 S: 32 O: 16	

[3 marks]

b What is the gram formula mass of magnesium chloride?

.... **[1 mark]**

4 Aluminium is extracted from aluminium oxide (Al_2O_3) by electrolysis.

a Why is aluminium not extracted by reacting its oxide with carbon? Put a tick (✓) in the box next to the correct answer.

i The melting point of aluminium oxide is high. ☐

ii Aluminium is a very reactive metal. ☐

iii Aluminium oxide is a very reactive compound. ☐

iv The density of aluminium is too low. ☐   **[1 mark]**

b Calculate the percentage of aluminium in aluminium oxide.

....

.... **[2 marks]**

c The equation for the reaction that happens at the positive electrode is: $2O^{2-} \longrightarrow O_2 + 4e^-$

i Explain why oxygen forms at the positive electrode.

....

....

.... **[2 marks]**

ii Write an equation to show what happens at the negative electrode.

....

.... **[2 marks]**

Metals and the environment

1 The table shows some metals and their properties.

Metal	Melting point (°C)	Electrical conductivity	Mass of 1 cm³ of the metal in g	Corrosion resistance	Cost
aluminium	660	good	2.70	does not corrode	high
iron	1535	fair	7.90	corrodes quickly	medium
copper	1083	excellent	8.90	corrodes quickly	high

a Copper is used to make electrical wiring for homes.

  i Give **one** reason why copper is a good choice for making electrical wiring.

.... [1 mark]

  ii Give **two** disadvantages of using copper for electrical wiring.

....

.... [2 marks]

b Aluminium is used to make overhead electrical cables.

  Explain why aluminium is a good choice for making overhead electrical cables.

....

.... [2 marks]

2 Read the information about mercury mining.

Mercury mining

Low-energy bulbs are used in the UK. They reduce carbon dioxide emissions from power stations because they use less energy than standard bulbs.

The bulbs contain mercury. Large amounts of mercury are mined in China.

People living near mercury mines need the jobs that the mines provide, but they complain that:

• children and animals are sick and die because water containing toxic mercury runs into the drinking-water supply

• extracting mercury from the ore gives off toxic gases that harm people living nearby.

Some environmental groups think that all mercury mining should be stopped.

a Use ideas about cost and benefit to explain why we mine mercury even though it is toxic.

....

....

....

....

.... [4 marks]

b Light bulbs contain metals other than mercury. Metals are used because they conduct electricity.

  Explain what happens when a metal conducts electricity.

....

....

.... [2 marks]

C5 Extended response question

Silicon dioxide and diamond are both giant covalent structures.

Silicon dioxide

Diamond

Discuss the similarities and differences between the properties and structures of silicon and diamond.

The quality of written communication will be assessed in your answer to this question.

B–A*

....

....

....

....

....

....

....

....

....

....

....

....

....

....

....

....

....

....

....

....

....

....

[6 marks]

Making chemicals, acids and alkalis

1 Jack uses petrol for his lawn mower. He buys the petrol in a can from a local garage. Petrol is very flammable.

   a What hazard symbol should be shown on the can? Put a ring around the correct answer.

   [1 mark]

G–E

   b What safety precautions should Jack take when he is handling the petrol?

   ....

   .... [2 marks]

2 Pure acid compounds have different states at room temperature and pressure.

   a Draw straight lines to connect each **pure acid compound** with its correct **state symbol**.

   pure acid compound   **state symbol**

   | (s) |

   sulfuric acid

   | (l) |

   citric acid

   | (g) |

   hydrochloric acid

   | (aq) | [3 marks]

D–C

   b Which acid in the table reacts with calcium to form a salt with the formula $CaSO_4$?

   .... [1 mark]

   c What other product is made in the reaction?

   .... [1 mark]

3 Tahira reacts copper carbonate with hydrochloric acid.

   a Complete the word and balanced symbol equation for the reaction.

   copper   +   hydrochloric $\longrightarrow$ +   +
   carbonate   acid

   $CuCO_3$ +   HCl $\longrightarrow$ $CuCl_2$ +   +   [3 marks]

   b Tahira puts a pH probe into the acid at the start of the reaction. She follows the pH changes as she adds the copper carbonate to the acid. She keeps adding copper carbonate until the reaction stops.

   Describe and explain the pH changes that Tahira sees during the reaction.

   ....

   ....

   ....

B–A*

   .... [3 marks]

   c Tahira makes some copper nitrate by adding copper carbonate, $CuCO_3$, to nitric acid, HNO_3.

   What is the formula for copper nitrate?

   .... [1 mark]

Reacting amounts and titrations

1 The table shows information about some compounds.

Complete the table. Use the Periodic Table to find any relative atomic masses that you need.

Name	Formula	Relative formula mass
magnesium oxide		24 + 16 = 40
sodium oxide	Na_2O	
	Na_2CO_3	

[4 marks]

2 Zinc reacts with hydrochloric acid to make zinc sulfate.

The balanced symbol equation for the reaction is: $Zn + H_2SO_4 \rightarrow ZnSO_4 + H_2$

a Explain why this equation is said to be 'balanced'.

....

.... [2 marks]

b What is the maximum mass of zinc carbonate that can be made from 130 g of zinc? (The relative atomic mass of zinc is 65.)

Use the Periodic Table to help you to find any relative atomic masses that you need.

....

....

.... [3 marks]

3 Rose wants to do a titration to find out how much sodium hydroxide is needed to neutralise 25 cm³ of hydrochloric acid. She has some dilute hydrochloric acid and an indicator. She also has some dilute sodium hydroxide in a burette.

a Describe how Rose should do the titration. Your answer should include details of what Rose should do to make sure that her results are as close to the true value as possible.

....

....

....

.... [4 marks]

Rose does further titrations using 25 cm³ samples of different concentrations of hydrochloric acid. She uses the same sodium hydroxide each time. Her results are shown below.

Acid	1	2	3	4
Volume of sodium hydroxide needed in cm³	50	25	20	25

b Which of the following statements about the results are true and which are false? Put a tick (✓) in one box in each row.

	True	False
a Acid 1 is the most concentrated acid.	☐	☐
b Acid 3 would react fastest with calcium carbonate.	☐	☐
c Acids 2 and 4 have the same concentration.	☐	☐
d Acid 4 is more dilute than acid 3.	☐	☐

[2 marks]

Explaining neutralisation & energy changes

1 Join the boxes to show which **acid** and **alkali** react together to make each salt.

Acid	Alkali	Salt
hydrochloric acid		
	calcium hydroxide	potassium chloride
citric acid		
	sodium hydroxide	calcium sulfate
nitric acid		
	potassium hydroxide	sodium nitrate
sulfuric acid		

[3 marks] D–C

2 Every reaction between an acid and an alkali makes a salt and one other product.

Give the name of this other product. .... **[1 mark]**

3 Neutralisation happens when ions from the acid react with ions from the alkali.

a Give the name and formula of the ion that is present in all acids.

Name: .... Formula: .... **[1 mark]**

b Give the name and formula of the ion that is present in all alkalis.

Name: .... Formula: .... **[1 mark]**

c Write an ionic equation for the neutralisation reaction between an acid and an alkali.

.... **[1 mark]**

4 Sulfuric acid reacts with sodium hydroxide to make sodium sulfate. The reaction is exothermic.

Complete the sentences by putting a ring around the correct words in **bold** in each line.

a During the exothermic reaction the temperature **increases** / **decreases**.

b Energy is **given out** / **taken in**.

c The reaction is an example of **neutralisation** / **combustion**. **[1 mark]**

G–E

5 This is the energy level diagram for the reaction between sulfuric acid and sodium hydroxide.

Energy

Reactants

Energy change

Products

Explain why the products are shown lower than the reactants.

....

.... **[2 marks]**

D–C

6 Explain why it is very important to control exothermic reactions when they are used in industry.

....

....

.... **[2 marks]**

B–A*

Separating and purifying

1 Fay reacts some calcium carbonate with some nitric acid in a beaker to make a salt.

— salt dissolved in water

— solid calcium carbonate

At the end of the reaction, the beaker contains the salt dissolved in water and some leftover solid calcium carbonate.

a What is the name of the salt that Fay has made? .... **[1 mark]**

b Describe how Fay can separate the solid calcium carbonate from the solution of the salt.

....

.... **[2 marks]**

c Which of the following statements about the experiment are true and which are false?

Put a tick (✓) in one box in each row.

	True	False
a Calcium carbonate is insoluble.	☐	☐
b The beaker contains a pure product.	☐	☐
c There is some acid left over at the end of the reaction.	☐	☐
d The beaker contains a solution.	☐	☐

[2 marks]

2 The diagram shows a flow chart for recrystallisation.

Step 1: dissolve crystals in small amount of hot water	→	Step 2: filter	→	Step 3: cool until crystals form	→	Step 4: filter	→	Step 5:

a Give **two** reasons why it is important to use hot water in Step 1.

.... **[2 marks]**

b Explain why the mixture must be filtered in Step 2 and in Step 4.

....

.... **[2 marks]**

c What is done to the crystals in Step 5?

.... **[2 marks]**

3 Ali makes some zinc chloride crystals by reacting zinc with hydrochloric acid:

zinc + hydrochloric acid → zinc chloride + hydrogen

Zn + $2HCl$ → $ZnCl_2$ + H_2

Ali uses 6.5 g of zinc in his experiment. The relative atomic mass of zinc is 65.

a What is the theoretical yield of zinc chloride in Ali's experiment? Use the Periodic Table to find any relative atomic masses that you need.

.... **[2 marks]**

b Ali weighs his product at the end of his experiment. He has made 10.2 g zinc chloride.

Calculate the percentage yield for his experiment.

.... **[2 marks]**

c Ali forgot to dry his crystals. What effect will this have on his percentage yield?

Explain your reasoning.

.... **[2 marks]**

Rates of reaction

1 Ray investigates the rate of reaction between solid calcium carbonate and dilute hydrochloric acid. He investigates the effect of changing the concentration of the dilute hydrochloric acid. He carries out four experiments using different concentrations of acid. He measures the time taken to collect 20 cm³ of gas.

The table shows his results.

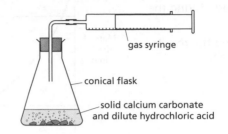

gas syringe

conical flask

solid calcium carbonate and dilute hydrochloric acid

Experiment	Concentration of hydrochloric acid in g / dm³	Time taken to collect 20 cm³ gas in s
1	30	10
2	15	20
3	7.5	40
4	1.5	60

a What factors should Ray control in the experiments?

....

.... **[2 marks]**

b Write a conclusion to summarise what the results show.

....

.... **[2 marks]**

c Ray thinks that the result for Experiment 4 is an outlier. Explain why he thinks this.

....

.... **[2 marks]**

---

2 Sara does an experiment to investigate the rate of reaction between magnesium and hydrochloric acid. She follows the rate of reaction by measuring the volume of hydrogen given off. The graph shows her results.

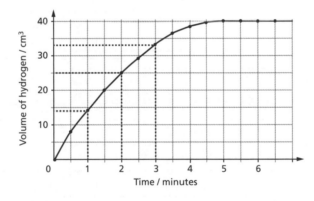

The reaction between magnesium and dilute hydrochloric acid.

a How often did Sara measure the volume of hydrogen? Circle the correct answer.

every minute   every 30 s   every second   every 6 minutes   **[1 mark]**

b How long did it take the reaction to finish? .... **[1 mark]**

c Use ideas about collisions to explain how and why the gradient of the line changes throughout the reaction.

....

....

....

.... **[4 marks]**

Joe did an experiment. He added excess solid calcium carbonate to a dilute acid in a flask.

He measured the mass of the flask during the reaction.

He plotted the results on a graph.

Mass of flask in g

Time in s

Explain how and why the shape of the graph changes during the reaction.

The quality of written communication will be assessed in your answer to this question.

[6 marks]

Speed

1 A cyclist travels 60 km in 3 hours. Calculate his speed in:

a kilometres per hour: .... [2 marks]

b metres per second: .... [2 marks]

G–E

2 How long would it take the cyclist in Question 1 to travel 100 km if he travels at the same average speed?

.... [2 marks]

D–C

3 The average speed of a train travelling from London to Birmingham is 150 km/h. Explain whether the train's average velocity will be lower or higher than its average speed.

....

.... [3 marks]

B–A*

4 Match these descriptions of motion with the distance–time graphs below, by writing the correct description next to each graph:

Constant high speed   **Constant slow speed**   **Speeding up**   **Stationary**

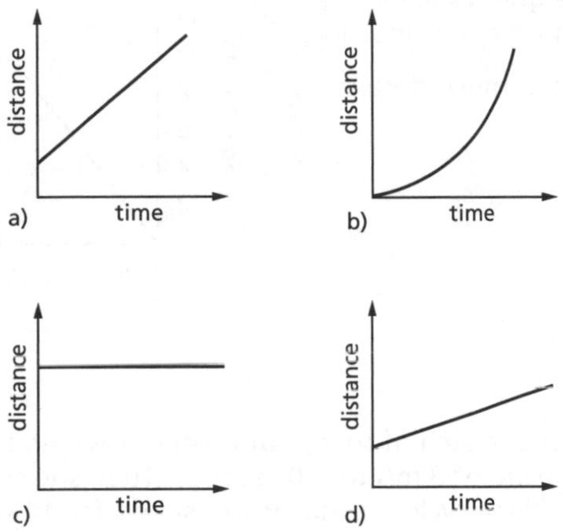

a)   b)

c)   d)

[4 marks]

G–E

5 Jennifer draws a distance–time graph for a walk to the local shop and back again (shown opposite).

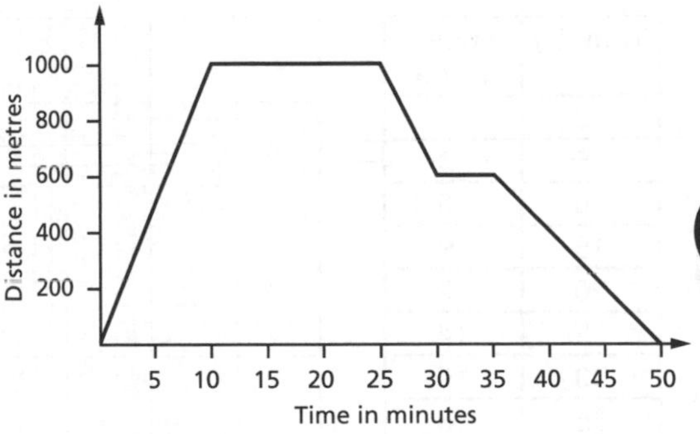

a How far does Jennifer travel in the first 10 minutes?

.... [1 mark]

b Describe the rest of Jennifer's journey.

....

....

....

.... [4 marks]

D–C

c Calculate Jennifer's speed in metres per second during the first 10 minutes.

.... [2 marks]

d What was the total distance for the whole journey? .... [1 mark]

e What was the total displacement for the whole journey? .... [1 mark]

f Calculate both the average speed and the average velocity for the whole journey, and comment on your answers.

....

....

....

.... [4 marks]

B–A*

Acceleration

1 Car A does 0 to 100 km/h in 12 seconds. Car B does 0 to 100 km/h in 10 seconds.

Which of these cars has the greatest acceleration? .... **[1 mark]**

2 Convert 100 km/h into metres per second and calculate the acceleration of the two cars above in metres per second squared.

....

....

....

.... **[5 marks]**

3 Joshi drops a ball out the window. It accelerates from rest at 10 m/s² for 1.6 seconds, then hits the ground. At what speed does the ball hit the ground?

....

.... **[2 marks]**

4 Diana cycles along a straight road. The graph shows how her speed changes during the first minute of her journey.

Describe how Diana's speed changes in as much detail as possible.

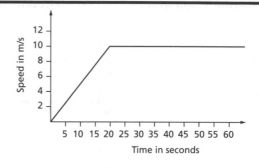

....

....

....

.... **[4 marks]**

5 The next day Diana cycles along the same stretch of road, but accelerates steadily for the first 20 seconds and then stays at a steady speed of 8 m/s for 20 seconds. Then she slows down steadily for 10 seconds until her speed is 5 m/s. Diana cycles at this slower speed for 10 seconds.

Draw a line on the graph in Question 4 to represent this journey.   **[4 marks]**

6 The table below shows how the speed of a vehicle changes over a minute.

Time (s)	Speed (m/s)
0	0
5	3
10	6
15	9
20	12
25	14
30	16
35	16
40	16
45	16
50	12
55	8
60	4

a Plot a speed–time graph on the axes.   **[4 marks]**

b Use the graph to calculate the acceleration during the first 20 seconds.

.... **[2 marks]**

c Calculate the deceleration in the last 15 seconds.

....

.... **[4 marks]**

Forces

1 Complete the passage below using words from this list:

a downwards   **a sideways**   **an upwards**   **space**   **pairs**

Forces always occur in .... When you lean on a table,

you exert .... force from your hand onto the table.

The table exerts .... force on your hand.   **[3 marks]**

G–E

2 For each the following situations, one force is described. Describe the second force that makes up the interaction pair.

a A tennis racket pushes a tennis ball forwards.

....   **[2 marks]**

b A horse pulls a cart forwards.

....   **[2 marks]**

D–C

c The Moon is attracted towards Earth.

....   **[2 marks]**

3 Add to the diagram below to show the two forces acting on this book lying on top of a table.
Label each force.

book
table

[4 marks]   B–A*

4 Joe did an experiment to find the maximum friction between a block of wood and three different surfaces. His results are shown in the table below.

Surface	Friction force (N)			Average force (N)
	Trial 1	Trial 2	Trial 3	
Glass	9.3	9.6	10.2	
Carpet	24.5	32.6	26.4	
Wood	15.0	18.1	16.4	

G–E

a Why did Joe repeat the experiment three times for each surface?

....   **[1 mark]**

b Identify an anomalous reading. ....   **[1 mark]**

c What did Joe have to keep the same for all his trials in this experiment?

....   **[1 mark]**

d Calculate the average value for force for each surface. Add your answers to the table.

....

....   **[3 marks]**

e Which surface has the least friction? Explain your answer.   D–C

....

....   **[2 marks]**

5 Friction can be wanted or unwanted. On a bicycle, give one example of useful friction and one example of unwanted friction. Explain your answers.

....   B–A*

....   **[4 marks]**

Effects of forces

1 The diagrams below show some forces acting on a spacecraft. For each diagram **a–d**, state whether the net force is **upward** or **downward** and give it a value.

100 N   200 N   100 N   200 N

150 N   150 N   200 N   100 N

a .... b .... c .... d ....

....   **[4 marks]**

2 Timothy drops a ball. Once the ball has left his hands there are two forces acting on the ball.

a Draw and label arrows on the ball opposite to show the direction of the two forces.

[2 marks]

b As the ball accelerates downwards, what can you say about the relative size of each force?

....   **[1 mark]**

c Describe what will happen to the speed of the ball if it falls far enough.

....

....   **[2 marks]**

3 When a vehicle crashes into a wall, large forces are involved. Tick (✓) to show which of the following will reduce the size of the forces.

a a larger (more massive) vehicle ☐   **c** increasing the time to stop ☐

b slower speed ☐   **d** decreasing the time to stop ☐   **[2 marks]**

4 Calculate the momentum of the following objects:

a A ball of mass 300 g travelling at a speed of 30 m/s.

....   **[2 marks]**

b A car of mass 1100 kg travelling at a speed of 15 m/s.

....   **[2 marks]**

c A lorry of mass 4000 kg travelling at a speed of 5 m/s.

....   **[2 marks]**

5 Class 11B are playing cricket. Nigel catches the ball when it is going fast and it makes his hands sting.

Using ideas about momentum, explain how Nigel could prevent his hands stinging.

....

....

....

....   **[4 marks]**

Work and energy

1 Complete the passage below, which explains a law about energy, using words from this list:

    generated   conservation   destroyed   transferred   devastation

Energy can never be created or .... It can only

be .... from one form to another or from one place to

another. This is called the .... of energy.   **[3 marks]**   G–E

2 Tick (✓) to show which activities are examples of work being done.

  a lifting a weight ☐   **c** kicking a football ☐

  b sitting on a chair ☐   **d** watching a film ☐   **[2 marks]**

3 Sophie lifted a ball weighing 3 N from the floor onto a shelf 1.3 m above the ground.

  a Calculate how much work Sophie did.

  .... **[2 marks]**

  b What sort of energy did the ball gain?

  .... **[1 mark]**   D–C

  c The ball then rolls off the shelf and falls to the floor. What is the maximum kinetic energy of the ball when it hits the floor?

  .... **[1 mark]**

  d What is the maximum speed at which the ball (of mass 0.3 kg) strikes the floor?

  ....   B–A*

  .... **[3 marks]**

4 A pendulum swings from side to side.

  a Mark on the diagram where the pendulum has maximum kinetic energy and where it has maximum potential energy. G–E

                                                   [2 marks]

  The pendulum bob has a weight of 1 N, and the height from which it is released is 0.2 m above its lowest point.

  b Calculate the initial gravitational potential energy.

  .... **[2 marks]**

  c At one point the pendulum is travelling at a speed of 1.2 m/s and its mass is 0.1 kg. What is its kinetic energy? D–C

  .... **[2 marks]**

  d At this point, what would be the gravitational potential energy of the pendulum bob?

  .... **[2 marks]**

5 **a** Describe the energy changes on a roller coaster.

  ....

  .... **[3 marks]**

  b Explain why the second peak must be lower than the starting point of the roller coaster, as shown in the diagram. B–A*

  ....

  ....

  .... **[3 marks]**

P4 Extended response question

Describe the safety features that car manufacturers build into their cars.

The quality of written communication will be assessed in your answer to this question.

[6 marks]

Electric current – a flow of what?

1 Atoms contain electrons, neutrons and protons. Draw lines to connect the position in the atom with the particle and its charge.

Position

Inside the nucleus

Orbiting the nucleus

Particle

Electron

Proton

Neutron

Charge

Positive

Neutral

Negative

[4 marks] **G–E**

2 Use words from this list to complete the paragraph below.

electrons   positive   negative   protons   neutrons   neutral   attracted   repelled

Lucy rubs a polythene rod with a duster. Some .... are transferred

from the duster to the rod. This makes the rod gain a ....

charge. When she puts the charged rod close to some small pieces of paper they are

.... to the rod.

[3 marks]

3 Explain why a metal rod will not become charged by rubbing.

.... [2 marks] **D–C**

4 Sadie is ironing some polyester shirts. When she has finished she switches off the iron at the wall socket and receives a small electric shock, which tickles her finger. Explain why.

....

....

....

....

....

B–A*

[5 marks]

5 John makes up a circuit to test some objects to find out whether they conduct electricity. He connects the objects between the crocodile clips in the circuit opposite.

Circle the objects in the list below that make the bulb light up.

2p coin   plastic ruler   steel scissors   eraser

G–E

[2 marks]

6 Explain how energy is transferred from the cell to the light bulb in this circuit.

....

....

....

D–C

[5 marks]

7 a In the circuit diagram for Question 6, draw an arrow to show the direction in which the charged particles flow when the current flows.

[1 mark]

b Explain why this happens.

....

.... [2 marks] **B–A***

c Explain how you could make the bulb glow brighter.

....

....

[2 marks]

Current, voltage and resistance

1 John is trying to measure the current and voltage in a simple circuit. He needs to add a voltmeter and an ammeter to the circuit.

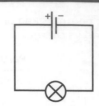

   a Add to the circuit diagram to show the correct symbols and positions for the voltmeter and ammeter.

                                                      [4 marks]

   b Describe what happens to the charged particles in the above circuit.

....

....

.... **[3 marks]**

2 What is meant by the term 'potential difference'?

....

....

.... **[3 marks]**

3 List the four circuits opposite in order of decreasing electrical current.

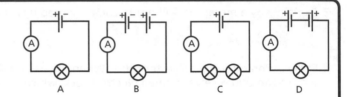

.... **[3 marks]**

4 A motor is connected to a power supply. When the voltage is 12 V the current is 0.5 A. Calculate the resistance (include the unit).

.... **[3 marks]**

5 John carried out an experiment to find the resistance of a piece of nichrome wire. He varied the potential difference across the piece of wire and recorded the current. The table shows his results.

Potential difference in volts	Current in amps
2	0.12
4	0.25
6	0.34
8	0.44
10	0.59
12	0.72

   a Plot these results on the axes opposite and draw in the straight line of best fit. **[2 marks]**

   b Calculate the average resistance of the wire.

.... **[2 marks]**

   c Suggest how John could improve his experiment.

....

.... **[2 marks]**

Useful circuits

1 Jody connects some bulbs in a circuit and measures the current. She first records the current with one cell and two bulbs in series.

  a What would happen to the current if Jody were to add another bulb in series?

.... [1 mark]

  b What would happen to the current if Jody used a second cell in series?

.... [1 mark]

  c What would happen to the current if one of the bulbs 'blew'?

.... [1 mark]

G–E

2 Abdul is investigating some sets of Christmas tree lights. One set has 20 bulbs connected in parallel and the other has 20 similar bulbs connected in series. Both are connected to a 24-V power supply.

  a How could Abdul easily tell which was which?

.... [1 mark]

  b What would happen to each set of lights if he unscrewed one bulb?

....

.... [2 marks]

D–C

3 In the circuit shown below, what would be the readings on ammeters A_1, A_2, A_3 and A_4?

  A_1: ....

  A_2: ....

  A_3: ....

  A_4: ....

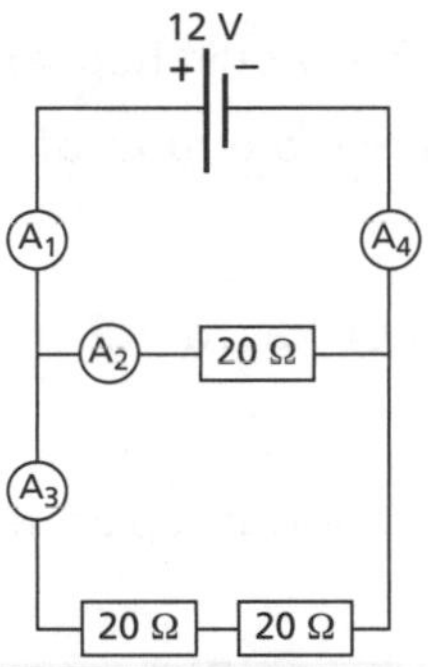

[4 marks]

B–A*

4 Describe the difference between an LDR and a thermistor.

....

.... [3 marks]

G–E

5 Thomas wants to use a thermistor as a thermometer. He needs to calibrate it before he can use it to measure temperatures. Describe how he could calibrate it.

....

....

....

.... [4 marks]

D–C

6 Rachel is investigating current in bulbs. She connects a bulb to a 12-V power supply and records the current flowing through the power supply as 0.7 A. She then connects a second identical bulb in series.

  a Why does Rachel expect the new current to be 0.35 A?

....

.... [2 marks]

  b The actual current is measured as 0.5 A. Explain why this current is higher than Rachel expected.

....

....

.... [4 marks]

B–A*

Producing electricity

1 Complete the paragraph below using words from this list:

sides   field   metal   magnetic   iron   wood   compasses   poles   force

A magnetic .... is the space around a magnet where

a .... material will experience a force. To see the shape of it we

can use .... filings or plotting ....

The force is stronger at the .... of the magnet.   **[5 marks]**

---

2 The diagram shows a simple generator.

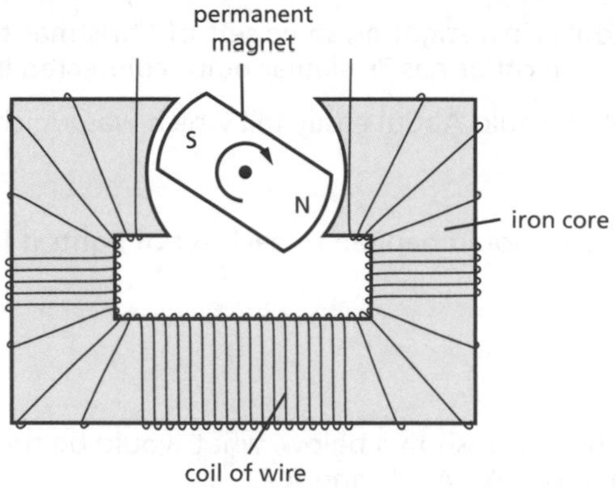

permanent magnet

iron core

coil of wire

(*Source: OCR (A182) Twenty First Century Science – Physics A*)

Describe what would happen to the output if:

a there were more turns of wire in the coil.

....   **[1 mark]**

b the magnet was rotated in the opposite direction.

....   **[1 mark]**

c the iron core was replaced by an aluminium one.

....   **[1 mark]**

---

3 Some electrical devices use d.c. and some use a.c. Complete the tick chart below to show which devices use which type of electricity.

	a.c. (✔)	d.c. (✔)			a.c. (✔)	d.c. (✔)
a Torch				c Computer		
b Transformer				d Electric iron		

[4 marks]

---

4 The mains voltage in the UK is 230-V a.c. A bar heater draws a current of 3.5 A when one bar is switched on.

Calculate the power used by the heater.

....

....   **[3 marks]**

---

5 Julian decides to replace his filament bulbs with compact fluorescent tubes because they are more efficient.

a The power rating for the filament bulb is 100 W. Calculate the current flowing in the bulb when connected to the mains at 230 V.

....

....   **[3 marks]**

b If the compact fluorescent tube uses a quarter the current of the filament to produce the same amount of light, calculate the power rating for the compact fluorescent tube.

....

....   **[2 marks]**

Electric motors and transformers

1 Add to this diagram to show the magnetic field lines around the coil. The field lines will go from left to right.

[2 marks]

D–C

2 This motor is a simple coil suspended in a magnetic field. The coil starts in the position shown. When a p.d. is applied to the connecting wires the coil rotates.

coil

brushes

split-ring commutator

Explain how the commutator and brushes enable the coil to experience continuous rotation.

....

....

....

....

.... [5 marks]

B–A*

3 Complete the paragraph using words from this list to explain how transformers work.

primary	voltage	iron	secondary	direct
current	alternating	copper	coils	magnets

Transformers are used to change the .... of electrical power

supplies. They only work with .... current. A transformer consists

of two .... wrapped around an ....

core. The input voltage is connected across the .... coil and the

new voltage will be output across the .... coil. [6 marks]

G–E

4 The diagrams below show four transformers (a–d), each with a different number of coils.

200 turns / 10 turns	20 turns / 1000 turns	20 turns / 10 turns	200 turns / 200 turns
230-V a.c. V_1	230-V a.c. V_2	230-V a.c. V_3	230-V a.c. V_4
a	**b**	**c**	**d**

i For each transformer, state whether it is a step up or step down transformer.

....

.... [4 marks]

ii List the output voltages (V_1–V_4) in order from lowest to highest.

.... [1 mark]

D–C

5 Explain why you cannot use a transformer to change the voltage from a battery.

....

....

....

.... [4 marks]

B–A*

P5 Extended response question

This diagram shows a simple generator. Describe how it produces electricity.

The quality of written communication will be assessed in your answer to this question.

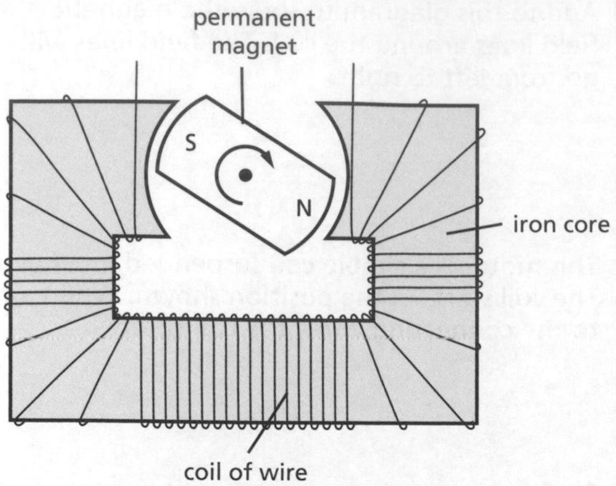

permanent magnet
S
N
iron core
coil of wire

(Source: OCR (A182) Twenty First Century Science - Physics A)

D–C

[6 marks]

Nuclear radiation

1 Complete the paragraph below about the structure of the atom with words from this list.

neutral   positively charged   mass   neutrons   protons

electrons   negatively charged   charge   nucleus

The atom has a .... containing neutrons and protons.

.... electrons orbit the nucleus. Overall the atom is neutral because

it has the same number of .... and ....

Most of the .... of the atom is concentrated in the nucleus, as the

electrons are very light.   **[5 marks]**

2 Rutherford and his team of scientists fired positively charged alpha particles at very thin gold foil, only a few atoms thick. The observations they made are shown below in the left-hand column. Draw arrows connecting the scientists' observations to the conclusions they made (in the right-hand column).

a Most alpha particles went straight through the gold foil.	**i** The nucleus is a very small but dense part of the atom.
b Very few particles were deflected through small angles.	**ii** The nucleus is positively charged.
c Some particles bounced straight back.	**iii** Most of the atom is empty space.

[3 marks]

3 Most carbon atoms are carbon-12. Carbon-14 is a radioactive isotope of carbon. What is an isotope?

....

....

....

[3 marks]

4 Ionising radiation causes atoms to become ions. What is the difference between an ion and an atom?

....

....

....

[3 marks]

5 The pie chart opposite shows the various sources of background radiation.

a What is meant by the term background radiation?

....

....

....   **[2 marks]**

gamma rays from ground and buildings 13.5%   medical sources 14%

radon gas from the ground 50%

cosmic rays 12%

food and drink 10%   nuclear power and nuclear weapons testing 0.5%

b Why do some parts of the country have more background radiation than others?

....   **[1 mark]**

6 Joanna and Michael are investigating the ionising radiation emitted from some rocks. Joanna suggests that they lower the temperature of the rocks because that will reduce the amount of radiation emitted. Michael disagrees with her – he thinks it will make no difference.

Discuss who you think is correct.

....

....

....

[3 marks]

G–E

D–C

B–A*

G–E

D–C

B–A*

Types of radiation and hazards

1 There are three types of ionising radiation: alpha, beta and gamma.
For each description below, indicate which type of radiation is being described.

a This type of radiation is an electromagnetic wave. ....

b This type of radiation can be stopped by a piece of paper. ....

c This type of radiation is a large, positively charged particle. ....

d This type of radiation is not deflected by electric or magnetic fields. ....

e This type of radiation is a fast-moving electron. .... **[5 marks]**

2 Sunita and Caroline are trying to find out which type of radiation is emitted from a gas mantle. They set up the source and the detector in a line as shown opposite, and they have some sheets of paper, aluminium foil and thin steel sheets.

gas mantle   radiation detector

Describe how Sunita and Caroline can find out which type of radiation is being emitted.

....

....

....

.... **[5 marks]**

3 What is an alpha particle?

....

....

.... **[3 marks]**

4 The bar chart below shows the typical annual radiation dose for a person in Britain from different sources.

a What is the total radiation dose a typical person in Britain would get from ground and buildings and medical scans in one year?

(*Source*: OCR(A182) *Twenty First Century Science – Physics A*)

.... **[2 marks]**

b In 2011 an earthquake damaged Fukoshima nuclear power station in Japan, causing a radiation leak. Explain how the bar graph would be different for a person in Japan in 2011.

....

.... **[2 marks]**

5 Explain how ionising radiation inside the body causes mutations.

....

....

.... **[3 marks]**

Radioactive decay and half-life

1 Why does the amount of radiation emitted from a radioactive sample decrease over time?

....

....

[3 marks]

2 When uranium emits an alpha particle it changes into Thorium, and when carbon-14 emits a beta particle it becomes nitrogen.

Explain why elements that emit gamma rays do not become a different element.

....

....

[3 marks]

3 Polonium-216 undergoes alpha decay to form a radioactive isotope of lead (Pb).

Complete the balanced equation to show this decay.

$$\frac{216}{84}\text{Po} \longrightarrow \boxed{}\boxed{} + \boxed{}\text{Pb}$$

[5 marks]

4 Different radioactive sources have different half-lives. The graph opposite shows the activity over time for three different radioactive sources (A, B and C).

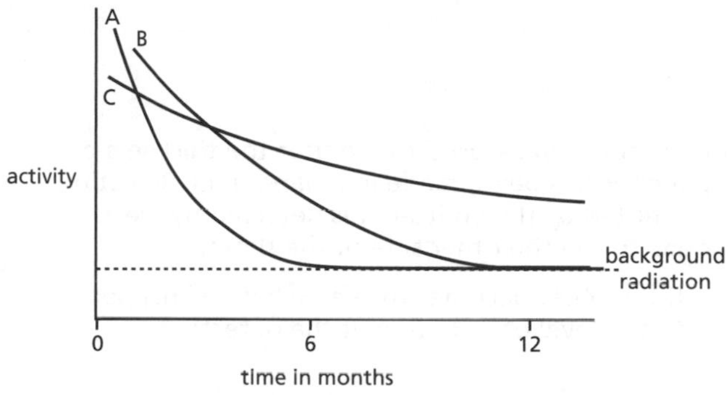

a Which radioactive source has the shortest half-life?

.... [1 mark]

b Which radioactive source has the most activity after 12 months?

.... [1 mark]

c Which radioactive source is likely to be a long-term storage problem? .... [1 mark]

5 John records the activity of a sample over a 30-minute period. His results are shown below.

Time (minutes)	0	5	10	15	20	25	30
Activity (counts per minute)	600	335	190	105	60	35	20

a Plot a graph of activity against time on the axes below.

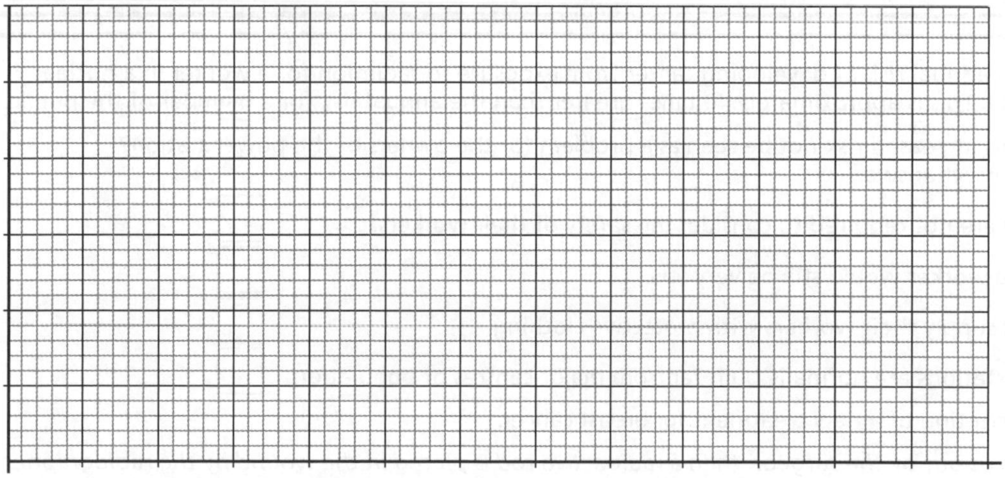

[3 marks]

b Use your graph to find the half-life of the sample.

....

....

[2 marks]

Uses of ionising radiation and safety

1 One of the three types of radiation is used for sterilising medical instruments, because the radiation destroys bacteria. An advantage of this method is that it does not require heat, which could damage the instruments.

G–E

  a Which of the three types of radiation is used for this? .... [1 mark]

  b Why is this type of radiation used?

  ....

  .... [3 marks]

2 The diagram shows radioactive sources being used to treat a deep-seated cancer.

  a Why is the radiation made to enter the body in a number of different and very carefully controlled directions?

radioactive source

D–C

  ....

  .... [2 marks]

  b Why is skin cancer treated with beta radiation and not gamma radiation?

  ....

  .... [2 marks]

3 Radiation can be used to monitor the thickness of paper as it is being made in a paper mill. Radiation is emitted by the emitter. It is detected by the detector on the other side of the sheet.

Source

Paper

Rollers

Radiation detector

  a If the sheet becomes thicker, what will happen to the level of radiation at the detector?

B–A*

  ....

  .... **[1 mark]**

  b Which of the three types of radiation could be used for this? .... [1 mark]

  c Why is this type of radiation used?

  ....

  .... [2 marks]

4 The maximum annual risk of developing cancer from exposure to radiation for a worker in a nuclear reactor is 0.1%. This is approximately 40 times greater than the annual risk for a member of the public.

Why might this increased risk not be seen as a problem for the owners of the power station? Tick (✓) the correct answer.

  a The owners are not required to consider the safety of their workers.

  b The risk to a worker would still be very low.

  c The owners supply their workers with protective clothing.

  d The power stations are normally built far from major centres of population.

D–C

5 Read this description of a new treatment for breast cancer.

The cancer is cut out by the surgeon then a radioactive rod is placed in the wound by the radiographer. Ionising radiation from the rod kills any cancer cells that the surgeon has missed. After a few hours the rod is removed and the wound is stitched up. No further treatment is needed.

Discuss the risks and benefits of the new treatment to all the people involved.

  ....

  ....

  .... [3 marks]

Nuclear power

1 Complete the passage below using words from this list:

fusion   fission   bomb   fuel   power

About a sixth of the electricity produced in the UK is from nuclear power stations.

The nuclear .... is either uranium or plutonium. The energy

is released by nuclear ...., when a large nucleus splits into two

smaller nuclei. At the moment nuclear .... is not commercially

used to generate electricity. **[3 marks]**

2 The difference in mass from fission reaction in a fuel pellet is 0.24 kg. Use Einstein's equation $E = mc^2$ to calculate the energy released from this reaction.

....

.... **[2 marks]**

3 Give **two** advantages and two disadvantages of using nuclear fuels to produce electricity.

....

....

....

.... **[4 marks]**

4 In most nuclear power stations uranium is the nuclear fuel. For nuclear fission to occur a neutron must be fired at a uranium nucleus.

a Explain how this leads to a chain reaction.

....

....

....

.... **4 marks]**

b What is done in a power station to control the chain reaction?

....

.... **[2 marks]**

5 In the Sun isotopes of hydrogen fuse together. What element is produced? Circle the correct response.

deuterium   helium   carbon   tritium **[1 mark]**

6 Currently nuclear fusion is not commercially used to produce power.

a Explain **two** advantages that nuclear fusion has over nuclear fission as a source of power.

....

.... **[2 marks]**

b Describe the difficulties that must to be overcome before nuclear fusion can be used commercially.

....

....

.... **[3 marks]**

A nuclear reactor produces radioactive materials for use in hospitals. The radioactive materials are used to treat patients.

Identify the different types of radioactive waste generated by the production and use of these radioactive materials, and describe how the waste should be dealt with.

The quality of written communication will be assessed in your answer to this question.

D–C

[6 marks]

B1 Grade booster checklist

I know that genes carry instructions that control how your body functions.	
I know that our characteristics are controlled by genes and by the environment.	
I understand that genes carry the instructions for us to make proteins.	
I know that we have 23 pairs of chromosomes, and that sex cells have 23 chromosomes (one from each pair).	
I know that the 23rd pair of chromosomes determines our sex (a female is XX; a male XY).	
I understand that a baby has a combination of genes from his or her parents, which leads to variation.	
I am familiar with disorders (Huntingdon's disease; cystic fibrosis) caused by faulty versions of a gene.	
I know that individuals of a clone have identical genes, and clones can occur in nature.	
I know that stem cells are unspecialised cells and are found in the human embryo and adults.	
I can explain that embryonic stem cells can develop into any cell type, but adult stem cells develop into fewer types.	
I am working at grades G/F/E	

I understand that each gene is a section of a molecule called DNA (deoxyribonucleic acid).	
I know that the DNA in our cells is coiled and packed into chromosomes.	
I can explain the difference between structural and functional proteins.	
I know that the different versions of a gene are called alleles.	
I know that for the chromosomes in a pair, the genes for a characteristic are in the same place.	
I understand that across the pair, two alleles can be the same or different.	
I understand how dominant and recessive alleles behave in combination with each other.	
I know how to use a Punnett square and a family tree to show inheritance of characteristics.	
I can describe how genetic testing and screening are used to check people, and embryos, for a disorder.	
I know that stem cells have the potential to treat illnesses.	
I am working at grades D/C	

I know and understand the terms genotype and phenotype, and homozygous and heterozygous.	
I can explain how the sex-determining gene on the Y chromosome triggers the development of testes.	
I know and understand why offspring are both similar and different to their parents.	
I know and understand the implications of testing and selecting embryos before implantation.	
I know that a clone can be produced by transferring an adult body cell to an empty, unfertilised egg.	
I am working at grades B/A/A*	

B2 Grade booster checklist

I understand how microorganisms cause the symptoms of infectious disease.	
I know that in the ideal conditions of the human body, microorganisms can reproduce rapidly.	
I know that white blood cells destroy microorganisms.	
I understand that a vaccine contains a safe form of the microorganism that causes a disease.	
I know that vaccines can never be completely risk-free, and some people show side effects.	
I know that antimicrobials are chemicals used to kill bacteria, fungi and viruses.	
I can explain that an antibiotic is an antimicrobial used to kill bacteria.	
I know that the heart is a double pump.	
I understand about heart rate and blood pressure, and how they can be measured.	
I know that the conditions inside our bodies are kept constant; this is called homeostasis.	
I understand that the kidneys help to balance water in the body by altering the concentration of urine.	
I am working at grades G/F/E	

I understand how to calculate the growth of a population of bacteria.	
I know that microorganisms carry antigens on their surface, and antibodies recognise these.	
I know that bacteria and fungi may become resistant to antimicrobials, and how this can be reduced.	
I understand about the use of placebos when testing new drugs, and the ethical implications of these.	
I know how the structure of arteries, capillaries and veins are related to their functions.	
I understand that high blood pressure increases the risk of heart disease (strokes and heart attacks).	
I know that lifestyle factors (poor diet, stress, cigarettes, drugs) can increase the risk of heart disease.	
I know that body control systems have receptors, processing centres and effectors.	
I am working at grades D/C	

I understand that after an infection, when antibodies have been produced, memory cells stay in the blood.	
I know that to prevent an epidemic, a high percentage of the population must be vaccinated.	
I can explain that antimicrobials are used to inhibit the growth of microorganisms, as well as killing them.	
I know that random mutations lead to microorganisms being less affected by antimicrobials.	
I am familiar with the use of 'open label', 'blind' and 'double blind' testing of new medical treatments.	
I know and understand the importance of long-term trials in investigating the effects of a drug.	
I know that epidemiological and genetics studies help us to understand factors involved in heart disease.	
I know and understand the principle of negative feedback in reversing changes in the body.	
I can explain how water in the body is controlled by ADH, and how drugs (alcohol and Ecstasy) affect this.	
I am working at grades B/A/A*	

B3 Grade booster checklist

I understand that a species is a group of organisms that can breed together and produce fertile offspring.	
I know that the adaptation of a species to its environment means that it can live and reproduce.	
I understand that nearly all species are dependent on energy from the Sun.	
I know that plants absorb only a small percentage of the Sun's energy for photosynthesis.	
I know how environmental change can be monitored using non-living and living indicators.	
I understand that life began on Earth about 3500 million years ago, and the first life was very simple.	
I know and understand the meaning of the terms biodiversity and sustainability.	
I am working at grades G/F/E	

I know that species living in a habitat are dependent on the environment and each other.	
I understand that there may be competition between the plants and animals in a habitat for resources.	
I know about food webs and the impact of removing a species from a food web.	
I know the factors that can lead to the extinction of a species.	
I can explain how energy is transferred between organisms when the organisms or their wastes are fed on.	
I know that energy is lost along a food chain, limiting its length.	
I can calculate the efficiency of energy transfer at different stages of a food chain.	
I know and understand how carbon and nitrogen are recycled in the carbon and nitrogen cycles.	
I know and understand the principles of natural selection (and selective breeding).	
I know that genetic variation (from sexual reproduction and mutation) can be passed on to offspring.	
I understand that evidence for evolution includes the fossil record and DNA analysis.	
I know that maintaining biodiversity is a key to sustainability.	
I am working at grades D/C	

I can explain the interdependence of organisms using food webs.	
I know that energy is transferred when partly decayed organisms are fed on by detritivores.	
I know and understand the processes of nitrogen fixation, protein synthesis and denitrification.	
I can explain how natural selection, with environmental change and isolation, produces new species.	
I understand why Darwin's model of evolution by natural selection is currently the best theory.	
I know that organisms are classified into groups according to similarities and differences.	
I am working at grades B/A/A*	

C1 Grade booster checklist

I know that the Earth's atmosphere was probably formed by gases from volcanoes.	
I know that power stations and vehicle use both add to air pollution.	
I can explain that hydrocarbons only contain carbon and hydrogen.	
I know that incomplete combustion produces carbon monoxide and carbon particulates.	
I understand that carbon monoxide, sulfur dioxide and nitrogen oxides are pollutants.	
I know that outliers are results that are different from all the others.	
I am working at grades G/F/E	

I understand that gas particles are very small with lots of empty space between them.	
I can recall that the Earth's atmosphere has changed over time.	
I know that solid particulates (soot) are released by both natural and man-made processes.	
I understand that sulfur dioxide and nitrogen oxides are pollutants that make acid rain.	
I know that hydrocarbon fuels burn in oxygen to form water and carbon dioxide.	
I know that oxidation is adding oxygen and reduction is removing oxygen.	
I understand that atoms are rearranged during chemical reactions to make new products.	
I know that nitrogen oxides are formed in hot car engines from nitrogen and oxygen in the air.	
I can explain that range is the difference between high and low results, and mean is the average.	
I can work out the true value from a set of results.	
I am working at grades D/C	

I know that explanations are evidence-based, but can change if new evidence is found.	
I understand that a correlation is a link between a factor and an outcome.	
I know that a causal link needs evidence showing that one factor always causes an outcome.	
I know that mass is always conserved in a chemical reaction.	
I can explain that NO is oxidised to NO_2, and jointly these are referred to as Nx.	
I know the benefits and problems of using biofuels and electric cars.	
I am working at grades B/A/A*	

C2 Grade booster checklist

I know that rubber, plastic and metals are useful materials that have different properties.	
I know that hydrocarbons only contain carbon and hydrogen atoms.	
I understand that crude oil is a mixture of thousands of different hydrocarbons.	
I know that crude oil can be separated by fractional distillation into fractions of similar-sized molecules.	
I know that monomers join up to make polymers.	
I understand that synthetic materials such as plastic have replaced some natural materials.	
I know that very small silver nanoparticles kill bacteria.	
I am working at grades G/F/E	

I understand that the properties of different materials need to be considered when choosing one for a job.	
I know that cotton, paper, silk, wool, iron ore and limestone are all natural materials.	
I know that synthetic materials are manufactured using raw materials from the Earth's crust.	
I understand that a fraction of crude oil contains similar-sized molecules with similar boiling points.	
I know that chain length, cross-linking and plasticisers alter the strength and flexibility of plastics.	
I know that nanotechnology makes molecules up to 100 nm in size, and that these have different properties to the same large-scale material.	
I am working at grades D/C	

I know that synthetic materials can be designed to provide the properties needed for a particular purpose.	
I know that the boiling point of a hydrocarbon molecule is linked to the intermolecular forces between molecules, and these forces increase with chain length.	
I understand that increasing crystallisation can be done by reducing the number of branches, giving a more regular pattern of aligned molecules, making a polymer stronger and increasing melting point.	
I know that carbon spheres containing 60 atoms are called 'buckyballs', which can be made into very strong nanotubes, and that this is a developing technology with possible unknown risks.	
I know that nanoparticles are very strong due to their high surface area to volume ratio.	
I am working at grades B/A/A*	

I understand that geological changes happened in Britain by the slow movements of tectonic plates.	
I know that coal, salt and limestone are important raw materials.	
I know that salt is used in the food industry as a flavouring and as a preservative.	
I understand that alkalis neutralise acid soils, are used for dying clothes, and for making glass and soap.	
I know that chlorine is used to kill microbes in drinking water.	
I know that electrolysis breaks up sodium chloride solution into chlorine, hydrogen and sodium hydroxide.	
I can explain that plasticisers are small molecules added to PVC to make it more flexible.	
I am working at grades G/F/E	

I know that Britain has experienced different climates and has rocks from different ancient continents.	
I understand how limestone, salt and coal formed.	
I know that solution mining extracts sodium chloride solution for use in industry, but that too much is bad.	
I know that a diet containing excess salt is linked to high blood pressure and heart failure.	
I can recall that the first alkali production created harmful by-products.	
I know that there are health concerns associated with plasticisers.	
I understand that a Life Cycle Assessment (LCA) measures the energy use and environmental impact over the life of a product, from cradle to grave.	
I am working at grades D/C	

I know that magnetic clues in igneous rocks can be used to track continental movements.	
I understand that fossils, shell fragments, grain shape and ripple marks give evidence about the conditions when sedimentary rocks formed.	
I know that salt extraction can cause subsidence and environmental problems.	
I can explain what the DH and Defra are, and what they do.	
I know that soluble hydroxides and carbonates are alkalis, and can predict the products when they react with acids.	
I understand that some people disapprove of chlorination, but the benefits outweigh the risks.	
I know the difference between perceived and calculated risk.	
I am working at grades B/A/A*	

P1 Grade booster checklist

I know that light travels through space at 300 000 km/s, and that a light-year is the distance travelled by light in a year.	
I know that we use light from distant stars and galaxies to detect them, and can measure distances to stars by comparing their brightness.	
I know that nuclear fusion is when two nuclei join together, forming a new element, and that this is the source of the Sun's energy.	
I understand that the Universe began about 14 thousand million years ago, and that distant galaxies are moving away from us.	
I can describe some rock processes taking place today and what they suggest about the past.	
I know that the Earth is about 4000 million years old.	
I know that continental land masses are moving very slowly and that this was described by Wegener in his theory of continental drift.	
I know that earthquakes produce P-waves and S-waves, and I can draw and label a diagram of the Earth's interior.	
I know that waves are disturbances that transfer energy. I can use the terms wavelength, frequency and amplitude, and can draw and interpret diagrams showing these.	
I am working at grades G/F/E	

I know the names, relative sizes and motion of different bodies in the solar system and the wider Universe, that the Sun is just one of the thousands of millions of stars in the Milky Way galaxy, and that the Universe contains thousands of millions of galaxies.	
I know how we can learn about distant stars and galaxies using their radiation, although light pollution and atmospheric conditions interfere with these observations.	
I know how relative brightness of stars and stellar parallax help us measure distances to stars.	
I understand that problems in measuring distances to and motion of distant objects mean that the future of the Universe cannot be accurately predicted.	
I can explain how Wegener's theory of continental drift was developed and modified; I understand that heating of the core causes convection in the mantle, and seafloor spreading.	
I know that earthquakes, volcanoes and mountain building generally occur at the edges of tectonic plates.	
I can describe the differences between P-waves and S-waves, and how they give evidence for the Earth's structure.	
I know how waves transfer energy, and can describe the difference between a transverse and longitudinal wave.	
I am working at grades D/C	

I know why the finite speed of light means we see distant objects in the Universe as they were in the past, and I understand why our observations of distant objects may be unreliable.	
I know that redshift shows more distant galaxies move away faster so space is expanding.	
I know that radiation emitted by the Earth has a lower principal frequency than radiation from the Sun, and that this radiation is absorbed or reflected back by some gases in the atmosphere.	
I can explain some problems with measuring the distances to and motion of distant objects and the mass of the Universe, and how this causes uncertainty in predicting its ultimate fate.	
I can explain the implications of the Earth being older than its oldest rocks.	
I know the implications of Wegener's theory and why geologists at first rejected it; I know the causes of seafloor spreading, and what the magnetisation of seafloor rocks can tell us.	
I know how moving tectonic plates cause earthquakes, volcanoes and mountain building.	
I understand how differences in P-waves (longitudinal waves) and S-waves (transverse waves) give evidence about Earth's structure.	
I am working at grades B/A/A*	

P2 Grade booster checklist

I know that a source emits electromagnetic radiation that is reflected, transmitted or absorbed by materials, and that it affects a detector when it is absorbed.	
I can list electromagnetic radiations in order of frequency and recall their speed through space.	
I know that absorbed electromagnetic radiation can heat and damage living cells.	
I know that some people worry about health risks from low-intensity microwave radiation.	
I know that ultraviolet radiation, X-rays and gamma rays are ionising radiation.	
I know that radioactive materials emit ionising gamma radiation continuously, and that exposure to ionising radiation can damage living cells, leading to cancer or cell death.	
I know that lead and concrete absorb X-rays, and that X-rays can produce shadow pictures.	
I know that some radiation from the Sun passes through the Earth's atmosphere, warming the Earth's surface.	
I know that carbon dioxide is a greenhouse gas and understand the causes of the greenhouse effect.	
I can interpret diagrams representing the carbon cycle.	
I know that global warming causes climate change and can describe some of these effects.	
I know that information superimposed onto an electromagnetic carrier wave creates a signal that can be transmitted.	
I know the features of analogue and digital signals, some advantages of digital signals over analogue signals, and that higher-quality sound or images use more digital information.	
I am working at grades G/F/E	

I know that energy from electromagnetic radiation is transferred by photons, and that higher frequency photons transfer more energy.	
I know that energy transferred by electromagnetic radiation depends on the energy and number of photons, and that electromagnetic radiation is less intense further from the source.	
I can relate the heating effect of radiation to the intensity of the radiation and its duration, and I know that water molecules strongly absorb microwave energy.	
I can explain why evidence for the health risk from microwaves is disputed.	
I know that photons of ionising electromagnetic radiations have enough energy to remove electrons from atoms or molecules when absorbed by substances.	
I know that sunscreen, clothing and the ozone layer absorb harmful ultraviolet radiation.	
I can apply understanding of the behaviour of X-rays to explain how images are produced.	
I know that all objects emit electromagnetic radiation, with a principal frequency that increases with temperature.	
I can use the carbon cycle to explain why the amount of carbon dioxide in the atmosphere was constant and is now increasing.	
I can explain and compare different ways in which electromagnetic radiation can transmit information.	
I know the advantages of digital signals over analogue signals, and that digital information is carried as pulses of an electromagnetic carrier wave, which are decoded when received.	
I am working at grades D/C	

I know that the intensity of electromagnetic radiation is the energy arriving per m^2 per s, and spreads over an increasing surface area and is partially absorbed as it moves away from the source.	
I know that ionised molecules can take part in chemical reactions.	
I know there are chemical changes in the atmosphere when ozone absorbs ultraviolet radiation.	
I know that radiation emitted by the Earth has a lower principal frequency than radiation from the Sun, and that this radiation is absorbed or reflected back by some gases in the atmosphere.	
I know that greenhouse gases include methane and water vapour, and that computer climate models provide evidence that human activities are causing global warming.	
I know that increased convection and water vapour in the hotter atmosphere can cause more extreme weather.	
I know why digital signals are less prone to noise than analogue signals, and how pulses of an electromagnetic carrier wave are created, are used to carry digital information and decoded.	
I am working at grades B/A/A*	

P3 Grade booster checklist

I know that power stations that burn fossil fuels emit carbon dioxide, and that growing energy demand raises issues about the availability and environmental effects of energy sources.	
I know that power is measured in watts, and that a more powerful appliance transfers energy more quickly. I can interpret information about energy use.	
I know that electric current passing through a component transfers energy to it and/or the environment, and a more efficient appliance transfers more of the energy to a useful outcome.	
I know that a domestic electricity meter measures energy use in kilowatt-hours (kWh).	
I can interpret and construct simple Sankey diagrams showing energy transfer.	
I can suggest how to reduce energy usage in personal and national contexts.	
I know that mains electricity is produced by generators, and that a generator produces a voltage across a coil of wire by spinning a magnet near it.	
I know that in many power stations a primary energy source heats water, producing steam which drives a turbine coupled to an electrical generator, and that some renewable energy sources drive the turbine directly. I can label a block diagram of the basic components of hydroelectric, thermal and nuclear power stations.	
I know that nuclear power stations produce radioactive waste, which emits ionising radiation.	
I know that electricity is convenient because it is easily transmitted and has many uses, and that the mains supply voltage to our homes is 230 volts.	
I know the advantages and disadvantages of different energy sources, and the effectiveness of methods of reducing energy use at home and work, and can interpret information about these.	
I am working at grades G/F/E	

I know that power in watts is the energy transferred each second, and can calculate how quickly an electrical device transfers energy using power = voltage × current.	
I can calculate energy transferred in joules or kWh using energy transferred = power × time.	
I can calculate the efficiency of an electrical device or power station using the equation efficiency = (energy usefully transferred ÷ total energy supplied) × 100%.	
I know how to calculate the cost of energy supplied by electricity.	
I can interpret and construct Sankey diagrams for various contexts, including electricity generation and distribution, and use them to calculate efficiency of transfer.	
I know how the voltage produced and current supplied by a generator can be increased, and that a generator uses more primary fuel per second when it supplies a bigger current.	
I can explain the difference between contamination and irradiation.	
I know how the distribution of electricity through the National Grid at high voltages reduces energy losses.	
I can discuss qualitatively and quantitatively the effectiveness of methods of reducing energy demand in a national context, and can interpret and evaluate information about different energy sources, considering efficiency, economic costs and environmental impact.	
I understand how different factors affect the choice of energy source for a given situation.	
I am working at grades D/C	

I know that power is the rate of energy transfer, and I can use and rearrange the equations power = voltage × current and energy transferred = power × time.	
I can use and rearrange the equation efficiency = (energy usefully transferred ÷ total energy supplied) × 100%.	
I can explain why contamination by a radioactive material is more dangerous than a short period of irradiation.	
I know that to ensure the security of national electricity supply, we need a mix of energy sources.	
I can interpret and evaluate information about different energy sources, also considering power output and lifetime.	
I am working at grades B/A/A*	

B4 Grade booster checklist

I know that respiration is the release of energy from food and that it occurs in all living cells.	
I know that respiration provides energy for chemical reactions, growth and movement.	
I understand that photosynthesis uses the Sun's energy; it makes food and energy available to food chains.	
I know the word equations for photosynthesis and respiration.	
I know that enzymes are proteins that speed up chemical reactions.	
I understand that enzymes have an optimum temperature and pH, at which they work best.	
I know how fieldwork techniques are used to investigate the effect of light on plants in the wild.	
I understand that aerobic respiration requires oxygen; anaerobic respiration takes place in the absence of oxygen.	
I know when organisms use aerobic and/or anaerobic respiration.	
I am familiar with the applications of anaerobic respiration.	
I know that the cell membrane regulates what enters and leaves cells.	
I know that oxygen and carbon dioxide move freely in and out of cells by diffusion.	
I am working at grades G/F/E	

I know that enzymes are assembled in the cytoplasm from instructions in genes in the nucleus.	
I know that an enzyme works on the substrate, and one enzyme works with one substrate only.	
I am familiar with the lock and key mechanism of enzyme action.	
I know that temperature increases enzyme activity, up to a certain point.	
I understand that energy for photosynthesis is absorbed by the green chemical chlorophyll in chloroplasts.	
I know that the products of photosynthesis are glucose and oxygen.	
I can explain how the rate of photosynthesis is affected by temperature, carbon dioxide and light intensity.	
I know that diffusion is the movement of molecules from an area of high to low concentration.	
I understand that osmosis is a special form of diffusion involving the movement of water only.	
I know that water is moved into plant roots by osmosis.	
I know that aerobic respiration releases more energy than anaerobic respiration.	
I can explain differences in cell structure between animals, plants, yeast and bacteria.	
I am working at grades D/C	

I understand how high temperatures and extremes of pH can denature enzymes.	
I know that active transport uses energy and is needed by plant roots to absorb nitrates.	
I know the symbol equations for photosynthesis and respiration.	
I am working at grades B/A/A*	

B5 Grade booster checklist

I know that in multicellular organisms, cells become specialised to do particular jobs.	
I know that in animals and plants, cells are grouped into tissues, and tissues into organs.	
I understand that sections from plant stems called cuttings will develop into new plants.	
I know that a type of cell division called mitosis is responsible for growth and repair.	
I understand that when a cell divides by mitosis, two identical daughter cells are produced.	
I know that a type of cell division called meiosis is used to produce gametes (sex cells).	
I know that chromosomes are found in the nucleus, and each is made from a DNA molecule.	
I understand that proteins are assembled in the cytoplasm, instructed by the genetic code in the nucleus.	
I know that only the genes required for a cell to carry out its function are switched on.	
I am working at grades G/F/E	

I know that a zygote (fertilised egg) divides by mitosis to form an embryo.	
I know that the cells in an embryo at the eight-cell stage are embryonic stem cells.	
I know that embryonic stem cells are unspecialised, and can become any type of specialised cell.	
I understand that adult stem cells can develop into certain cell types.	
I know that in plants, the only cell division by mitosis occurs in regions called meristems.	
I know that plants from cuttings are clones of the parent and will have identical features.	
I understand that a plant grows towards the light, and this response is called phototropism.	
I know how growing towards the light will increase a plant's chances of survival.	
I know and understand why, in gamete production, the chromosome number of the cell is halved.	
I understand the structure of DNA (a double helix) and how the four bases pair up.	
I understand that in embryonic stem cells, any gene can be switched on to produce any specialised cell.	
I know that embryonic and adult stem cells have the potential to repair damaged tissues.	
I know that because of ethical issues, the use of embryonic stem cells is government-regulated.	
I am working at grades D/C	

I understand that the growth of roots on plant cuttings can be promoted by plant hormones called auxins.	
I know that auxin is redistributed, and the effects of this, when a plant is lit from one side.	
I understand the events during the growth phase and mitosis phase of the cell cycle.	
I know that the order of bases in a gene defines the order of amino acids assembled into a protein.	
I know that a copy of the gene is carried into the cytoplasm as messenger RNA.	
I know that it is now possible to switch on genes in some body cells to produce required cell types.	
I am working at grades B/A/A*	

B6 Grade booster checklist

I understand the definition of a stimulus, and that stimuli are detected by receptors.	
I know that the nervous and hormonal systems coordinate our responses to stimuli.	
I know that the cerebral cortex is connected with traits that make us human.	
I know that nerve cells, or neurons, transmit electrical impulses when stimulated.	
I am familiar with the structure of a neuron (cell membrane; cytoplasm; nucleus; an extension called an axon).	
I know the path followed by a nerve impulse in a reflex arc.	
I know that reflex actions enable 'automatic' responses to aid survival, and can give examples.	
I understand that nerve impulses are transmitted across gaps between nerves called synapses.	
I know that chemical transmitter substances transmit an impulse across a synapse.	
I know that behaviours in simple animals are instinctive and depend on reflexes.	
I know that as humans interact with their environment, new neuron pathways are formed.	
I understand that memory is the storage and retrieval of information, and that there are two forms of memory.	
I am working at grades G/F/E	

I know that the central nervous system (CNS) is made up of the brain and spinal cord.	
I know that the peripheral nervous system (PNS) is made up of the nerves.	
I understand that the axon of a nerve is covered with a fatty (myelin) sheath, which has gaps.	
I am familiar with how conditioning works and can give two examples of conditioned reflexes.	
I am familiar with the techniques used to map the brain.	
I know that neuron pathways can be strengthened by repetition.	
I understand how models can be used to describe memory, including the multi-store model.	
I am working at grades D/C	

I know about and can compare the responses of the nervous and hormonal systems.	
I know that the myelin sheath insulates the nerve and speeds up the transmission of nerve impulses.	
I know that a transmitter binds to a receptor and initiates the nerve impulse in a second nerve.	
I understand that some drugs affect the transmission of nerve impulses across a synapse.	
I know how Ecstasy affects the concentration of a transmitter called serotonin.	
I know how, in certain circumstances, a reflex action can be overridden.	
I know that conditioning can develop in response to a new stimulus, introduced with the primary stimulus.	
I understand that because of the huge number of potential neuron pathways, humans are able to adapt.	
I know about evidence to suggest that children may only acquire skills at a certain age.	
I can explain how models can be used to describe memory, and their limitations.	
I am working at grades B/A/A*	

C4 Grade booster checklist

I understand that elements are arranged into patterns in the Periodic Table.	
I know that an atom consists of a nucleus containing protons and neutrons, with electrons arranged in shells around the outside.	
I know that the first electron shell holds 2 electrons and the second shell holds 8 electrons.	
I know that a horizontal row across the Periodic Table is called a Period.	
I know that a vertical column in the Periodic Table is called a Group.	
I know the colours and states of the halogens (Group 7) at room temperature and as gases.	
I know that the halogens react with Group 1 metals and iron.	
I understand that ionic compounds contain charged particles and conduct electricity when they are molten or dissolved in water.	
I am working at grades G/F/E	

I know that Döbereiner, Newlands and Mendeleev were three scientists who had different ideas about how to arrange elements into patterns.	
I know that each element has a unique flame colour and line spectrum.	
I understand that the proton number of an atom gives the number of electrons for the atom.	
I know that for the first 20 elements, the third electron shell in an atom holds 8 electrons.	
I know that proton numbers, numbers of electrons and properties change across a period.	
I know that Group 1 metals have trends in their physical properties.	
I can explain how Group 1 metals react with water and chlorine.	
I know that the halogens have trends in their physical properties.	
I understand that halogens contain diatomic molecules (molecules that contain two atoms).	
I know that the halogens get less reactive down the group.	
I know that more reactive halogens can displace less reactive halogens from their compounds.	
I understand that ionic compounds conduct electricity when their ions are free to move.	
I am working at grades D/C	

I can use the Periodic Table to work out the number of protons, neutrons and electrons in an atom.	
I know that the electron arrangement of the atoms in an element is linked to its position in the Periodic Table.	
I know that the electron arrangement in atoms is linked to the reactivity of the element.	
I know that Group 1 elements are more reactive when they have more electron shells and Group 7 are less reactive when they have more electron shells.	
I understand that positive ions form when atoms lose electrons and negative ions form when they gain electrons, and that the formula of an ionic compound contains positive and negative ions with a balance of charges.	
I am working at grades B/A/A*	

C5 Grade booster checklist

I know that air contains elements and compounds that contain non-metal atoms, and I know the main gases in the air.	
I know that the Earth's hydrosphere contains water with dissolved ionic salts.	
I know that ionic salts can be identified by testing with dilute sodium hydroxide, dilute acid, dilute silver nitrate and dilute barium chloride.	
I understand that the lithosphere contains rocks, minerals and ores.	
I know that metals can be extracted from metal oxides by reduction (taking away oxygen).	
I understand that electrolysis breaks down electrolytes when an electric current passes through.	
I know that metals are strong, malleable, have high melting points and conduct electricity.	
I am working at grades G/F/E	

I know that molecules in the air can be shown in either 2D or 3D.	
I know that molecules in the air have low melting points and boiling points because there are weak attractions between the molecules.	
I know that in ionic compounds there are strong forces between positive and negative ions called ionic bonds, and that these determine the properties of the compound, such as melting point and electrical conductivity.	
I understand that tests for positive and negative ions depend on the formation of insoluble precipitates.	
I know that large amounts of rock have to be mined to produce small amounts of metals.	
I know that diamond, graphite and silicon dioxide are giant covalent structures with similar properties.	
I know that relative atomic mass can be used to work out relative formula mass and relative gram mass.	
I understand that, during electrolysis, metals form at the negative electrode and non-metals form at the positive electrode.	
I know that metal processing harms the environment because of toxic waste and large amounts of waste rock.	
I am working at grades D/C	

I know that molecules in the air are held together by sharing electrons in covalent bonds.	
I know that the positive and negative charges on ions balance in the formula of an ionic compound.	
I know that precipitation reactions can be represented using ionic equations.	
I know that redox reactions involve both reduction and oxidation.	
I know that the bonding and structure of diamond and graphite are linked to their properties.	
I know that relative gram mass can be used to work out the mass of metal in a mineral.	
I know that ionic equations can be used to show electrode reactions during electrolysis.	
I know that metals contain positive ions in sea of free-moving electrons.	
I am working at grades B/A/A*	

C6 Grade booster checklist

I know that chemicals have hazard symbols to show that they are hazardous.	
I know that acids react with metals, metal oxides, metal hydroxides and metal carbonates to give a salt and other products.	
I know how to work out the relative formula mass of a compound by adding up the relative atomic masses of the atoms.	
I understand that neutralisation reactions happen when an acid reacts with an alkali.	
I know that some reactions are exothermic and some reactions are endothermic.	
I know that filtration separates solids from liquids or solutions.	
I understand that the rate of reaction is usually measured in amounts per second.	
I know that a catalyst speeds up a reaction without being used up.	
I am working at grades G/F/E	

I know that pure acids can be solids, liquids or gases, and I can name examples of common alkalis.	
I know that reactions of acids can be shown using word and symbol equations with state symbols.	
I know that balanced equations can be used to work out the masses of reactants and products in a reaction.	
I understand that titration results vary, and that the mean is an indicator of the true value of a result.	
I know that acids contain hydrogen ions (H^+) and alkalis contain hydroxide ions (OH^-), and that these react to form water (H_2O) during neutralisation reactions.	
I know that energy-level diagrams can be used to show exothermic and endothermic reactions.	
I know that crystallisation purifies solid crystals and I can list the steps involved.	
I understand that there are several ways of measuring rates of reactions, and how surface area, temperature and catalysts affect the rate of reaction.	
I am working at grades D/C	

I know that the pH scale is used to show the strength of acids and alkalis.	
I know that the formulae of salts can be worked out using the charges on the ions, and that symbol equations balance if the numbers of each type of atom are the same on both sides.	
I can use relative formula masses to work out reacting masses in equations.	
I know that the titration results can be compared if the same concentrations of the same solutions are used.	
I understand how to work out theoretical and percentage yields.	
I know that the gradient on a rate of reaction graph changes as the rate changes.	
I know that the rate of reaction depends on the frequency of collisions of particles.	
I am working at grades B/A/A*	

I know how to calculate the average speed of a moving object.	
I can draw and interpret distance–time graphs showing objects that are stationary or moving at constant speed.	
I know how to calculate the acceleration of moving objects.	
I know that forces between objects occur in 'interaction pairs', and I am able to identify the partner to forces.	
I understand that friction is a force between two moving surfaces, which acts to slow down the movement.	
I know that many forces can act on an object and I can find the resultant force.	
I know that falling objects accelerate towards Earth, but are slowed down by drag forces.	
I know about the safety features used in vehicles to reduce the forces in collisions.	
I understand that work is done when a force moves an object.	
I know the law of conservation of energy.	
I am working at grades G/F/E	

I understand the difference between the instantaneous speed and the average speed of a moving object.	
I know how to calculate the speed of an object from a distance–time graph.	
I can draw and interpret both distance–time graphs and speed–time graphs showing objects that are stationary, moving at constant speed and increasing speed.	
I know that when the forces on an object are balanced there is no acceleration of the object.	
I can give examples of situations where friction is wanted and unwanted.	
I can explain what is meant by terminal velocity.	
I can calculate the momentum of a moving object.	
I know that a resultant force causes a change in momentum.	
I know that energy is transferred when work is done.	
I can calculate the gravitational potential energy of an object.	
I know that in all energy transfers, some energy is always dissipated as heat energy.	
I am working at grades D/C	

I know what is meant by the terms displacement and velocity.	
I understand that acceleration can cause a change in direction as well as a change in speed.	
I can calculate acceleration from speed–time graphs.	
I can explain friction and reaction forces.	
I can draw force diagrams.	
I can explain how the forces on moving vehicles affect their speed.	
I can calculate the forces involved when there is a change of momentum.	
I can calculate the kinetic energy of moving objects.	
I am working at grades B/A/A*	

P5 Grade booster checklist

I know that rubbing insulators causes them to become charged, and that like charges repel and unlike charges attract.	
I know that electrons are negatively charged.	
I know that electric current is a flow of charge and is measured in amps.	
I know that the larger the voltage in a circuit the higher the current, and the larger the resistance in a circuit the lower the current.	
I know that LDRs are semi-conductors whose resistance changes with light intensity, and that thermistors are semi-conductors whose resistance changes with temperature.	
I understand current and resistance in series circuits.	
I know that electromagnetic induction produces a voltage, and that a current will flow if there is a circuit.	
I can describe a transformer.	
I know the difference between d.c. and a.c., and I know that the domestic electricity supply in the UK is 230-V a.c.	
I know that electric motors are used in many different appliances.	
I am working at grades G/F/E	

I know that metals (conductors) contain lots of electrons which are free to move, and that when a battery is connected in a circuit it causes the free electrons to move around the circuit.	
I understand that work is done by a battery when it causes an electric current to flow, and that energy is transferred to the component (usually as heat).	
I can calculate electrical power.	
I can calculate resistance and I know how the current in a resistor varies as the voltage is increased.	
I know that the potential difference (or voltage) is a measure of the work done by the charges in a circuit.	
I can explain currents in parallel circuits, and potential differences in series circuits.	
I understand that if a wire experiences a changing magnetic field, a voltage will be induced.	
I can describe how a generator (or dynamo) works and explain how you can increase the size of the voltage produced.	
I know that a wire carrying a current in a magnetic field will experience a force, and that this is the principle of the motor effect.	
I am working at grades D/C	

I can explain why resistance changes with temperature in metals.	
I can explain currents and potential differences in series and parallel circuits.	
I can explain clearly how a transformer works, and know that the voltage ratio equals the number of turns ratio.	
I can explain why a.c. is used for domestic electricity.	
I can explain how motors work.	
I am working at grades B/A/A*	

P6 Grade booster checklist

I know that radioactive elements emit ionising radiation, and that some radioactive elements are naturally occurring and contribute to background radiation.	
I know that an atom has a positively charged nucleus, made of protons and neutrons, which is surrounded by electrons.	
I know that the behaviour of radioactive materials cannot be changed by chemical or physical processes.	
I can describe the penetration properties of the three types of ionising radiation (alpha, beta, gamma).	
I understand that over time, the activity of radioactive materials decreases.	
I know that ionising radiation can damage living cells and may cause cancer, and that it is used to treat cancer, sterilise food and surgical instruments, and as a tracer.	
I know that radiation dose (in Sieverts) is a measure of the possible harm done to your body, and that people are exposed to risk by contamination or irradiation.	
I know that nuclear fission and fusion can produce a lot more energy than chemical reactions.	
I understand that nuclear power stations use nuclear fuels, such as uranium, and produce radioactive waste.	
I know that some people, such a radiographers and nuclear power workers, are regularly exposed to radiation, and that their exposure to radiation must be monitored.	
I am working at grades G/F/E	

I understand that the results from the alpha-scattering experiment give evidence for the structure of the atom.	
I know that two hydrogen nuclei can fuse to form helium in a nuclear fusion reaction.	
I know the relative mass and charge of alpha and beta particles, and can relate this to their ionising power.	
I understand the half-life of radioactive materials, and know that there is a very wide range of values for half-life.	
I know that ionising radiation causes atoms to become ions, and I can list some uses of ionising radiation.	
I can interpret data on risk related to radiation dose, and know that we are exposed to radiation all the time from background radiation.	
I can relate half-life to the amount of time it takes for a radioactive source to become safe.	
I understand the difference between nuclear fission and fusion reactions.	
I know that radioactive waste is categorised as high level, intermediate level and low level, and I can relate this to disposal methods.	
I am working at grades D/C	

I know that protons and neutrons are held together in the nucleus by a strong nuclear force, which overcomes the electrostatic repulsion of the protons.	
I know that Einstein's equation $E = mc^2$ is used to calculate the energy released during nuclear fission and fusion reactions.	
I can explain what isotopes are.	
I know that an alpha particle is a helium nucleus (2 protons and 2 neutrons), and that a beta particle is a fast-moving electron.	
I can complete nuclear equations for alpha and beta decay, and I know that radioactive nuclei are unstable. I can carry out simple calculations involving half-life.	
I know that ions formed by ionising radiation can take part in chemical reactions.	
I can explain the uses of ionising radiation, using ideas about the properties of the ionising radiation and half-life.	
I understand that nuclear fission involves a neutron hitting a large unstable nucleus, which then splits into two smaller nuclei, releasing neutrons and energy.	
I understand that nuclear fission needs to be controlled in nuclear power stations, and can use the terms chain reaction, fuel rod, control rod and coolant.	
I am working at grades B/A/A*	

Answers

B1 You and your genes

Page 146 What genes do

1 a Nucleus

  b DNA (deoxyribonucleic acid)

2 a Collagen or keratin

  b Enzyme, e.g. amylase or antibody or hormone

3 a The project identified the location of genes on the different chromosomes, enabling scientists to improve their understanding of the way in which genes control our development and function; OR providing scientists with an opportunity of understanding diseases that have a genetic cause OR diseases where genes provide protection against the disease, and an opportunity to control or screen for the disease

  b In identifying genes linked with disease: some drugs' companies have tried to patent/claim ownership of genes, limiting research; OR some insurance companies might withhold insurance from someone likely to get a genetic disease

4 a Blue eyes; dimples; straight hair *(Any 2)*

  b Pink hair; decayed tooth; pierced ear; scar *(Any 2)*

  c Skin colour. Skin colour is controlled by genes, but is affected by exposure to the Sun (an environmental factor)

5 Eye colour or height; skin colour/height will show a continuous range across the population

6 Genotype refers to the genetic description of the person for the characteristic, written as letters, e.g. for a person with dimples, DD. Phenotype is a description of the features of the organism, e.g. for dimples, has dimples or does not have dimples. The phenotype will depend on the person's genes, but may also be affected by how these interact with the environment

7 c

Page 147 Genes working together and variation

1 a; c

2 a A mistake occurs when producing sex cells; so that the individual has a different number of chromosomes to the normal 46

  b Down's syndrome (or another suitable example); the person has an extra chromosome 21/ three chromosome 21s instead of two

3 You are similar because: you inherit genes from your parents (half from your mother and half from your father); these genes control your characteristics; some behavioural characteristics will be learned from parents

You are different because: sex cells contain half the genetic material from each parent; combinations of genes/ chromosomes from each parent will make you different; you may inherit genes not seen (expressed) in either parent; the environment will also produce differences, e.g. diet and exercise

4 Different forms of a gene

5 The term homozygous means that both alleles for a characteristic are the same (on each chromosome of the pair). So the combinations DD and dd are homozygous. The term heterozygous means that the alleles for a characteristic are different (on each chromosome of the pair). So the combination Dd is heterozygous

Page 148 Genetics crosses and sex determination

1 Identical positions; each chromosome; alleles

2 a Dominant; the genotype is written in upper case letters

  b Tt; tt

  c The phenotype for Tt is tongue roller. The phenotype for tt is non tongue roller (cannot roll tongue)

3 a The couple could be TT; or Tt

  b i The genotype of both the man and women must be Tt; without the presence of a t allele in both parents, all the children would be tongue rollers

    ii

		Mother possible alleles in eggs	
		T	t
Father possible alleles in sperm	T	TT Tongue roller	Tt Tongue roller
	t	Tt Tongue roller	tt Non tongue roller

In the Punnett square:
- 1 mark for the different alleles that could be passed on to the offspring from the mother and father (the alleles in the egg cells and sperm cells)
- 1 mark for the possible combinations (genotypes)
- 1 mark for the phenotypes produced

4 The presence of the sex determining gene on the Y chromosome in the cells of the embryo means that it will be male. It is thought that the sex-determining gene triggers the development of testes in the embryo

5 The genes for some traits such as haemophilia and colour blindness are found on part of the X-chromosome not present in the Y-chromosome. If a child inherits a defective allele for blood clotting or normal colour vision, a male baby will have the condition because they have just a single X-chromosome. But for a female baby inheriting a defective allele, it's likely that a normal allele is present on their other X-chromosome. Although there is a small chance that a female baby will inherit two recessive genes from their parents

Page 149 Gene disorders, carriers and genetic testing

1 a Dominant; Huntingdon's disease; concentrating

  b Recessive; cystic fibrosis; digesting food

2 Write out your working stage by stage.

There are 700 000 babies born.

So, if there are 1 in every 10 000 babies born with the allele, there are

$\dfrac{700\,000}{10\,000}$ in every 700 000 babies born

$= 70$ born in the UK every year

Or: the proportion of babies born $= \dfrac{1}{10\,000} = 0.00001$

So, if there are 700 000 babies born each year, the number born

$= 0.00001 \times 700\,000 = 70$

3 a So that people can get treatment for the disorder (if it is treatable); allow people to plan for the future (if treatment of the disorder is less successful, or it is untreatable)

  b Amniocentesis; Chorionic villus sampling

4 a i Used to detect five different conditions, so identification; may make treatment possible

    ii Has ethical implications, because identification of a disorder may cause employer to change work role; an insurance company to withhold insurance

  b Embryo testing detects genetic disorders; only embryos free of genetic disorders are then implanted

Page 150 Cloning and stem cells

1 a Successful characteristics can be passed to offspring; advantage when plants or animals live in isolation *(Any 1)*

  b As they are genetically identical; changes in conditions could result in the population being wiped out

2 The nucleus from a body cell is extracted; and inserted into an egg cell; that has had its nucleus removed (giving the egg cell a full set of genes without having been fertilised); the embryo is implanted into a suitable surrogate mother

3 Embryonic; embryo; any; unspecialised; diseases; certain

4 a Embryonic stem cells are found in the five-day-old embryo. They have the potential to develop into any cell in the human body. Adult stem cells are found in adults in a few places in their bodies. They are able to develop into fewer cell types than adult stem cells

  b Embryonic stem cells are extracted from embryos. In the process, the embryo is destroyed. There is no destruction of an embryo / potential human life / ethical implications in the use of cells from a person's skin

5 Stem cells are unspecialised so can be used to produce different types of body cells. They could be used in the testing of new drugs. Understanding how cells become specialised in the early stages of the person's development by the switching on and off of particular genes. Renewing damaged or destroyed cells in spinal injuries, heart disease, Alzheimer's disease and Parkinson's disease

Page 151 B1 Extended response question

5–6 marks

Explains why the genotypes of the parents must be Pp, the cross involved and produces a suitably detailed and accurate Punnett square. Explains that there will always be a 1 in 4 (25%) chance of producing a baby with PKU because which 'type' of sperm produces which 'type' of egg will always be random. All information in answer is relevant, clear, organised and presented in a structured and coherent format. Specialist terms are used appropriately. There are few, if any, errors in grammar, punctuation and spelling

3–4 marks

States that the parents are carriers, and draws an incomplete or not fully-detailed Punnett square for the cross. States that fertilisation is a random event but does not fully explain how this relates to probability of the offspring having PKU. For the most part the information is relevant and presented in a structured and coherent format. Specialist terms are used for the most part appropriately. There are occasional errors in grammar, punctuation and spelling

Answers

B2 Keeping healthy

Page 152 Microbes and disease

1 Bacteria make us feel ill by releasing toxins. Microorganisms that cause disease are called pathogens

2 The number of bacteria after 3 hours would be 2 621 440. So the area of flesh eaten would be 2.62 cm^2

3 a Lag phase; exponential (logarithmic) phase; stationary phase; death phase

  b Phase 1: Lag phase – the bacteria are not reproducing. They are copying DNA and producing proteins

   Phase 2: Exponential (logarithmic) phase – the bacteria are rapidly growing and dividing

   Phase 3: Stationary phase – food is running out. The bacteria are dying at the same rate as they are reproducing

   Phase 4: Death phase – the bacteria are dying as toxins build up

4 Skin; oil; saliva; tears; stomach; acid

Page 153 Vaccination

1 Memory cells are left in the blood after vaccination. An infecting microorganism is destroyed very quickly by antigens in the blood

2 a Builds up immunity

  b To protect all children against diseases that are preventable

  c Antigens on the surface of the flu virus change over time/do not stay the same, giving new strains. Antibodies produced after exposure to one vaccine would not work against new antigens. New vaccines are produced regularly from different strains of the virus

3 William is correct in that the widespread use of vaccines has already eradicated one disease from the world – smallpox

   In response to Xavier's point, although it may be impossible to vaccinate the whole of the world's population, a high percentage would do

   Yvonne is incorrect. Vaccines are very safe. But some people would be reluctant to have their children vaccinated, as there is a very small chance of them causing harm

   Zak is correct. We do not yet have vaccines for all infectious diseases. Also, some microorganisms, such as the flu virus, change continuously, so flu would be difficult to eradicate completely using a vaccination programme

4 a Side effect

  b Because of genetic variation

5 Effective antimicrobials either kill bacteria; or inhibit their growth

Page 154 Safe protection from disease

1 a The normal population of bacteria contains some bacteria that are resistant to the antibiotic. These survive the treatment with the antibiotic. They reproduce (by splitting into two/binary fission). The genes for antibiotic resistance are passed on. The genes for antibiotic resistance spread through the population

  b Reduce their use/only prescribe when absolutely necessary. Complete the course. Never pass on antibiotics prescribed for you to anyone else

2 a The resistance of the bacterium to three antibiotics increases; from 1985/6 to 2000 (for penicillin) and 2001 for erythromycin and tetracycline; then shows a fall. Almost no resistance was found in the bacterium to ofloxacin until 1995. Resistance since has been more or less stable, at around 1%

  b There is a decrease in total antibiotic use between 1997 and 2007

  c There is a correlation between the decrease in antibiotic resistance from 2001 and the decrease in antibiotic use. But this does not prove the reduction of antibiotic use was the cause

3 One type is carried out on healthy volunteers, to check for safety. A second type is carried out on people with the illness, to test for safety and effectiveness

4 In the double-blind test, neither patient nor researcher knows who is receiving the new drug, existing drug or placebo. That way, researchers can have more confidence that any difference between the new treatment/existing treatment/placebo is real. Also, it means that the research is not affected by perceptions of doctors, patients or data analysts

Page 155 The heart

1 a Oxygen; nutrients; wastes

  b Double pump; half; lungs

  c Heart; blood vessels; blood

2 Arteries transport blood away from the heart under high pressure; walls are very thick, elastic and muscular to withstand the pressure

   Capillaries link arteries and veins; walls are one-cell thick to allow the transfer of substances to and from cells

   Veins collect blood and return it to the heart; walls contain elastic, muscular tissue, but are thinner than those of arteries; the blood is under low pressure and veins have valves to prevent the backflow of blood

3 Build-up of fatty material in the coronary arteries, reducing or preventing blood flow. These are arteries that supply the heart muscle with oxygen

4 a There is a fall in deaths from CHD in all age groups; in the early and late 1970s, the death rate rose in the 45–54 and 35–44 age groups

  b The 45–54 age group

  c Smoking decreased; more exercise; improvement in diet / less obesity / less saturated fat in diet (Change in lifestyle = 1 mark only)

  d Review statistics / carry out research on smoking, exercise and dietary habits

Page 156 Cardiovascular fitness

1 Stroke/ heart attack; damage to the organs, e.g. the kidneys

2 Stress; poor diet; smoking; misuse of drugs (Any 3)

3 Pulse rate is measured in beats per minute. Blood pressure is measured using a sphygmomanometer

4 a i In nurses not exposed to cigarette smoke = 17

    ii In nurses exposed to any amount of cigarette smoke = 135

  b How many times more likely to get CHD were the nurses who were exposed to smoke, compared with those exposed to no smoke = 7.9 / 8 times

  c The data suggest that occasional exposure to cigarette smoke is more likely to cause CHD than regular exposure. With regular exposure to cigarette smoke, there is a greater chance of the CHD being fatal: a 1 in 4.5 chance (or 22%) of the CHD being fatal, compared with 1 in 5.7 chance (or 18%) for occasional exposure

  d Studies are difficult to carry out on large samples/numbers of volunteers. Scientists can't expose large numbers of volunteers to cigarette smoke. If the study is carried out on a normal sample, people in the sample do other things in their lifestyles that might affect the result/it is difficult to control other factors

Page 157 Keeping things constant

1 The process by which internal factors in the body are kept constant

2 Temperature

3 Flow is from: receptor; to processing centre; to effector

4 a Hypothalamus; effectors cause vasodilation; heat lost from skin surface; body temperature falls; receptors detect body temperature

  b The receptors detect this and pass the message to the processing centre. The effector (the blood vessels) return to their normal size

  c Negative feedback

5 a Blood plasma is diluted; the kidneys produce a large volume of urine

  b Blood plasma becomes concentrated; the kidneys produce a small volume of urine

6 a ADH acts on the kidneys and causes them to reduce the amount of water lost in urine

  b Alcohol suppresses the release of ADH. ADH increases the amount of urine produced. Alcohol therefore causes us to produce a greater volume of more dilute urine

Page 158 B2 Extended response question

5–6 marks
Provides a detailed discussion responding to the points in the article, referring to the data in the Vaccination Green Book and explaining how this refutes the article's claims:

- The number of deaths from the different diseases was not negligible in the year before the vaccine was introduced. In 1938, over 2000 people in the UK died from diphtheria.
- Introduction of the vaccine caused a marked decrease in numbers of deaths / reduced deaths from diphtheria, measles and polio to zero in 2003. However, in Hib, the number of deaths was still 58% of what they were before it was introduced in 1991. But we do not know the percentage of children vaccinated for Hib.

Answers

- The article refers to 'waves of epidemics' and 'many deaths'. We cannot comment on this, because the table does not provide information on the years immediately after the vaccine – just 2003 data. And we do not have data on other diseases; just six (although these are some of the most serious).

Overall, the data suggest that vaccinations are beneficial, and not harmful, although we do know that a very small minority of people are affected severely by vaccines. All information in answer is relevant, clear, organised and presented in a structured and coherent format. Specialist terms are used appropriately. There are few, if any, errors in grammar, punctuation and spelling

3–4 marks
There is some discussion of the number of deaths from the different diseases prior to their introduction, and the recognition that these were not negligible in the year before the vaccine was introduced. There is comment on the reduction of deaths after vaccination, and the lesser lack of effectiveness against Hib. Limited discussion, or just one point, regarding the limitations of the data, e.g. data only covers six diseases. For the most part the information is relevant and presented in a structured and coherent format. Specialist terms are used for the most part appropriately. There are occasional errors in grammar, punctuation and spelling

1–2 marks
Limited discussion of patterns in data, but concludes that vaccinations have reduced deaths. Answer may be simplistic. There may be limited use of specialist terms. Errors of grammar, punctuation and spelling prevent communication of the science

0 marks
Insufficient or irrelevant science. Answer not worthy of credit

B3 Life on Earth

Page 159 Species adaptation, changes, chains of life

1 a A species is a group of organisms that are able to breed together and produce fertile offspring

   b Ability to accumulate and store water in the stem; thick, waxy cuticle to reduce water loss over the surface; spines instead of leaves to reduce water loss; a stumpy shape that gives a small surface area in relation to its volume; a shallow, spreading root system that stores water when it rains (Any 3)

2 i The foxes feed on shrews, whose population will decrease because of the lack of insects and spiders. The foxes also eat rabbits, however, so the population of foxes is unlikely to be affected, unless the population of rabbits is small

   ii The insects and spiders form the main food for the birds, so food will be in very short supply. As the birds can fly, it is likely that they will move to another area where the food is more plentiful

   iii The insects and spiders form the main food for the lizards, so food will be in very short supply. The lizards are likely to starve and die. They are less likely than the birds to be able to move to another area where the food is more plentiful

   iv The insect-eating birds will either move to another area or die. The owls will have to hunt for food in another area to survive

3 A change in the environment, or an example of a change in the environment, e.g. loss of habitat, climate change, predated on by newly-introduced species, or disease

4 a 15% (111 – 94/111 × 100 = 17/111 × 100 = 15%)

   b The energy is transferred to decomposers as they feed. Decomposers and detritivores use the grasshoppers' bodies to obtain their energy by respiration; some of this energy is lost as heat

Page 160 Nutrient cycles, environmental indicators

1 a Carbon dioxide; air; photosynthesis; fixed

   b Respiration; decomposition; bacteria; fungi; soil

2 Nitrogen-fixing involves bacteria (in the soil or in root nodules) converting atmospheric nitrogen into nitrates; denitrification involves converting nitrates and other nitrogen-containing compounds into atmospheric nitrogen

3 Mayfly larvae are good biological indicators because they can only live in clean water. Monitoring carbon dioxide gives information on climate change

4 a Carbon dioxide levels in the air show a steady increase from around 315 ppm in 1958 to 370 ppm in 2010. The increases are the result of pollution from the burning of fossil fuels. But this does not give a full explanation of the data. The graph also shows yearly peaks and troughs. These seasonal changes are the result of plant activities, and the removal of carbon dioxide from the air for photosynthesis

   b Carbon dioxide in water (the oceans and freshwater); carbon dioxide in air trapped in polar region ice (ice core data)

Page 161 Variation and selection

1 Million; simple; complex; evolution; variation

2 Fossils of eggs that have been found are where the eggs have been turned into rock

3 A mutation is the change in the genetic information in a cell that can occur at random; a mutation will result in a change in the characteristics of an organism; if a mutation occurs in a sex cell, the characteristic will be passed on to the next generation; most mutations are harmful, but those that give the individual an advantage will be passed on, and lead to changes in the population as the gene spreads through it

4 Sometimes genes change because of mutations. Most mutations are harmful, but some will be beneficial. The beneficial changes will spread throughout the population. So the gene pool – the alleles of genes that occur in a population of organisms – will change; some alleles will become more common, others will disappear. Over millions of years, with the change in frequencies of alleles, a new species will arise

5 The cattle breeder selects a female animal/cow that produces milk with a low fat content (but also large volumes of milk) and breeds this with a male. (Be careful! Usually in selective breeding, the scientist would select a male and female with the best qualities, but only females produce milk!) The cattle breeder breeds these from the offspring, then selects the female that produces the milk with the lowest fat and largest volume. The selective breeding process is continued over several generations

6 a In a population, some rats will be resistant to the rat poisons and survive their use. These rats will pass on their resistant genes. These will spread through the population as the rats breed, until most of the rats become resistant

   b Stadtlohn. This town has the highest percentage of poison-resistant rats, because the rat poison has been overused

Page 162 Evolution, fossils and DNA

1 b, a, d, c, e

2 The Galapagos Islands are volcanic, so Darwin realised that ancestors of the organisms on the islands must have arrived, at some point, from the mainland. He observed that the organisms there, such as mockingbirds, were similar but had slight differences to those on the mainland. The mockingbirds were also different from one island to the next. This led to Darwin's idea that species were not fixed, but could change over time

3 Classified; groups; appearance; DNA; evolved; ancestor

4 Fossil evidence: shows that the simplest organisms are found in the earliest rocks, with more complex ones appearing in younger rocks; more recent fossils have features that look like adaptations or developments of those of older organisms. DNA analysis: of today's organisms has confirmed predictions made from the fossil record, including when branches in the tree of life occurred, and how long ago common ancestors were shared and these branches occurred

5 Characteristics that change during the lifetime of an organism are not usually passed on to the offspring unless they occur by mutation in the sex cells; so changes to a giraffe's neck during its lifetime would not be passed on to the next generation; Darwin suggested that new species of organisms evolved by natural selection; differences in the inherited features of organisms that were advantageous would be passed on; over time, a new species would evolve

Page 163 Biodiversity and sustainability

1 A plantation of tropical oil palm trees has very high biodiversity

2 Meeting our current needs without depriving future generations. Recycling and re-using items instead of throwing them away

3 Each organism in an ecosystem is interdependent/dependent on each other. One organism feeds on another, and will itself be food for another. The removal of one organism will mean that others starve, unless they have another source of food, and may mean that the numbers of others, that are preyed upon, increase

4 a Biodiversity is reduced by: growing only one crop in the field; removing hedgerows (along with the organisms that live in the hedgerows); spraying herbicides and pesticides that kill plants and animals

   b Replanting hedgerows; creating beetle banks

5 Use of less, and sustainable packaging materials; considering the materials used; considering the energy used; considering the pollution created (Any 3)

6 They still use energy in their transport; they will decompose slowly in oxygen-deficient landfill sites

Answers

Page 164 B3 Extended response question

5–6 marks
Carries out thorough analysis of the data. Appreciates that yields are in tonnes per hectare, and number of plants per hectare varies, so calculates average yields of bananas per plant (Plot 1 – 13.8 kg per plant; Plot 2 – 12.0 kg per plant; Plot 3 – 10.4 kg per plant). Describes trends across plots 1–3: average height of plants increases from plot 1 to 2 (by one-fifth); there is no further increase from plots 2 to 3; time to reach harvest increases; average mass of a bunch of bananas decreases; total yield of bananas per plot increases; total yield of bananas per plant decreases. Gives detailed suggestions for differences: the plants are competing for resources, including light, water and minerals; as the density of plants in the field increases, the availability of these resources to each plant decreases; so the growth/ yields of each plant is reduced. Makes a suggestion for the plants' increased height, e.g. the plants are growing taller to try to 'reach'/'capture' light. All information in answer is relevant, clear, organised and presented in a structured and coherent format. Specialist terms are used appropriately. There are few, if any, errors in grammar, punctuation and spelling

3–4 marks
Reasonable analysis of the data, but no or little in the way of mathematical comparison. States that there is a reduction in growth of the banana plants, and suggests that competition is a factor in this reduction. For the most part the information is relevant and presented in a structured and coherent format. Specialist terms are used for the most part appropriately. There are occasional errors in grammar, punctuation and spelling

1–2 marks
Makes simple comments and limited or no suggestions for the trends in data. Answer may be simplistic. There may be limited use of specialist terms. Errors of grammar, punctuation and spelling prevent communication of the science

0 marks
Insufficient or irrelevant science. Answer not worthy of credit

C1 Air quality

Page 165 The changing air around us

1 a Oxygen; nitrogen
  b Mixture
  c Clouds are not made of gases / contain solids and liquids
2 Ratio of 4 nitrogen to 1 oxygen; filling all available space / well spread out
3 Suitable apparatus, e.g. two syringes and a tube containing (excess) copper connected together. Continually pass air over heated copper for a few minutes. Note change in volume and calculate percentage oxygen
4 a Carbon dioxide, water vapour
  b (Earth's) first atmosphere
5 iv before iii; iii before ii; ii before i; i before v
6 Evidence; that is consistent / reliable; little / no evidence to refute the theory; consensus about the explanation of the evidence *(Any 2)*

Page 166 Humans, air quality and health

1 a Factories; power stations; for transport; in homes *(Any 2)*
  b Carbon dioxide
  c Air with few pollutants in it / low pollution levels
2 a Goes up / increases each year; more rapidly over last 10 years, consistent / 15 ppm rise between 1960–80; rise is 1990 less than previously *(Any 3)*
  b i Volcanoes   ii Open fires / vehicle exhausts
  c Sulfur dioxide; nitrogen oxides
  d Asthma; heart disease; lung disease *(Any 2)*
3 a Stands for parts per million; means 1 gram of pollutants in 1 million grams of air
  b To look for trends; to find areas where pollution is a problem; so people at risk can be warned
4 a Both factors lead to the same outcome
  b Can be caused by other factors/carbon particulates; cannot be certain which one is the cause

Page 167 Burning fuels

1 a Carbon; hydrogen
  b Carbon + oxygen; $\longrightarrow$ carbon dioxide
2 a Oxygen; water; carbon dioxide
  b Oxidation
3 Visual representation to show: $2C_2H_2 + 5O_2 \longrightarrow 2H_2O + 4CO_2$ *(1 mark each for: correct representation of molecules; correct equation; balanced)*

4 a Molecules   b Rearranged; new
5 a Reactants; products   b The same
6 Atoms cannot be destroyed; they can react / become new substances
7 a Colour change; new substance / gas formed
  b Sulfur is yellow and insoluble; sulfur dioxide colourless and soluble
8 a Contained in plants and animals; so present when fossil fuels forms
  b $S + O_2 \longrightarrow SO_2$ *(1 mark for reactants; 1 mark for products)*
9 a Acid rain forms when sulfur (and nitrogen oxides) dissolves in water vapour; it damages forests by lowering soil pH; lower pH kills aquatic animals
  b It doesn't affect humans directly

Page 168 Pollution

1 a Power stations; transport (accept examples, e.g. cars)
  b Carbon monoxide   c Sulfur
2 Nitrogen and oxygen (both needed for mark)
3 a Visual representation to show: $2NO + O_2 \longrightarrow 2NO_2$ *(1 mark each for correct representation of molecules, correct equation, balanced)*
  b Harmful to humans or the environment; any two from: damage buildings; make acid rain; cause breathing / lung problems
4 a Sulfur dioxide   b Carbon dioxide
  c Nitrogen monoxide
5 a Deposited on surfaces   b Acid rain formation
  c Photosynthesis / dissolving in oceans
6 a Reading 4 (8 units)   b 12 to 15
  c $15 + 12 + 14 + 14 + 12 = 67 \div 5 = 13.4$
7 Result might be correct; caused by unusual event (e.g. one very polluting car on a quiet road, gust of wind, etc.)

Page 169 Improving power stations and transport

1 Gas contains less sulfur / makes less sulfur dioxide; easier to remove sulfur from gas before burning; no solid waste products / doesn't make ash; more energy efficient *(Any 2)*
2 Using flue gas desulfurisation; using alkaline lime slurry; using sea water *(Any 2)*
3 Switching off devices when not in use / not leaving on stand-by; using newer, more efficient products; accept changing lifestyle *(Any 2)*
4 Idea of fossil fuels running out / fossil fuels are finite non-renewable resource
5 Wood chips; palm oil; coconut husks; biodiesel; alcohol from sugar *(Any 2)*
6 a Biofuels are grown as plants; carbon dioxide is trapped in plant as it grows and released when burnt; they are carbon neutral if another plant is grown for every one used
  b Take up land needed to grow food; cannot produce enough to replace fossil fuels
7 Use cars less / walk; use cleaner fuels; remove pollutants from exhaust fumes; use more public transport, make public transport cheaper / more assessable / more frequent *(Any 2)*
8 a Part of the exhaust system
  b Nitrogen monoxide; carbon dioxide; nitrogen
  c Carbon monoxide gains oxygen and is oxidised to carbon dioxide; nitrogen monoxide loses oxygen and is reduced to nitrogen
9 a Very expensive; not many charging points currently available; takes a long time to charge; limited distance range per charge; dangerous as too quiet for people to hear *(Any 3)*
  b Most electricity used to charge them is produced at power stations; burning fossil fuels releases carbon dioxide

Page 170 C1 Extended response question

5–6 marks
A detailed description of the patterns shown is given, along with suitable science ideas about causes and why the levels change – an example might be: PM10 levels are lower than NOx and both vary day to day. A correlation exists: when NOx levels rise and fall, so do levels of PM10; level rise might be linked to traffic levels / more people travelling weekdays / less at weekends; changes may be due to environmental conditions / wind / rainfall. All information in answer is relevant, clear, organised and presented in a structured and coherent format. Specialist terms are used appropriately. There are few, if any, errors in grammar, punctuation and spelling

3–4 marks
Limited description of the graph and explanation, or a good explanation of one part. For the most part the information is relevant and presented in a structured and coherent format. Specialist terms are used for the most part appropriately. There are occasional errors in grammar, punctuation and spelling

Answers

C2 Material choices

Page 171 Using and choosing materials

1 a Rubber – hard and elastic – car tyres; Plastic – can be moulded – washing-up bowls; Fibres – can be woven – making clothes (*Any 2*)

   b When talking about materials, a property is something that makes it suitable for the job it does

2 a Melting point

   b Compressive strength

   c Hardness

   d Density

3 a i Climbing ropes need to be dynamic (stretchy) to absorb fall energy and not snap

      ii Climbing ropes need to be dry treated so they do not get heavy when wet and pull the climber off

   b Thicker / 11-mm rope contains more woven fibres; increasing tensile strength

4 a 5.3

   b 5.6 to 5.8

   c To find the true value

   d 5.7

   e An outlier can be discarded if an error occurs in measurement; if one measurement is very different from all the rest

Page 172 Natural and synthetic materials

1 Good conductor; hard; malleable

2 Clay; glass; cement

3 Polymers

4 a Cotton / paper

   b Silk

   c Limestone / iron ore

5 Synthetic materials are made by a chemical reaction; using raw materials from the Earth's crust

6 Natural materials are in short supply; they can be designed to give particular properties; they can be cheaper; they can be made in the quantities needed (*Any 3*)

7 Carbon, hydrogen

8 a Fuels

   b Equal

9 a C_4H_{10}

   b C_6H_{14}

10 Due to differences in the marine organisms; that decomposed to make it

11 Diagrams showing correct representation for each molecule; diagram balanced showing two oxygen molecules on the left, and two water molecules on the right

Page 173 Separating and using crude oil

1 a Distillation

   b Evaporates

   c Condenses

   d Fractions

2 The larger the hydrocarbon molecule the higher the boiling point (*Accept reverse*)

3 Petrol is a smaller molecule; so it has less attractive forces between molecules; so less energy is needed to change it from liquid to gas

4 a Separate paper clips are labelled monomer and joined-up paper clips are labelled polymers

   b i Carbon fibre

      ii Plastic / polythene / polyethene

5

   (*Ignore how many – marks given for repeated units; all joined by single bonds*)

6 A PET polymer is strong; has low density; does not shatter

7 The properties of polymer chains can be changed by replacing hydrogen atoms; with other groups of atoms

Page 174 Polymers: properties and improvements

1 a i   b i

2 a Low melting point; weak; flexible; soft; insulator (*Any 3*)

   b Little or no side branches; long chains; strong forces between molecules; high crystallinity (*Any 3*)

3 Crystallinity holds molecules in regular patterns; the more molecules present the stronger the polymer; and the higher the melting point

4 a Stronger   b More   c Higher

5 Plasticisers are small molecules; that fit between polymer chains; weakening the force between them increases flexibility

6 Thermoplastics have no or few cross-links; so they melt when heated; allowing them to be moulded

7 Natural rubber can be made harder by increasing molecule chain length; and by having more cross-linking

8 a Crystallinity can be increased by drawing the polymer though a small hole to align molecule chains

   b Increased crystallinity makes the plastic brittle

Page 175 Nanotechnology and nanoparticles

1 Microscope

2 10

3 a Salt in sea spray

   b Carbon particulates

   c Designed / made in labs

4 Large molecule

5 One thousand million / one billion

6 Buckyballs are spheres; of 60 carbon atoms. Nanotubes are made from (graphene) sheets; folded into tubes

7 Nanoparticles are effective catalysts because they have a larger surface area; so there are more sites for reactions to occur

8 a Graphene sheets are only one atom thick; but millions of atoms long

   b Volume is the space taken up by a substance. A volume of $1\ cm^3$ stays the same when cut up. Surface area increases when cut up. The surface area starts at $6\ cm^2$. When cut in half, it becomes $8\ cm^2$. When cut into four it becomes $10\ cm^2$

Page 176 The use and safety of nanoparticles

1 To kill bacteria

2 a 80.5 cm

   b 77 to 79 cm

   c 75 cm after 3 months

   d The range overlaps in each set of results so the true value might be the same; but the means do show it is slightly decreasing. So it is possible it might be getting less bouncy

   e Composites

3 Nanotechnology means the science of making, using and controlling nanoparticles

4 Carbon nanotubes are held together by very strong carbon bonds / bonded like graphite; lightweight as carbon has a low atomic mass; more ductile as while each small tube is stiff, they are not very long, and adjacent ones can flex

5 The silver nanoparticles can be washed into sewage works; and kill the bacteria that clean the water

6 Nanoparticles are small enough to pass through the skin into the body; and possibly lodge in body organs; long-term effects are not known

7 Little actual evidence (so far) of any risk; natural nanoparticles have been around forever without noticed risk; nothing is ever completely safe; people can decide themselves whether to take the risk of using them; ideas that many nanoparticles have useful applications (*Any 3*)

Page 177 C2 Extended response question

5–6 marks
Answer states at least three properties needed (flexible, lightweight / low density, strong in compression / strong in tension, hard and stiff). Also refers to patterns in the table, e.g. weight and cost links to material used: graphite reduces mass but adds to the cost; nano graphite increases mass as spaces filled in to increase strength and stiffness. All information in answer is relevant, clear, organised and presented in a structured and coherent format. Specialist terms are used appropriately. There are few, if any, errors in grammar, punctuation and spelling

Answers

3–4 marks
Answer states three or four properties only, or some properties and a pattern. For the most part the information is relevant and presented in a structured and coherent format. Specialist terms are used for the most part appropriately. There are occasional errors in grammar, punctuation and spelling

1–2 marks
Answer gives one or two properties, and only directly quotes information from the table. Answer may be simplistic. There may be limited use of specialist terms. Errors of grammar, punctuation and spelling prevent communication of the science

0 marks
Insufficient or irrelevant science. Answer not worthy of credit

C3 Chemicals in our lives – risks and benefits

Page 178 Moving continents and useful rocks
1 Geologists
2 Colliding; pulling apart
3 Pangea was formed when many continents collided; forming mountain ranges
4 Over the last 600 million years Britain has moved; weather is linked to (latitude) position / warmer nearer equator
5 Lava erupts; containing magnetic materials; that line up along Earth's magnetic field; their direction shows Earth's magnetic field when formed
6 a The Industrial Revolution started in the north-west of Britain because the industries needed to be where all the raw materials were found
   b Coal; limestone; salt
7 In warm shallow seas shellfish died; compacted and hardened; forming limestone
8 Salts dissolve in water; enclosed shallow salty sea water evaporates; wind-blown sand becomes mixed with the salt; and buried over time
9 a Ripple marks in the rock; indicate river or wave action
   b Fossils in coal show it is made from plant material; shell fragments in limestone show it is made from sea creatures; shaped grains in rock salt show both water and wind erosion took place *(Any 1)*

Page 179 Salt
1 You can obtain salt by mining it; or by evaporating sea water
2 Salt is put onto icy roads to prevent ice forming; by lowering the freezing point of water; and giving better grip
3 Salt is not obtained from sea water in the UK because it is too expensive to evaporate the water
4 Subsidence could result in large holes underground causing the ground / people's homes to sink into the earth; evaporating salt blown into the environment can damage habitats / pollute water supplies; mining can allow water in mines, which may let salt leach out and contaminate water supplies *(Any 2)*
5 As flavouring; and a preservative
6 As salt concentration increases; the bacteria numbers drop; drop increases at a greater rate after 48%
7 a Salt can cause high blood pressure / heart failure; the risk is estimated by measuring salt intake / by adding totals from labels
   b People can reduce the risk from salt by changing diet to a lower intake; or using 'low salt' alternatives
8 The government regulates food safety through agencies; DH and Defra; the agencies carry out risk assessments; and provide advice to the public

Page 180 Reacting and making alkalis
1 c; d
2 a Potassium sulfate
   b Sodium chloride
3 Stale urine; burnt wood
4 There were few sources; only from burnt wood or seaweed; the demand was greater than the supply
5 a Visual representation to show: $CaCO_3 \longrightarrow CaO + CO_2$ *(1 mark each for correct representation of molecules, correct equation, balanced)*
   b Lime only reacts when acid is present; so any excess just remains in the soil; too much or too little alkali has adverse effects on plant growth; adding the correct amount of alkali is hard to judge; as the amount of acid in a patch of soil is likely to vary *(Any 4)*
6 A mordant sticks dye to fabric

7 a Hydrogen chloride gas; hydrogen sulfide gas
   b Find a use for unwanted products; e.g. as raw materials for a new process
8 a Sodium sulfate + water + carbon dioxide
   b Ammonium nitrate + water
9 Alkalis must be soluble hydroxides; providing hydroxide ions when dissolved

Page 181 Uses of chlorine and its electrolysis
1 a Any number between 90 and 98
   b In Spain the drinking water; is treated with chlorine
2 They may have been infected on holiday; or by drinking untreated water
3 a Chlorine reacts with organic materials in water; forming toxic or carcinogenic compounds / disinfectant by-products (DBPs)
   b Amounts of DBPs are very small; so the risks are small; compared to the risk of cholera or typhoid; in this case the benefits outweigh calculated risks;
4 a Electrolysis
   b Hydrogen; chlorine; sodium hydroxide
   c All the products are useful (so there is no waste)
   d Making these products requires large amounts of electricity or energy; to melt the electrolyte; and to separate using electrolysis
   e Chlorine is used for PVC / medicines / crop protection; hydrogen is used for margarine / rocket fuel / fuel cells; sodium hydroxide is used for paper making / in domestic cleaners / for refining aluminium
5 Environmental impacts plus potential solutions:
   • Chlorine is linked to ozone depletion; it was used in fridges and aerosols; but has now been banned. Collection points have been set up by local authorities so fridges can be disposed of safely
   • Chlorine is linked to dioxins in paper bleaching; which are linked to cancer. Manufacturers try to find alternative for bleaching; or ways to find to prevent dioxins escaping
   • Mercury is linked to toxic poisoning; which can build up over time in animals higher up the food chain. Laws are in place in some countries to ban the mercury method of electrolysis. You could also use an alternative, more efficient membrane cell method *(Any 1)*

Page 182 Industrial chemicals and LCA
1 a Lead could spread to the environment from dumping; escapes during manufacture or mining; is released into the air if batteries are burnt *(Any 2)*
   b Recycling lead batteries reduces the need for processing or mining new lead
   c A cumulative poison is when something builds up; slowly over time; in body tissues
2 People *perceive* a greater risk with a less familiar named chemical; long-term studies to measure actual risk are not possible; the *actual* risk may be very small or non-existent; the EU believe that if there might be a risk it should be avoided
3 a Carbon
   b Chlorine
   c Hydrogen
4 a It will not break down naturally
   b Plasticisers
5 Idea that they leach out easier; unknown health risk
6 An LCA measures energy use and environmental impact; from when something is made; until its disposal
7 The LCA is better for the table: though more energy is used to make a table; it has a longer life and may never be disposed of; the table has less environmental impact; than burning wood which releases carbon dioxide
8 A full LCA is not always possible because of insufficient data; it is too hard to measure some aspects; e.g. a product life / the disposal method is often unknown

Page 183 C3 Extended response question
5–6 marks
LCA defined as the energy needed to make it, use it, dispose of it. Answer contrasts the two charts effectively using examples, e.g.:
• large amount of energy needed for extracting the oil, refining it, cracking into monomers and making the polymer;
• energy needed to collect plastics disposed off, and to clean and shred, but less than making from new;
• LCA during the life of an individual item the same, but recycling has better LCA as it extends the life and reduces the need to burn or go to landfill.

Answers

All information in answer is relevant, clear, organised and presented in a structured and coherent format. Specialist terms are used appropriately. There are few, if any, errors in grammar, punctuation and spelling

3–4 marks

LCA defined with some contrast of the LCA and some explanation. For the most part the information is relevant and presented in a structured and coherent format. Specialist terms are used for the most part appropriately. There are occasional errors in grammar, punctuation and spelling

1–2 marks

'Cradle to grave' stated with some reference to differences, although not explained. Answer may be simplistic. There may be limited use of specialist terms. Errors of grammar, punctuation and spelling prevent communication of the science

0 marks

Insufficient or irrelevant science. Answer not worthy of credit

P1 The Earth in the Universe

Page 184 Our solar system and the stars

1 Comet, star, planet, moon, dwarf planet
2 a Mercury; Earth; Uranus; Jupiter
  b Mercury; Earth; Jupiter; Uranus
3 Light has a speed of 300 000 km/s; a light-year is how far light goes in a year; this is $300\,000 \times 60 \times 60 \times 24 \times 365 = 9.46 \times 10^{12}$ km
4 The Milky Way is one of many *galaxies* in the Universe. Each galaxy is made of many *stars*
5 Look at radiation from stars; and compare it with radiation from the Sun. If they are different, then hypothesis is disproved
6 Radiation from a distant galaxy takes a long time to reach us; but it can only tell us about the galaxy when it left. So if the galaxy is a billion light-years away, we see it as it was a billion years ago
7 Earth; Sun; Solar System; Milky Way
8 Stars which are further away have a lower apparent brightness than ones which are closer; so the relative brightness of two stars allows you to work out their relative distance; if you know the distance to one star, then you can work out the distance to the other
9 Only close stars have a large enough parallax to be measured accurately; most stars do not appear to move relative to the others as the Earth moves around its orbit

Page 185 The fate of the stars

1 The energy comes from fusion of hydrogen
2 Nuclear fusion in stars; forces hydrogen atoms together to make helium. Big enough stars can use fusion to make helium into heavier elements. The heaviest elements are made when a star explodes at the end of its lifetime
3 Solid planets are made out of elements heavier than hydrogen or helium. Scientists believe that when the Universe was created it only contained hydrogen and helium. Other elements were only created when the first stars ran out of fuel and exploded as supernovae. The dust and gas from these explosions then condensed to make our planets
4 Most galaxies are moving away from us. This increases the wavelength of the light we receive from them
5 The Sun is 5000 million years old; the Earth is 4500 million years old; the Universe is 14 000 million years old
6 The future expansion of the Universe depends on its mass. If there is enough mass in the Universe, gravity will eventually reverse the expansion and the Universe will end up at a single point. If there is too little mass, the Universe should continue to expand forever. But finding the mass of the Universe is difficult because we can only see those bits of the Universe which emit radiation

Page 186 Earth's changing surface

1 Fossils are the remains of plants and animals in rock. Mountains are made by folding the rocks of the Earth's crust. Volcanoes are new mountains made from lava. Sediments are materials from the erosion of mountains
2 Mountains are broken down into small bits by the weather. The bits are carried into the sea where they fall to the bottom and form layers of sediments. Each new layer buries and crushes the previous layers to make rock
3 The Earth's crust is always moving. Where bits of the crust collide head-on, the sedimentary rocks made under the sea from material eroded from mountains are folded and pushed up to make new mountains. Where bits of the crust are moving apart, liquid rock can rise up from the centre of the Earth to make new mountains called volcanoes out of lava. So although erosion is always scraping the tops off mountains and putting them into the sea, new mountains are made all the time to replace them

4 b; e
5 Wegener's theory was very different to the other theories which explained similar rocks and fossils where the continents fitted together. Wegener had no science qualifications. Nobody could detect the motion of the continents. Nobody could explain why the continents should move
6 Lava comes up from the mantle at the oceanic ridge down the centre of the seafloor. As the lava solidifies, it is magnetised by the Earth's field and then gradually moves away from the oceanic ridge as it makes new rock. Every so often the Earth's magnetic poles swap over, changing the magnetisation direction of the new rock made at the ridge. So magnetisation of the rocks suddenly changes as you move away from the ridge

Page 187 Tectonic plates and seismic waves

1 Earthquakes; volcanoes; mountains
2 Where two tectonic plates meet. Tectonic plates are pieces of the Earth's crust
3 Liquid magma comes up where plates are moving apart, forming volcanoes. Where plates are moving together, one plate may be forced under the other. The plate melts as it moves down; and the liquid rock is forced up through cracks in the other plate to form volcanoes
4 As the plates move past each other, tension built up at the boundary is suddenly released as an earthquake
5 Most of the original rocks were eroded and formed sediments in the sea. Some of these sediments were pushed into the mantle and melted; as one tectonic plate was pushed under another. The material was then fed back up to the surface through volcanoes to make fresh rock
6 Core – liquid iron; crust – solid rock; mantle – semi-liquid rock
7 Tectonic; seismic; S-waves; P-waves; faster
8 The evidence comes from observations of seismic waves; around the world. The time of arrival of the waves allows the speed of the waves to be calculated. This tells us about the density of the material that the waves have passed through. There are places where S-waves don't arrive, suggesting a liquid core which S waves can't pass through

Page 188 Waves and their properties

1 Speed – how far the energy of the wave travels in a second; Frequency – the number of vibrations of the wave source in one second; Amplitude – maximum value of the disturbance in one wave; Wavelength – distance along wave from one zero disturbance to the next
2 Energy
3 Time delay for light is almost nothing, so distance = $300 \times 1.5 = 450$ m
4 The frequency of a wave is the number of vibrations per second produced by it
5 Speed – m/s; frequency – Hz; wavelength – m
6 Frequency = 20 000 Hz, wavelength = 0.40 m; Speed = $20\,000 \times 0.40 = 8\,000$ m/s
7 a Frequency = speed / wavelength = $2.0 \times 10^8 / 0.44 \times 10^{-6}$ = 4.5×10^{14} Hz
  b Assuming speed is still 2.0×10^8 m/s; wavelength = speed / frequency = $2.0 \times 10^8 / 6.7 \times 10^{14} = 3.0 \times 10^{-7}$ m

Page 189 P1 Extended response question

5–6 marks

Answer includes the majority of relevant points for both theory and observations, as follows:
- Relevant scientific points about theory of an expanding Universe:
  - Every galaxy should appear to be moving away from every other galaxy.
  - This should redshift the light from each galaxy.
  - Redshift increases the wavelength of the light from it received by another galaxy.
  - The further away the galaxies are, the greater the redshift should be.
- Relevant scientific points about observations of the Universe:
  - Observations of the spectra; of galaxies seen from Earth shows a redshift.
  - A few nearby galaxies do not have a redshift.
  - Fainter galaxies have a larger redshift than brighter ones.
  - This suggests that redshift increases with distance between galaxies.

All information in answer is relevant, clear, organised and presented in a structured and coherent format. Specialist terms are used appropriately. There are few, if any, errors in grammar, punctuation and spelling

Answers

P2 Radiation and life

Page 190 Waves which ionise

1 b; a; d; c
2 300 000 km/s
3 Radio waves ➔ infrared ➔ ultraviolet ➔ gamma rays
4 Energy; photons; light; frequency
5 b; c; d
6 The intensity of a wave decreases as it moves away from its source. This is because the photons are spread over an increasing area as the wave moves
7 Intensity = energy / (time × area) = 4.8×10^{-3} / ($8.0 \times 2.0 \times 10^{-6}$) = 300 J/s/m^{-2}
8 The atom becomes positively charged. This is because the electron which was removed has negative charge, and the atom started off with no charge at all
9 a Gamma rays; X-rays; ultraviolet
   b The photons have enough energy; to knock an electron out of the atom
10 Photons absorbed by a molecule in a cell; can ionise it; starting off a chemical reaction; which might damage a cell's DNA; and cause cancer

Page 191 Radiation and life

1 a ii
   b iii; iv
2 Absorbed; transmitted; shields; decreasing
3 a; b; c
4 The microwaves are only absorbed; by water molecules in food
5 They select a large number of people who use mobile phones; and another control group of people who don't; but are otherwise the same. They then record how many people in each group develop cancer; over many years
6 a Ultraviolet
   b Give you skin cancer
   c Put on sunscreen; and cover up with clothes
7 a You can get sunburn; and skin cancer
   b You get a tan; and it improves your health (by making vitamin D)
8 As the ozone absorbs the radiation; it is chemically changed

Page 192 Climate and carbon control

1 b; f
2 All; greatest; increasing
3 A – ii; B – i; C – iii; D – iv
4 Plants have absorbed carbon through photosynthesis; at the same rate; as it has been released by respiration; of living organisms
5 Methane has the most effect; then carbon dioxide; and finally water vapour. There is a lot of water vapour, so it has most effect. There is very little methane, so its effect is small
6 Farmland could be flooded by the sea; it could become too hot for some crops; the weather could become too stormy
7 The model is a computer program; which uses laws of science; to work out future climate from today's climate. A good model will be able to predict today's climate from that of some period in the past

Page 193 Digital communication

1 Radio, microwaves, visible and infrared
2 They are not absorbed by air; so can travel from the transmitter to receiver
3 It can travel through optical fibres; with very little absorption by the glass
4 It changes either the frequency; or the amplitude; of the wave
5 C

6 a They are the two values for the carrier wave; which can be on or off
   b The value of the sound is coded as a string of 1s and 0s; many time a second; at the transmitter. The strings are used to switch the carrier wave on and off. The receiver uses the strings of code to construct a copy of the original sound
7 The noise signal is there all the time; so can be usually distinguished from the digital signal; which turns the carrier wave on and off
8 2; 1; 0; 8
9 The player receives a large number of samples in each second. They arrive one after the other. Each sample is a string of binary digits; which is used to set the value; of the sound at that instant
10 The information is easily stored electronically; it can be processed in a computer; it can be music, speech or picture; it doesn't get lost or altered during transfer

Page 194 P2 Extended response question

P3 Sustainable energy

Page 195 Energy sources and power

1 Coal, oil and gas
2 Biofuels; wind; solar; waves; geothermal; hydroelectric; tidal (Any 3)
3 It has to be transferred from another source of energy
4 Most of our electricity comes from burning fossil fuels; and these are non-renewable so they will run out one day
5 Burning gas makes carbon dioxide. This is a greenhouse gas; so it increases the temperature of the atmosphere. This impacts on the environment by causing floods and stormy weather
6 Energy; watts; joules; seconds
7 250 W is 0.25 kW, so energy = 0.25×24 = 6.0 kWh
8 The energy in the circuit – flows from the supply to the kettle
   The current in the circuit – transfers energy from the supply
   The power of the kettle – is the energy transfer per second
   The voltage of the supply – provides the current with energy
9 Power; volts; current; amperes
10 230×3.0 = 690 watts
11 Power = 1150 W so current = 1150 / 230 = 5.0 A

Page 196 Efficient electricity

1 One kilowatt-hour is – 3 600 000 joules. The meter readings are in – units of kilowatt-hours. An electricity meter records – the energy transferred into a house.
2 Power = 0.80 kW and time is 1.5 h, so energy used is 0.8 kW × 1.5 h = 1.2 kWh; the cost is 1.2 kWh × 15 p/kWh = 18 p
3 a How much of the electrical energy entering the TV is transferred to various other sorts of energy
   b Light 10 J

Answers

4 An arrow splits – where there is an energy transfer. Electrical energy flows – in from the left. Waste heat energy flows – out downwards. The electrical energy input is – equal to the sum of the energy outputs. The thickness of each arrow is – proportional to its energy. Useful transferred energy flows – out to the right

5 Efficiency = 20/50 = 0.4 or 40%

6 Put very expensive taxes on inefficient cars; subsidise house insulation; replace old power stations with more efficient ones

7 The world's population is increasing; so each of those extra people will need a certain amount energy. At the moment, a lot of the energy is used by the few people in rich countries; and it is likely that everyone else will also want to have the same energy-rich lifestyle

Page 197 Generating electricity

1 Generators; magnet; coil

2 Voltage

3 e, a, d, b, c

4 **a** Coal; gas; nuclear fuel; oil; biofuels such as wood *(Any 3)*
   b Wind; water behind dams; tidal

5 It spins; the magnet inside the generator

6 Exhaust from burning gas; the air as wind; and water in rivers and the sea

7 The fuel is burnt; and transfers energy to water; which boils to make steam; which passes through the turbine

8 The gas is burnt; to make a high pressure gas; which passes through the first turbine. When the gas leaves the turbine it is hot enough to boil water into steam; which is sent through the second turbine

9 The rods of uranium; fuel are stacked close together; in the reactor. Nuclear reactions transfer energy in the rods, heating them up. This thermal energy is carried away by high pressure water; pumped through the stack of rods to a heat exchanger where it is used to boil water to make steam; for the turbine

Page 198 Electricity matters

1 It is radioactive; so emits radiation

2 **a** The radiation from nuclear waste is ionising; so it damages cells in your body
   b Stay well away from it; or put thick shields of metal and concrete in the way
   c When you are contaminated some of the waste has got inside your body. This means that you are exposed to the radiation for a much longer time; and it is difficult to isolate and remove the material

3 They compile statistics of deaths caused by each technology; taking into account the amount of electricity produced and how many years it has been used. They can then decide which technology is low risk (smaller number of deaths per MWh of electricity)

4 Hydroelectric; tidal; wind

5 Hydroelectricity can be switched on and off quickly, so is useful for meeting surges in demand; and it can be used to store excess electricity from elsewhere by pumping water back behind the dam. However, the dams cost a lot to build and the lakes behind them flood a lot of land

6 You can put wind turbines out at sea; so that they can't visually pollute the landscape. You can use the dams of hydroelectric schemes; to prevent the sudden flooding of rivers downstream when there is a lot of rain

7 230 V

8 The most important wasteful energy transfers occur in the power station. Whenever the energy is being transferred as heat, some of it escapes into the environment. Some energy has to be used to provide current for the electromagnet in the generator; and finally some is lost as heat in the cables of the National Grid

Page 199 Electricity choices

1 Fossil fuel – produces greenhouse gas; Wind power – noise and visual pollution; Nuclear power – produces radioactive waste; Hydroelectricity – floods large areas of land

2 Coal; oil; gas; nuclear

3 Geothermal; nuclear; solar; wind

4 They could install 40 wind farms, but this would take up a lot of land and upset a lot of people who think they are ugly. Some could go out at sea where they would work better and not be seen. They could build 20 nuclear power stations, but that would mean a lot of radioactive waste to deal with in the future. I think they should go for a mixture of both; so that they don't have to rely on just one technology

5 People in the USA and the UK will both be affected by the global warming caused by the carbon dioxide

6 There will be more people on Earth in future which means less energy per person to keep the total energy the same. Most people around today use less energy than we do in the UK, so future people will need to use more energy to adopt our lifestyle. So we will need to use a lot less energy in the future

7 They need to make sure that they can always generate enough electricity to meet demand. To avoid over-dependence on just one energy source, a variety of technologies should be used to generate electricity. Since fossil fuels will inevitably run out, they need to invest in a range of renewable energy sources. They could also make sure that old power stations are replaced by more efficient ones to make better use of the energy source

Page 200 P3 Extended response question

5–6 marks
Answer includes nearly all of the relevant points, as follows:
• Is the cost of the energy source low enough?
• Is the energy source near enough to be reliable?
• Does the energy source contribute to global warming?
• How long will the energy source last for?
• Will the power station be efficient?
• How easy is it to dispose of waste from the power station?
• How much it will cost to run and maintain the power station?
• What is the public's perception of risks from the power station?

All information in answer is relevant, clear, organised and presented in a structured and coherent format. Specialist terms are used appropriately. There are few, if any, errors in grammar, punctuation and spelling

3–4 marks
Answer includes at least half of the relevant points (see above). For the most part the information is relevant and presented in a structured and coherent format. Specialist terms are used for the most part appropriately. There are occasional errors in grammar, punctuation and spelling

1–2 marks
Answer includes at least two relevant points (see above). Answer may be simplistic. There may be limited use of specialist terms. Errors of grammar, punctuation and spelling prevent communication of the science

0 marks
Insufficient or irrelevant science. Answer not worthy of credit

B4 The processes of life

Page 201 The chemical reactions of living things

1 c, d

2 A chemical that speeds up the rate of a chemical reaction (but is, itself, unchanged)

3 Proteins; amino acids; genes; substrate; active site; lock and key

4 **a** Stomach: pH 1.5; small intestine pH 8.0
   b Enzyme activity depends on the substrate (protein) being able to fit into the active site of the enzyme. Each enzyme has an optimum pH. At other pHs, the structure of the active site changes, and the substrate is unable to fit as well. At a certain point, the structure of the enzyme is changed permanently (it is denatured) and is no longer able to work

5 **a** The graph only shows that papain is unaffected by pH between 4.0 and 8.5. We do not know how its activity is affected at lower or higher pHs
   b The shape of the active site of the enzyme is unaffected by pHs from 4.0 to 8.5

Page 202 How do plants make food?

1 Reactants are carbon dioxide and water; products are oxygen and glucose. The reaction is driven by light energy – in the box above the arrow

2 The Sun; chlorophyll; chloroplasts; sugar; glucose; respiration; starch; cellulose; proteins *(The last three terms can appear in any order)*

3 Reactants are $6CO_2$ and $6H_2O$; products are $6O_2$ and $C6H_{12}O_6$. The reaction is driven by light energy – in the box above the arrow

4 **a** Cell membrane: controls what enters and leaves the cell (it allows gases and water to pass in and out of the cell freely, but is a barrier to other chemicals)
   b Cell wall: gives the cell support. It lets water and other chemicals pass through freely
   c The cytoplasm is the jelly-like material that fills the cell, and is where most of the chemical reactions in the cell occur
   d A mitochondrion is a structure in the cell responsible for the release of energy by aerobic respiration/ responsible for the release of most of the energy by the cell
   e Nucleus: contains DNA, which stores the genetic code. The genetic code carries information the cell uses to make enzymes and other proteins

Answers

5		Bacteria	Yeast
	Outer layer of cell	Cell wall	Cell wall
	Genetic material	As circular DNA in the cytoplasm	In the nucleus (as chromosomes)
	Respiration	Enzymes for respiration associated with cell membrane	In mitochondria (with some in the cytoplasm)

Page 203 Providing the conditions for photosynthesis

1 The movement of molecules from an area of high concentration to an area of low concentration

2 Carbon dioxide; into; diffusion; oxygen; out of; passive; increase

3 a The chip in the distilled water increased in mass, and the chip in the concentrated sucrose solution decreased in mass. The potato cells contain a dilute solution (in their vacuoles), and the water concentration in the cells is lower than in the distilled water; so the water moves in by osmosis. In the concentrated sucrose solution, the water concentration is lower than in the potato cells; so water is lost by osmosis

b Cut a number of potato chips and place them in a range of solutions with known concentrations of sucrose. Calculate the percentage change in mass. Repeat the experiment; and calculate the mean change in mass. Plot a graph of percentage change in mass over sucrose concentration. Read off the sucrose concentration where the line crosses the x-axis

4 The concentration of nitrates is higher in the root cells than in the soil, so they cannot be taken up by diffusion. They must be taken up against a concentration gradient; by active transport, which requires energy

5 a As the light intensity increases, the rate of photosynthesis increases; as light energy is required to drive the process. At a certain point, the graph levels off, so any further increase in light intensity will result in no further increase in photosynthesis. At this point, some other factor must be limiting, e.g. carbon dioxide

b i The graph for the high carbon dioxide concentration has an identical gradient, but reaches a greater height, i.e. photosynthesis reaches faster rate, before levelling off)

ii In a higher concentration of carbon dioxide, the graph will continue to a higher point (i.e. a higher rate of photosynthesis); until it levels off. At this point (with light and carbon dioxide being available), some other factor (e.g. temperature); must be preventing any further increase in the rate of photosynthesis

Page 204 Fieldwork to investigate plant growth

1 By examining the leaf and answering a sequence of yes or no questions; e.g. does the leaf have needles/is the leaf a typical shape/ are the leaves in groups; Ruby will be able to place the leaf in smaller groups; until she identifies the tree it is from

2 a i 5 cm $\times$ 5 cm; ii 0.5 m $\times$ 0.5 m; iii 0.5 km $\times$ 0.5 km

b 6 dandelions per m² / 5.6 dandelions per m² (the total number of dandelions over the 10 quadrats is 14, so the mean is 1.4 per quadrat; each quadrat is 0.25 m², so the distribution is 5.6 per m²) *(1 mark for answer; 1 mark for units)*

c When the plants show an obvious change in distribution across a location

3 a Light is needed for photosynthesis; and products of photosynthesis are required to synthesise the molecules required for growth; in low light intensities, plants will not be able to photosynthesise; but the tolerance of low light intensities will vary from plant to plant, so some are better able to live in shade than others

b This evidence supports/increases confidence in the hypothesis; and a mechanism relating to photosynthesis could account for these; but correlation does not prove cause; other factors could also contribute, e.g. competition among the plants for water and minerals

Page 205 How do living things obtain energy?

1 c; e

2 Glucose, oxygen; carbon dioxide, water, energy

3 a 1 mark for reactants; 1 mark for products

$C_6H_{12}O_6 + 6O_2; \longrightarrow 6CO_2 + 6H_2O + energy$

b The reaction takes place as a series of stages/the equation is a summary; with energy being released in stages

4 Human muscle cells during vigorous exercise; plant roots in waterlogged soil; bacteria in deep puncture wounds

5 a Ethanol/alcohol; carbon dioxide; lactic acid

b Yeast – beer/wine/other alcoholic drink/bread or bacteria – yogurt

6 Aerobic respiration requires the presence of oxygen; anaerobic respiration takes place in the absence of oxygen, or in very low oxygen concentrations. The products of aerobic respiration are carbon dioxide and water; the products of anaerobic respiration vary/ products include alcohol, carbon dioxide, lactic acid, but not water. The energy released by aerobic reaction is much greater than that released by anaerobic respiration

Page 206 B4 Extended response question

5–6 marks

Explains how the sucrose is an exact fit to the enzyme in terms of the active site, protein structure and sequence of amino acids. Explains enzyme specificity and that maltose will not be an exact fit to the active site, and uses an accurate diagram to illustrate these principles. Explains the importance of pH in enzyme action, and the effects of an inappropriate pH on the structure of the active site. Recognises that the optimum pH for invertase must be around 4.5, but the pH optima for other enzymes involved in brewing must also be around this pH. All information in answer is relevant, clear, organised and presented in a structured and coherent format. Specialist terms are used appropriately. There are few, if any, errors in grammar, punctuation and spelling

3–4 marks

Explains enzyme action in terms of the enzyme and lock and key mechanism, but explanation is incomplete, not fully detailed, or related to sucrose. States that enzyme action is specific to one substrate and that enzymes work best at a specific pH, and relates these to enzyme shape, but the answer does not go beyond this. For the most part the information is relevant and presented in a structured and coherent format. Specialist terms are used for the most part appropriately. There are occasional errors in grammar, punctuation and spelling

1–2 marks

Limited description of enzyme action. States that the enzyme works on sucrose only and at a certain pH, with little or no explanation of this. Answer may be simplistic. There may be limited use of specialist terms. Errors of grammar, punctuation and spelling prevent communication of the science

0 marks

Insufficient or irrelevant science. Answer not worthy of credit

B5 Growth and development

Page 207 How organisms develop

1 a Multicellular; specialised

b Tissue; nerve cells; nervous tissue

c Tissues; organs; brain

2 After the eight-cell stage of the embryo, cells become specialised. This is called differentiation. In these specialised cells, only the genes needed to enable the cell to function as that type of cell remain switched on; other genes are switched off

3 Plant tissue: phloem; xylem. Plant organ: flower; root; stem

4 a, b

5 Meristem at tip of shoot/stem: division of meristem cells, followed by enlargement of one of the daughter cells, produces an increase in height/length of stem, or growth of new leaves or flowers. Meristem in side bud: division of meristem cells, followed by enlargement of one of the daughter cells, produces side growth, or growth of new leaves or flowers. Meristem along the length of the stem/shoot and root: division of meristem cells, followed by enlargement of one of the daughter cells, produces an increase in girth/thickness of the stem and root. Meristem in tip of root: division of meristem cells, followed by enlargement of one of the daughter cells, produces an increase in the length of the root *(1 mark for each)*

Page 208 Plant development

1 Roots grow at the base of the stem; while the shoot continues to grow. The technique enables people to produce many new plants from a single plant

2 Description of taking a cutting or tissue culture:
- Taking a cutting – cut a small length of plant stem which includes a meristem; dip the cut end into hormone rooting powder; put the end of the stem into damp compost
- Tissue culture – remove a small piece of tissue, or a few cells from a plant; place on agar; containing nutrients and plant hormones

3 Cell division; cell enlargement

4 Light is coming from one direction, so the plant grows towards the light to expose more surface to the light. This helps the plant's survival by enabling it to photosynthesise. Without photosynthesis, the plant would not be able to grow (as it produces glucose, from which the molecules needed for growth are produced)

Answers

5 One direction; towards; phototropism; photosynthesis

6 There is no significant difference between the amount of auxin in the plants kept in the dark or light, or total auxin in plants illuminated on one side; so light has no effect on the *production* of auxin. About 71% of the auxin in the plant illuminated from one side is on the dark side; so as the total auxin was unaffected by light, the auxin must have been *redistributed* from the light to dark side

Page 209 Cell division

1 a A type of cell division; that produces two cells that are genetically identical; and have identical numbers of chromosomes as the parent cell

 b During growth; and when cells divide to repair tissues

2 Percentage of total time spent in mitosis: stomach – 2.2%; small intestine – 3.9%; large intestine – 1.3%

3 a The cell increases in size; the number of organelles increases; the DNA in each chromosome is copied

 b Rat intestine (30/2000 = 0.15 or 1.5%)

 c The developing fruit fly egg; because the egg is developing, so is undergoing rapid cell divisions

4 a, c

Page 210 Chromosomes, genes, DNA and proteins

1 Chromosomes; DNA; genetic; amino acids; cytoplasm

2 a Phosphate (green circle); bases (white rectangles); sugar (yellow pentagon)

 b An alpha-helix / like a twisted ladder

3 The genetic code carries the instructions for protein synthesis

4 c, b, e, d

Page 211 Cell specialisation

1 a, d

2 Damaged / diseased; diseased / damaged; adult; limited

3 Michael's first sentence is correct. So for many people, their use is unethical, and is sufficient to prevent their use. Ahmed's first statement is also correct, but many people think that any individual – even an early embryo – has the right to life. Beatrice's statement is correct in that most embryonic stem cells currently come from embryos surplus to IVF treatments, but it is very controversial. Many consider that embryo use is justifiable under any circumstances, but work with, and use of stem cells, is subject to legislation in many countries. Maia is incorrect. While adult stem cells have the potential to replace some cell types, this is much less than that of embryonic stem cells

4

Source of stem cells	One advantage	One disadvantage
Embryo	Can be used to produce any cell type	Removal of stem cells involves destruction of an embryo
Adult	Can be removed from the patient	Used to produce a limited number of cell types only
Therapeutic cloning	Stem cells are genetically identical to those of the patient, so won't be rejected	The 'embryo' produced is still destroyed as stem cells are extracted
Transformed body cells	Potentially, could be used to produce any cell type	None (although the technique is only in its early stages of development)

Page 212 B5 Extended response question

5–6 marks

States that auxins are plant hormones that regulate plant growth, and explains that auxin promotes cell elongation (and cell division) in a plant, so is involved in the plant's growth response to light (phototropism). States that auxin is produced by the tip of the shoot and produces growth below the tip. Describes how, when a plant is exposed to light from one side, auxin is redistributed away from the light to the shaded side, where it produces growth. The shoot therefore grows towards the light. Describes how this is an advantage to the plant because the plant needs light energy for photosynthesis, in order to produce the materials for growth (and energy). All information in answer is relevant, clear, organised and presented in a structured and coherent format. Specialist terms are used appropriately. There are few, if any, errors in grammar, punctuation and spelling

3–4 marks

Describes that the plant grows towards the light because there is more auxin on the shaded side, but the explanation is incomplete and not fully detailed. States that the plant grows towards the light, and describes how, as light is needed for photosynthesis, this is important for the plant to stay alive, but the answer does not go beyond this. For the most part the information is relevant and presented in a structured and coherent format. Specialist terms are used for the most part appropriately. There are occasional errors in grammar, punctuation and spelling

1–2 marks

States that the plant grows towards the light, but there is limited or no description of the action of auxin. States that light is essential for the plant to live, with little or no explanation of this. Answer may be simplistic. There may be limited use of specialist terms. Errors of grammar, punctuation and spelling prevent communication of the science

0 marks

Insufficient or irrelevant science. Answer not worthy of credit

B6 Brain and mind

Page 213 The nervous system

1 a Nervous; hormonal

 b The nervous system uses electrical impulses/messages, the hormonal system uses chemical messages. The nervous system produces a quick, short response, while the hormonal system produces a slower response, but the response is longer-lasting. The nervous system sends messages using nerve cells or neurons, while in the hormonal system, hormones are transported in the blood *(Any 2)*

2 Sensory, receptors, stimuli, central; Motor, central, effectors, muscles

3 Insulin – produced by the pancreas; Oestrogen – produced by the ovaries. It is a sex hormone that controls the development of the adult female body at puberty and the menstrual cycle

4 a Dendrite; Myelin (fatty) sheath; Cell body; Nucleus; Axon

 b Arrow is from left to right

5 Temperature – a higher temperature (but not higher than the body temperature of mammals and birds) speeds up transmission. The diameter of the axon – the wider the axon, the faster the speed. The myelin sheath – the presence of the sheath increases the speed of transmission

Page 214 Linking nerves together

1 b, c

2 a Work in different areas of the body; work between nerves and other nerves, and nerves and muscles; some excite nerves or muscles, different ones inhibit them

 b The receptors on the second nerve or muscle are a specific shape to receive each type of chemical transmitter

3 Stimulus; receptors; organ; ear/eye; eye/ear

4 a Central

 b Hormones, enzymes

 c Contraction, stimuli, heartbeat

Page 215 Reflexes and behaviour

1 a Eye/ receptor; Sensory neuron; Relay neuron in CNS/ brain; Motor neuron

 b It helps to protect the eye from damage if it's suddenly exposed to a bright light

2 Picking up a hot object normally sets up a reflex action where you would drop the plate; when a message reaches the brain that you have picked up the plate, the brain sends a message to motor neurons; which instead of causing you to release the plate, make you hold on to it

3 a 20%; 80%; 70%; 30%

 b Woodlice move towards dark; and wet places

 c Instinctive

 d It assists their survival; since they cannot learn from experience

4 As it has happened many times, the dog has learned to associate the ringing of the doorbell with the arrival of a stranger/someone at the door, so will bark, anticipating the arrival of the stranger

5 A predatory bird, at some stage, will have tried to eat a wasp and will have been stung/harmed in the process. The bird will have come to associate the yellow and black pattern of the wasp with danger; so will avoid insects with similar patterns, such as hoverflies

Page 216 The brain and learning

1 a Cerebral cortex; cerebellum; medulla/brain stem

 b Intelligence; memory; language; consciousness

Answers

2 Different types of MRI can investigate the structure and activity; of the brain by scanning it. Other techniques involve stimulating the brain using electrodes and monitoring how the person responds. This requires access to the brain by removing part of the skull; and is usually carried out during brain surgery; so opportunities will be more limited

3 Intelligence; memory; language; consciousness

4 Neurons, links, neuron pathway; repeating, impulses, stimuli

5 It is easier to establish new neuron pathways in children than adults

6 Babies are born with simple instinctive behaviours, e.g. the rooting and sucking reflexes; but then need to learn behaviours. This requires exposure to new, appropriate stimuli. These enable many neuron pathways to develop in the brain. There is evidence to suggest that these pathways, and therefore learning, only develop at a certain age; feral children brought back into society, for instance, develop only limited language skills

Page 217 Memory and drugs

1 a The storage; and retrieval; of information

  b Long-term

  c Arranging them so as to produce a pattern; repetition (of reading, reciting, etc.); other sensible answer (Any 2)

2

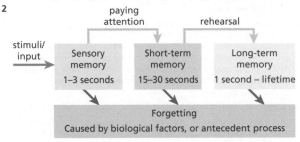

Stimulus; Paying attention; Rehearsal; Sensory, short-term, long term; 1-3 seconds, 15-30 seconds, 1 second-lifetime (1 mark for stimulus; 1 mark for each arrow; 1 mark for each box)

3 a Justine; the word (mnemonic) that she makes up is the stimulus

  b Lucas; he is storing information **and** retrieving it

4 Drugs; toxins

5 Bungarotoxin blocks the acetylcholine receptors of the muscles, so the transmitter substance acetylcholine is unable to bind with the receptors on the muscles affected. These muscles are therefore unable to contract, and the person is paralysed

6 When a nerve impulse reaches the end of a neuron, a chemical transmitter is released, which causes the next nerve to fire. After the nerve impulse has passed, the transmitter is taken back up into the neuron and therefore removed from the synapse. MDMA prevents this re-uptake and therefore increases the levels of serotonin in the synapse. The person's mood is lifted. Some time after taking MDMA, the brain becomes depleted of serotonin, and the person feels irritable and tired

Page 218 B6 Extended response question

5–6 marks
Describes in detail the experiments that Pavlov carried out. Uses the terms primary (the food) and secondary stimulus (the bell). Describes that, after a period, the dogs will come to salivate in response to the ringing of a bell, and defines this as a conditioned reflex, where the response has no direct connection with the stimulus. All information in answer is relevant, clear, organised and presented in a structured and coherent format. Specialist terms are used appropriately. There are few, if any, errors in grammar, punctuation and spelling

3–4 marks
Describes in outline the work that Pavlov carried out, and defines this work as being on conditioned reflexes, but misses out some points, e.g. that salivation is the normal response of dogs to the smell, sight or taste of food. States that the dogs came to associate the ringing of the bell with food, but the explanation and terminology used are incomplete. For the most part the information is relevant and presented in a structured and coherent format. Specialist terms are used for the most part appropriately. There are occasional errors in grammar, punctuation and spelling

1–2 marks
There is limited description of the experimental work, and states that the dogs linked the ringing of the bell with food, but with little or no explanation of conditioned reflexes. Answer may be simplistic. There may be limited use of specialist terms. Errors of grammar, punctuation and spelling prevent communication of the science

0 marks
Insufficient or irrelevant science. Answer not worthy of credit

C4 Chemical patterns

Page 219 Atoms, elements and the Periodic Table

1 Relative atomic masses / properties / specific examples of properties, e.g. reactivity / melting points / density (Any 2 = 1 mark each)

2 a $(7 + 39) \div 2 = 22.5$

  b The RAM of sodium is roughly the mean of the other two

3 a They have similar properties

  b It left gaps for undiscovered elements; made predictions about the properties of elements

4 Looking at light from the Sun; splitting light using a prism; looking at the patterns of lines; idea that the pattern is unique to an element / element can be identified from the pattern (Any 3)

5 Protons / neutrons in either order; electrons; shells

6

Proton number	Relative atomic mass	No. of protons	No. of neutrons	No. of electrons
9	19	9	10	9
13	27	13	14	13
3	7	3	4	3

(1 mark for each correct row)

Page 220 Electrons and the Periodic Table

1 a 2 dots in first shell, 8 in second, one in third; electron arrangement: 2.8.1

  b Both have 1 electron in their outer shell; potassium has more electrons than sodium / 19 instead of 11; potassium has more electron shells than sodium / 4 shells instead of 3

2 a

Name of element	Symbol	Proton number
Neon	Ne	10
Fluorine	F	9
Lead	Pb	82

(1 mark for each correct row)

  b Aluminium

3 c; d

4 The number of electrons in the outer shell is the same as the group number; elements with 3 or fewer electrons in the outer shell are usually metals; elements with 5 or more electrons in the outer shell are non-metals

Page 221 Reactions of Group 1

1 a Decreases down the group

  b 850–1200°C (must have units); boiling point increases down the group

2 a a to iii; b to i; c to ii

  b Both fizz / produce hydrogen; both turn pH indicator blue / make an alkali / make a hydroxide; potassium reaction is faster / produces a flame; produces potassium hydroxide rather than sodium hydroxide

3 a $2Li + Cl_2 \longrightarrow 2LiCl$ (Cl_2 = 1 mark, balancing = 1 mark)

  b Lithium has 2 electron shells / electron arrangement 2.1. Sodium has 3 electron shells / electron arrangement 2.8.1. Potassium has 4 electron shells / electron arrangement 2.8.8.1. Reactivity increases down the group; number of electron shells also increase down the group (Any 4 points)

Page 222 Group 7 – The halogens

1 Solid – iodine – dark grey; liquid – bromine – red-brown; gas – chlorine – pale green (States all correct = 1 mark, only two colours correct = 1 mark, all three colours correct = 2 marks)

2 Cl_2; contains two atoms in each molecule

3 a Gases at the top, then a liquid, then solids at the bottom

  b Boiling points increase; melting points increase; density increases; colour becomes darker; reactivity decreases (2 marks for any one answer)

4 a Iron chloride

  b Rose should do the experiment in a fume cupboard; chlorine is a toxic gas

  c Rate of the reaction is slower; bromine is less reactive

5 a $Cl_2 + 2KAt \longrightarrow 2KCl + At_2$ (At_2 = 1 mark, balancing = 1 mark)

  b Chlorine has fewer electron shells; Group 7 / non-metals are more reactive when they have fewer electron shells

Page 223 Ionic compounds

1 b; e

2 a C

  b A

Answers

3 a Below melting point it does not conduct; above melting point its conductivity increases / it does conduct

b Ions cannot move in the solid; when the solid melts the ions can move

4 a Both have the same number or 11 protons; both have same number or 12 neutrons; ion has fewer electrons / loses an electron / has 10 not 11 electrons

b A sodium atom loses an electron; a chlorine atom gains an electron

5 Sodium ions have a charge of +1; chloride ions have a charge of –1; idea that charges balance one to one in sodium chloride; the charge on a calcium ion must be +2; need two chloride ions to balance the charge in calcium chloride *(Any 4 points)*

Page 224 C4 Extended response question

5–6 marks
Answer makes clear comparisons and discusses both lithium and chlorine and gives information about electron arrangement, metal/non-metal character and reactivity. Possible marking points:
- lithium has 1 electron in the outer shell
- fluorine has 7 electrons in the outer shell
- lithium is a metal
- chlorine is a non metal
- lithium is less reactive than other elements in Group 1
- fluorine is more reactive than other elements in Group 7.

All information in answer is relevant, clear, organised and presented in a structured and coherent format. Specialist terms are used appropriately. There are few, if any, errors in grammar, punctuation and spelling

3–4 marks
Answer discusses both lithium and chlorine and discusses at least two from: electron arrangement, metal/non-metal character and reactivity. For the most part the information is relevant and presented in a structured and coherent format. Specialist terms are used for the most part appropriately. There are occasional errors in grammar, punctuation and spelling

1–2 marks
Answer gives some properties of lithium and/or chlorine but comparison is not clear. Answer may be simplistic. There may be limited use of specialist terms. Errors of grammar, punctuation and spelling prevent communication of the science

0 marks
Insufficient or irrelevant science. Answer not worthy of credit

C5 Chemicals of the natural environment

Page 225 Molecules in the air

1 O_2 – oxygen – 21%; Ar – argon – about 1%; CO_2 – carbon dioxide – 0.04%; N_2 – nitrogen – 78%

2 b; c

3 a Oxygen

b The forces between molecules are very weak; so a small amount of energy can overcome the forces

c Nitrogen has the lowest boiling point; but it does not have the lowest melting point. Oxygen has the lowest melting point; but oxygen does not have the lowest boiling point. The pattern is not true for any of the three gases

Page 226 Ionic compounds: crystals and tests

1 a Positive and negative ions; are very strongly attracted together; and need large amounts of energy to overcome the forces between them

b Ions cannot move

c Ions move about; become further apart / are not arranged regularly/ spread through the water

2 Formula for calcium chloride: $CaCl_2$. Potassium sulfate: positive ion – K^+ and negative ion – SO_4^{2-}; formula – K_2SO_4

3 a Calcium carbonate

b Similarity – still get a white precipitate; difference – this time it dissolves in excess

c Insoluble

d $Ca_2+(aq) + 2OH–(aq) \longrightarrow Ca(OH)_2(s)$ *(correct equation = 1 mark, state symbols = 1 mark)*

Page 227 Giant molecules and metals

1 a 12%

b Silicon and oxygen *(both = 1 mark)*

c Reduction

2 Only very small amounts of copper are found in the rock; so to make a small amount of copper, large amounts of waste rock are produced

3 a Simple molecular structure contains small molecules; idea that only a few atoms are joined in each molecule; idea that many atoms are joined in a giant covalent structure; idea that the atoms are joined in 3 dimensions *(Any 3 points)*

b i False **ii** True **iii** True **iv** False *(All 4 correct = 2 marks; 2 or 3 correct = 1 mark)*

Page 228 Equations, masses and electrolysis

1

	iron oxide	carbon monoxide	iron	carbon dioxide
reactant (✓)	✓	✓		
product (✓)			✓	✓

2 Copper; carbon dioxide and CO_2

3 a RAM of Mg: 58.5; RFM of NaCl: 24; RFM of $CaSO_4$: 136

b 95 g *(must have g)*

4 a b

b 54% *(Uses correct atomic masses 27 and 16 = 1 mark, correct answer = 1 mark)*

c i Oxygen ions are negatively charged; negatively charged ions are attracted to the positive electrode

ii $Al^{3+} + 3e^- \longrightarrow Al$ *(3e = 1 mark, rest correct = 1 mark)*

Page 229 Metals and the environment

1 a i Excellent electrical conductivity

ii Expensive; corrodes quickly

b Excellent conductivity; very light in weight

2 a Reducing the energy needed for lighting is a benefit; it is impossible to stop mining mercury altogether because we need mercury to make light bulbs; people buy low-energy light bulbs because they think it helps the environment; carbon dioxide causes climate change; if the mines were closed people would lose their jobs; the costs of the mines include the toxic water; the costs include the toxic gases that the production produces; it might be possible to reduce the environmental impact of the mercury mine *(Any 4)*

b The sea of electrons in the metal; can move

Page 230 C5 Extended response question

5–6 marks
Discusses and compares similarities and differences for both diamond and silicon dioxide and gives details of the structure and state (solids) of both. Possible points:
- both solids / high melting points
- both hard
- both regular arrangement or lattice
- both covalently bonded; both giant / 3-D structures
- diamond is an element / only contains carbon atoms
- silicon dioxide contains both silicon and oxygen atoms / it is a compound.

All information in answer is relevant, clear, organised and presented in a structured and coherent format. Specialist terms are used appropriately. There are few, if any, errors in grammar, punctuation and spelling

3–4 marks
Covers some points about the structure of both diamond and silicon dioxide. Gives at least one similarity and one difference between the two. For the most part the information is relevant and presented in a structured and coherent format. Specialist terms are used for the most part appropriately. There are occasional errors in grammar, punctuation and spelling

1–2 marks
Gives some information about either silicon dioxide or diamond but does not make a comparison clear. Answer may be simplistic. There may be limited use of specialist terms. Errors of grammar, punctuation and spelling prevent communication of the science

0 marks
Insufficient or irrelevant science. Answer not worthy of credit

Chemical synthesis

Page 231 Making chemicals, acids and alkalis

1 a Flammable hazard symbol

b No naked flames; keep the top on the can; be careful not to spill any *(Any 2)*

2 a Sulfuric acid (l); citric acid (s); hydrochloric acid (g)

b Sulfuric acid

c Hydrogen

Answers

3 a Copper chloride, carbon dioxide and water; $2HCl$; CO_2 and H_2O

  b pH starts off low / at pH 1; pH rises when copper carbonate is added; because the acid is neutralised

  c $Cu(NO_3)_2$

Page 232 Reacting amounts and titrations

1 MgO; $(23 \times 2) + 16 = 62$; sodium carbonate; $(2 \times 23) + 12 + (3 \times 16) = 106$

2 a Same number of atoms idea; on both sides of the equation

  b Relative atomic masses: S = 32 and O = 16; relative formula mass $ZnSO_4$ = 161; mass of zinc sulfate = $161 \times 2 = 322$ g

3 a Add sodium hydroxide until indicator changes colour; with shaking or stirring; add drop by drop near the end-point; do repeats; check that repeats are close together; take an average / mean of the repeats *(Any 4)*

  b a = True; b = False; c = True; d = False *(All correct = 2 marks; 3 correct = 1 mark)*

Page 233 Explaining neutralisation and energy changes

1 Hydrochloric acid, potassium hydroxide, potassium chloride; nitric acid, sodium hydroxide, sodium nitrate; sulfuric acid, calcium hydroxide, calcium sulfate

2 Water

3 a Hydrogen, H^+ b Hydroxide, OH^-

  c $H^+ + OH^- \longrightarrow H_2O$

4 Increases; given out; neutralisation

5 Energy is given out; the reactants have more energy than the products

6 Idea of very large temperature rise; causing explosion / fire / hazard

Page 234 Separating and purifying

1 a Calcium nitrate

  b Filter; the solid stays on the filter paper / does not go through

  c a = True; b = False; c = False; d = True *(All correct = 2 marks; 3 correct = 1 mark)*

2 a More soluble / more dissolves / dissolves faster *(Any 2)*

  b Step 2 takes out insoluble impurities / impurities that do not dissolve; Step 4 filters off the crystals

  c Drying; in an oven or dessicator

3 a RFM of $ZnCl_2$ = 65 +(2×35.5) = 136. Theoretical yield = 13.6 g

  b $10.2 / 13.6 \times 100$; = 75%

  c It will be too high; due to extra mass of water

Page 235 Rates of reaction

1 a Volume of acid; mass of calcium carbonate; temperature *(Any 2)*

  b As the concentration increases; the rate of reaction increases

  c Idea that is it too short a time; it should be about 200 s / ten times slower than Experiment 2 / it does not fit the pattern

2 a Every 30 s   b 5 minutes

  c The gradient becomes less steep; because the reaction slows; it becomes level when the reaction stops; because the reactants are being used up; collisions become less frequent; less frequent collisions lead to a slower reaction *(Any 4)*

Page 236 C6 Extended response question

5–6 marks

Links shape of graph to rate of reaction and gives explanation for the changes, e.g. a gas is made and the acid is used up. Points to include:

• mass decreases during the reaction; due to loss of a gas

• (describes shape of graph) graph changes fast at first and then levels out

• (links shape to rate of reaction) reaction is fastest at first and then slows

• reactions stops when line goes flat

• reaction slows when acid concentration falls

• reaction stops when acid is used up.

All information in answer is relevant, clear, organised and presented in a structured and coherent format. Specialist terms are used appropriately. There are few, if any, errors in grammar, punctuation and spelling

3–4 marks

Links shape of graph to rate of reaction and discusses reaction slowing and stopping. For the most part the information is relevant and presented in a structured and coherent format. Specialist terms are used for the most part appropriately. There are occasional errors in grammar, punctuation and spelling

1–2 marks

Describes how the shape of the graph changes over time. Answer may be simplistic. There may be limited use of specialist terms. Errors of grammar, punctuation and spelling prevent communication of the science

0 marks

Insufficient or irrelevant science. Answer not worthy of credit

P4 Explaining motion

Page 237 Speed

1 a Speed = distance ÷ time = 60 ÷ 3 = 20 km/h

  b Speed = distance ÷ time = 60000 ÷ (3 × 3600) = 5.5 m/s

2 Time = distance ÷ speed = 100 ÷ 20 = 5 hours

3 The average velocity will be less than 150 km/h; because the displacement of the journey will be lower than the total distance; the railway line will not be a straight line

4 a Constant high speed   b Speeding up

  c Stationary   d Constant slow speed

5 a 1000 m

  b Speed = 1000 m ÷ (10 × 60) = 0.17 m/s

  c She stopped for 15 minutes; walked back 400 m in the next 5 minutes; stopped for 5 minutes; then went back 600 m in the next 15 minutes

  d 2000 m

  e 0 m

  f Average speed = 2000 ÷ (50 × 60) = 0.67 m/s; average velocity = 0 m/s; average velocity is zero because she went back to the original starting point

Page 238 Acceleration

1 Car B

2 100 km/h = 100 000 m ÷ 3600s = 27.8 m/s; Car A's acceleration = change in speed ÷ time = 27.8 ÷ 12 = 2.3 m/s²; Car B's acceleration = 27.8 ÷ 10 = 2.8 m/s²

3 Change in speed = acceleration × time = 10 × 1.6 = 16 m/s. Answer = 16 m/s

4 Speed increases steadily; for first 20 seconds; up to maximum speed of 10 m/s; then remains at this constant speed for 40 seconds

5 Graph needs to show: a straight line from the origin to 8 m/s at 20 s; a horizontal line at 8 m/s from 20 to 40 seconds; a straight line from (40, 8) to (50, 5); a horizontal line at 5 m/s from 50 to 60 seconds

6 a All points correctly plotted *(3 marks)*; continuous line drawn between points *(1 mark)*

  b Acceleration = gradient of line = 12 ÷ 20 = 0.6 m/s²

  c Deceleration = (16 - 4) ÷ 15 = 0.8 m/s² (ignore negative sign)

Page 239 Forces

1 Pairs; a downwards; an upwards

2 a The ball pushes on the tennis racket backwards

  b The cart pulls the horse backwards

  c The Earth is attracted to the Moon

3

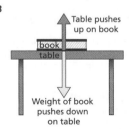

4 a To make the results more reliable

  b 32.6

  c Same block of wood

  d Average values 9.7; 25.5 or 27.8; 16.5 *(1 mark each)*

  e Glass; it is much smoother than wood and carpet *(1 mark each)*

5 Useful friction on brakes/tyres; friction is used to slow the wheels down/friction is needed to grip the road; there is unwanted friction on axles/gears/chain, etc.; friction on moving parts causes heat/wear/energy loss

Page 240 Effects of forces

1 a 50N downwards   b 50 N upwards

  c 100 N downwards   d 100 N upwards

2 a Diagram needs to include: downwards arrow labelled weight/gravity; upwards arrow labelled air resistance/drag

  b Downwards force must be bigger than upwards force

  c Speed eventually becomes constant; when drag force is equal in size to weight

3 b and c

Answers

4 a Momentum = 0.3 × 30 = 9 kg m/s

b Momentum = 1100 × 15 = 16 500 kg m/s

c Momentum = 4000 × 5 = 20 000 kg m/s

5 The ball has a lot of momentum; when he catches it the momentum drops to zero. If he moves his hands towards his body as he catches it the time to stop the ball increases; this reduces the force

Page 241 Work and energy

1 Destroyed; transferred; conservation

2 a, c

3 a Work done = force × distance = 3 × 1.3 = 3.9 J

b Gravitational potential energy

c 3.9 J

d KE = ½ mv², so v² = 2 × KE ÷ m; v² = 2 × 3.9 ÷ 0.3 = 26; v = $\sqrt{26}$ = 5.1 m/s

4 a The diagram should show: maximum kinetic energy at the lowest point; maximum potential energy at either high point

b GPE = weight × height = 1 × 0.2 = 0.2 J

c KE = ½ × mass × v² = ½ × 0.1 × 1.2² = 0.072J

d GPE = 0.2 – 0.072 = 0.128 J

5 a The roller coaster starts with gravitational potential energy (GPE); which is converted to kinetic energy (KE) on the way down the track. As it goes up again the KE is converted back to GPE again

b Some of the energy is dissipated as heat; due to friction and air resistance; so it does not have as much GPE as it started with

Page 242 P4 Extended response question

Relevant points include:

- safety devices that protect in the event of a crash: crumple zones; seat belts; air bags
- how they work: increase time taken for person to slow down; slowing down momentum change; reducing force on the person

5–6 marks

Mentions all three devices (see above). Clearly links reduction in force during a collision to the increase in time needed to change momentum. All information in answer is relevant, clear, organised and presented in a structured and coherent format. Specialist terms are used appropriately. There are few, if any, errors in grammar, punctuation and spelling

3–4 marks

Mentions at least two devices that protect passengers during a crash (see above). Includes two out of three points about how they work. For the most part the information is relevant and presented in a structured and coherent format. Specialist terms are used for the most part appropriately. There are occasional errors in grammar, punctuation and spelling

1–2 marks

Mentions at least two devices that protect passengers during a crash (see above). Includes one relevant point about how they work. Answer may be simplistic. There may be limited use of specialist terms. Errors of grammar, punctuation and spelling prevent communication of the science

0 marks

Insufficient or irrelevant science. Answer not worthy of credit

P5 Electric circuits

Page 243 Electric current – a flow of what?

1 Inside nucleus – proton – positive; Inside nucleus – neutron - neutral; Orbiting nucleus – electron – negative

2 Electron; negative; attracted

3 Metals are conductors; they contain free electrons

4 The movement/friction of the iron charges up the shirt; electrons are transferred between Sadie/the iron and the shirt. Sadie becomes charged up. When she touches the wall she discharges/electrons move away. An electric current causes the small shock

5 2p coin; steel scissors

6 The cell gives energy to the electrons/charged particles; the electrons are free to move in the connecting wires; electrons/charged particles carry the energy to the bulb; at the bulb the energy is transferred to light; and heat

7 a Arrow showing flow of charged particles towards positive terminal of cell

b Electrons are negatively charged; so they will be attracted towards the positive terminal

c Put another cell (in series with the first one); so more energy will be supplied to each electron

Page 244 Current, voltage and resistance

1 a Circuit diagram should be amended to show correct voltmeter symbol; correct ammeter symbol; voltmeter placed in parallel to either cell or bulb; ammeter placed in series (either side of the bulb)

b The charged particles carry the energy from the cell to the bulb; the charged particles in this circuit are electrons; the electrons/charged particles are given energy in the cell; the elctrons/charged particles flow all the way round the circuit; at the bulb the energy carried by the electrons/charged particles is converted to heat and light (Any 3)

2 Potential difference is the scientific term for voltage and is measured in volts; it is a measure of how much energy is given to/converted from the charged particles; 1 volt means that 1 joule of energy is given to/converted for each unit of charge

3 B, A, C, D

4 Resistance = voltage ÷ current = 12 ÷ 0.5 = 24 Ω

5 a All points correctly plotted; line of best fit drawn accurately

b Either: calculate gradient of line; then calculate 1/gradient, Or: calculate a value for resistance from table values; calculate two or more values for resistance and work out average (value for resistance should be in the range 16 Ω < R < 18 Ω)

c He should repeat the experiment; and work out average values of current at each voltage

Page 245 Useful circuits

1 a Current decreases

b Current increases

c Current would be zero

2 a The series circuit would have dimmer bulbs

b Series circuit – all bulbs would go out; parallel circuit – all the other bulbs would stay on with the same brightness

3 A_1 = 0.9 A; A_2 = 0.6 A; A_3 = 0.3 A; A_4 = 0.9 A

4 The resistance of an LDR varies with light intensity; the resistance of a thermistor varies with temperature; Either: the lower the light intensity the higher the resistance (or vice versa), Or: the lower the temperature the higher the resistance (or vice versa)

5 He records the resistance at a known temperature; he records the resistance at a second known temperature; the lower known temperature could be melting ice; the higher known temperature could be boiling water; he makes a scale between the two known temperatures; it is not a linear temperature scale; record values of resistance at several different known temperatures and plot a graph; read the temperature off the graph (Any 4)

6 a With two components you add the resistances; twice as much resistance would give half the current

b Bulbs do not have constant resistance; at lower currents the bulbs are cooler; so they have lower resistance; so more current will flow

Page 246 Producing electricity

1 Field; magnetic; iron; compasses; poles

2 a The output voltage would be higher

b The voltage produced would be in the opposite direction

c The output voltage would become very low

3 a Torch – d.c.   **b** Transformer – a.c.

c Computer – d.c.   **d** Iron – a.c.

4 Power = V × I = 230 × 3.5; = 805; W

5 a Current = power ÷ voltage = 100 ÷ 230; = 0.43; A

b New current = 0.43 ÷ 4 = 0.11; power = 230 × 0.11 = 25 W

Page 247 Electric motors and transformers

1 Magnetic field lines drawn similar to those shown; arrows on field lines from South towards North marked on diagram

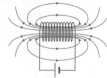

2 The force on the sides of the coil/wire carrying current is at right angles to both the magnetic field; and to the current. The forces on either side of the coil produce a turning effect; once every half turn the current in the coil has to change direction; in order for the turning effect to be in the same direction

3 Voltage; alternating; coils; iron; primary; secondary

4 i a Step-down **b** Step-up **c** Step-down **d** Neither

ii V_1, V_3, V_4, V_2

5 The output from a battery is d.c.; d.c. will create a constant magnetic field; for a voltage to be induced on secondary coil; there must be a varying magnetic field

Answers

Page 248 P5 Extended response question

5–6 marks
Includes all main details and some additional details. Main details:
- rotate magnet
- to alter magnetism/magnetic field of iron/coil
- voltage across/current in coil.

Additional details:
- process is called (electromagnetic) induction
- voltage keeps on changing / a.c. / not d.c.
- current in components connected to ends of coil
- work done turning magnet transfers to electrical energy.

All information in answer is relevant, clear, organised and presented in a structured and coherent format. Specialist terms are used appropriately. There are few, if any, errors in grammar, punctuation and spelling

3–4 marks
Includes some of the main details and some additional details (see above). For the most part the information is relevant and presented in a structured and coherent format. Specialist terms are used for the most part appropriately. There are occasional errors in grammar, punctuation and spelling

1–2 marks
Includes at least one main detail and at least one additional detail (see above). Answer may be simplistic. There may be limited use of specialist terms. Errors of grammar, punctuation and spelling prevent communication of the science

0 marks
Insufficient or irrelevant science. Answer not worthy of credit

P6 Radioactive materials

Page 249 Nuclear radiation

1 Nucleus; negatively charged; protons; electrons; mass
2 a to iii; b to i; c to ii
3 An atom of the same element; with different number of neutrons in nucleus; and the same number of protons
4 An atom is neutral / has same number of protons and electrons; an ion has an overall charge; less protons than electrons / more protons than electrons
5 a Background radiation is ionising radiation which is all around us; from cosmic rays / rocks / man-made sources
   b Some rocks are more radioactive than others
6 Michael is correct; radioactivity is not affected by physical conditions; it only depends on the amount of radioactive material present in the sample

Page 250 Types of radiation and hazards

1 Gamma; alpha; alpha; gamma; beta
2 Put the different materials in between the mantle and the detector. Find out if the reading on the detector changes; if the radiation is stopped by paper it is alpha; if it is stopped by aluminium or steel it is beta; if it is not stopped by any of the materials it is gamma
3 A helium nucleus; with two protons and two neutrons; a large, positively charged particle
4 a 900; microSieverts
   b The bar from nuclear power and weapons would be much higher; because a lot more radiation was released as the power station was damaged
5 Ionising radiation ionises (water) molecules in the body; the ions can then react with other molecules in the body; such as DNA

Page 251 Radioactive decay and half-life

1 Once a radioactive nucleus (atom) has emitted some radiation; it can't do it again; so the number of radioactive nuclei decreases
2 Alpha and beta particles are made up from parts of the nucleus; so the number of protons in the nucleus changes; gamma rays are electromagnetic waves carrying energy and do not change the nucleus's structure
3 4; 2; He; 212; 82
4 a A   b C   c C
5 a All points plotted correctly; smooth curve drawn (1 mark for each)
   b One value of half-life read off; average of more half-life values calculated (1 mark for each)

Page 252 Uses of ionising radiation and safety

1 a Gamma
   b The instruments can be sealed in packaging; and the radiation will pass through the packaging
2 a So radiation is concentrated on the cancer tissue; and is not going to damage healthy cells around the cancer
   b Beta radiation will just pass though skin; and won't damage deeper tissue
3 a It will get less
   b Beta
   c Alpha will not get through any thickness of paper; gamma passes through the paper too easily, so too little variation as paper changes thickness
4 b
5 Health/cancer risk for all participants due to irradiation by the rod; this risk is greatest for the radiographer who will repeat the procedure many times; patient will benefit if their existing cancer is cured; but the risk of patient and/or radiographer developing a new cancer may outweigh the benefits of the procedure (Any 3)

Page 253 Nuclear power

1 Fuel; fission; fusion
2 $E = 0.24 \times (3 \times 10^8)^2$; $= 2.16 \times 10^{16}$ J
3 Advantages: no CO_2 given off; no contribution to greenhouse effect; large amounts of energy from small amounts of fuel; fuel is not running out (Any 2)

   Disadvantages: employees at risk of radiation; radioactive waste produced; risk of nuclear accident (Any 2)
4 a The neutron is absorbed by the uranium nucleus, making it unstable; it then splits into two smaller nuclei; and gives off two or three neutrons; which then go on to hit more uranium nuclei causing them to fission
   b Control rods (boron); are used to absorb some of the neutrons
5 Helium
6 a Larger quantities of energy could be available; ready source of fuel/hydrogen; no radioactive waste (Any 2)
   b The fuel needs to be contained; at very high temperatures and pressures. This uses up more energy than is created from the reaction

Page 254 P6 Extended response question

Relevant points include:
- high level only produced in reactor; is very radioactive; so is stored in ponds of water; until it becomes intermediate waste / less radioactive
- hospital produces mostly intermediate waste; intermediate waste is encased in concrete / glass; and stored in metal drums; under guard / in secure conditions
- low level waste is produced at both hospital and reactor; is put in landfill; with waterproof linings; to keep radioactivity out of ground water
- all radioactive waste is harmful / cancerous; becoming less harmful as time goes on

5–6 marks
Evaluates production and use of the radioactive materials, and correctly identifies sources for all three types of waste. Suggests how to dispose of them safely. Gives a valid reason why waste needs to be stored carefully. All information in answer is relevant, clear, organised and presented in a structured and coherent format. Specialist terms are used appropriately. There are few, if any, errors in grammar, punctuation and spelling

3–4 marks
Evaluates production and/or use of the radioactive materials, and correctly identifies sources for at least two types of waste, perhaps omitting some important details. For the most part the information is relevant and presented in a structured and coherent format. Specialist terms are used for the most part appropriately. There are occasional errors in grammar, punctuation and spelling

1–2 marks
Refers to at least one type of waste and valid disposal method for it; may not give a reason for the need for careful disposal. Answer may be simplistic. There may be limited use of specialist terms. Errors of grammar, punctuation and spelling prevent communication of the science

0 marks
Insufficient or irrelevant science. Answer not worthy of credit